APPLIED NATURAL SCIENCE

Environmental Issues and Global Perspectives

APPLIED NATURAL SCIENCE

Environmental Issues and Global Perspectives

Mark D. Goldfein, PhD, DSc, and
Alexey V. Ivanov, PhD

AAP | APPLE ACADEMIC PRESS

Apple Academic Press Inc.
3333 Mistwell Crescent
Oakville, ON L6L 0A2
Canada

Apple Academic Press Inc.
9 Spinnaker Way
Waretown, NJ 08758
USA

International Standard Book Number-13: 978-1-77188-272-9 (Hardcover)

International Standard Book Number-13: 978-1-77188-273-6 (eBook)

Library and Archives Canada Cataloguing in Publication

Goldfein, Mark D., author
Applied natural science : environmental issues and global perspectives / Mark D. Goldfein, PhD, DSc, and Alexey V. Ivanov, PhD.

Includes bibliographical references and index.
Issued in print and electronic formats.
ISBN 978-1-77188-272-9 (hardcover).--ISBN 978-1-77188-273-6 (pdf)

1. Science. I. Ivanov, A. V. (Alekse™i Viktorovich), author II. Title.

Q158.5.G64 2016 500 C2016-901111-9 C2016-901112-7

Library of Congress Cataloging-in-Publication Data

Names: Goldfein, Mark D. | Ivanov, Alexey V.
Title: Applied natural science : environmental issues and global perspectives /
Mark D. Goldfein, PhD, DSc, and Alexey V. Ivanov, PhD.
Description: Toronto : Apple Academic Press, 2016. | Includes bibliographical references and index.
Identifiers: LCCN 2016006427 (print) | LCCN 2016015765 (ebook) | ISBN 9781771882729 (hardcover : alk. paper) | ISBN 9781771882736 ()
Subjects: LCSH: Ecology--Philosophy. | Nature and civilization. | Environmental protection.
Classification: LCC QH540.5 .G65 2016 (print) | LCC QH540.5 (ebook) | DDC 508--dc23
LC record available at https://lccn.loc.gov/2016006427

Apple Academic Press also publishes its books in a variety of electronic formats. Some content that appears in print may not be available in electronic format. For information about Apple Academic Press products, visit our website at **www.appleacademicpress.com** and the CRC Press website at **www.crcpress.com**

ABOUT THE AUTHORS

Mark D. Goldfein, PhD, DSc

Mark D. Goldfein, PhD, DSc, is currently a Professor and Head of the Chair of Environmental Protection and Life Safety, Biological Faculty at Saratov State University (SSU) in Saratov, Russia. He has authored more than 300 educational and research papers and several textbooks and monographs in the field of chemical physics, the physico-chemistry of polymers, ecology, environmental protection, and life safety. He also holds 20 patents. He studied at Saratov State University (SSU), named after N.G. Chernyshevsky.

Alexey V. Ivanov, PhD

Alexey V. Ivanov, PhD, is currently a Professor and Dean of the Faculty of Ecology and Service at Saratov State Technical University in Saratov, Russia. He has authored over 300 scientific papers, 20 monographs, and 15 textbooks in the field of geology, ecology, paleontology, and global processes in nature and society. He graduated from Saratov State University, named after N.G. Chernyshevsky, with a PhD in geology and mineralogy.

CONTENTS

List of Abbreviations .... *ix*

Introduction........ *xiii*

PART 1

1. **Interrelation Between the Natural-Scientific Knowledge and the Human Knowledge........ 1**
2. **Some General Trends in the Development of Modern Natural Science........ 17**
3. **Role of Mathematics in Solving Scientific and Practical Tasks of Modern Natural Sciences........ 85**
4. **The Application of Thermodynamics to the Open Systems........ 99**
5. **Global Problems in the Biospheric Studies........ 117**
6. **Phenomenon of the Logistic Equation as a Basic Model for Describing Global Processes in Society........ 193**

PART 2

7. **Theory and practics of organic paramagnetics........ 229**
8. **Problems of Environmental Pollution with Polymers........ 329**
9. **Polymers and Human Environment........ 407**

Index........ 429

LIST OF ABBREVIATIONS

2,4-D	2,4-dichlorophenoxyacetic acid
2,4,5-T	2,4,5-trichlorophenoxyacetic acid
AAD	α,α′-azo-isobutyric acid dinitrile
ABS	Acrylonitrile butadiene styrene
ABTS	2,2'-azine-bis-3-ethylbenzothiazoline-6-sulfonic acid
AIDS	Acquired immunodeficiency syndrome
AL	Artificial leather
ANN	Artificial neural network
ATR	Attenuated total reflection
BA	Butyl acrylate
BAPM	Biologically active polymeric material
BIPO	Butadiyn-*bis*-2,2,6,6-tetramethyl-1-oxyl-4-oxi-4-piperidyl
CERN	Conseil Européen pour la Recherche Nucléaire
CHXDTPA	Trans-1,2-cyclohexyldiethylenetriamine pentaacetic acid
CMB	Cosmic microwave background
CMS	Carboxymethyl starch
CS	Coherent structures
DPPH	1,1-diphenyl-2-picrylhydrazyl
DTPA	Diethylenetriamine pentaacetic acid
EDTA	Ethylenediaminetetraacetic acid
ELISA	Enzyme-linked immunosorbent assay
EMIT	Enzyme-multiplied immunoassay technique
EPR	Electron paramagnetic resonance
EW	Evanescent wave biosensors
FIA	Flow injection analysis
FILIA	Flow-injection liposome immunoanalysis
FITC	Fluorescein isothiocyanate
FMIAs	Fluorescence microbead immunosorbent assays
FOBs	Fiber optic biosensors
GFRP	Glass fiber-reinforced plastic
GRP	Glass-reinforced plastic
HDPE	High-density polyethylene

HELC	High-efficiency liquid chromatography
HIPS	High-impact PS
IAC	Immunoaffinity chromatography
IEA	Immunoenzymatic analysis
IR	Infrared
IT	Information technologies
IUPAC	International Unity of Pure and Applied Chemistry
LAPS	Light-addressable potentiometric sensor
L-DOPA	L-dihydroxyphenylalanine
LIA	Liposome immunoaggregation
LLDPE	Linear low-density polyethylene
MA	Methyl acrylate
MDPE	Medium-pressure polyethylene
MFI	Melt flow index
MIF	Mean information field
MPC	Maximum permissible concentration
MWD	Molecular weight distribution
NIT	New IT
NR	Natural rubber
OEL	Occupational exposure limit
PA	Polyamides
PAA	Polyacrylamide
PAGs	Polyalkylene guanidines
PAN	Polyacrylonitrile
PC	Polycarbonate
PCs	Personal computers
PCBs	Polychlorinated biphenyls
PE	Polyethylene
PET	Polyethylene terephthalate
PFIA	Polarization fluoroimmunoassay
PHMG	Polyhexamethylene guanidine
PRFIA	Phase-resolved fluoroimmunoassay
PM	Polymeric material
PMQ	Permissible migration quantity
PP	Polypropylene
PPE	Personal protective equipment
PS	Polystyrene
PVC	Polyvinylchloride

QCM	Quartz crystal microbalance
RP	Rubber products
SAW	Surface acoustic wave
SET	Substance energy technologies
SFE	Supercritical fluid extraction
SHA	Sterically hindered amines
SLDPE	Secondary LDPE
SPR	Surface plasmon resonance
STI	Scientific and technical information
STW	Surface transverse wave
TCNE	Tetracyanoethylene
TCNQ	7,7,8,8-tetracyan-p-quinodimethane
TEMPOP	2,2,6,6-tetramethyl-4-oxypiperidyl-1-oxyl phosphite
TIRF	Total internal reflection fluorescence
TLV	Threshold limit value
TNCs	Transnational corporations
TREC	Trans-Mediterranean Renewable Energy Cooperation
UFI	Ultrafine interaction
VA	Vinyl acetate

INTRODUCTION

The natural sciences as a collection of scientific knowledge about natural objects, phenomena, and processes have been playing a crucial role in the development of mankind and its culture over several millennia. Convergence of the philosophical and methodological positions of people of rather different professions and businesses is the most important function of the natural sciences. As the human environment has a variety of levels of organization and properties, its in-depth study is only possible by using a multidisciplinary approach, that is, by applying the knowledge of various scientific fields.

The history of the development of religion, philosophy, and various sciences is connected with the problem of the relation between the natural-scientific and humanitarian cultures of mankind. In recent years, due to new general-scientific views on system analysis, self-organization processes and the mechanisms of evolutionary processes in animate and inanimate nature, the scientific culture and the human culture have begun to converge. In particular, at the present time, such theoretical and practical methods of natural sciences as mathematical and physical modeling, new information technologies, self-organization processes, and so forth are already used in many fields of social and spiritual life. All this justifies the necessity of the formation of a broader scientific outlook and an increased level of the culture of thinking. The main purpose of this book is to provide the reader with the most complete insight into the natural-scientific pattern of the world, that is, the acquisition of knowledge of most important historical stages of the development of various areas of science, methods of natural-scientific research, general scientific and philosophical concepts, and the fundamental laws of nature. Of particular importance is our discussion of some problems associated with the place of man in the biosphere, issues of the globalization of science and technology, new ideas about the Universe, and the concept of universal evolutionism. At the same time, the book discusses more specific issues related to solving major global and regional environmental problems (particularities of organic paramagnetic materials, the influence of polymers on the man and environment, etc.). All this leads to the fundamental conclusion of the unity of animate and inanimate nature, as well as the improvement of the process of cognition of the real world, which consists in objective and natural changing of worldviews.

However, the existing forms of science development are increasingly a matter of concern for public opinion for the results and perspectives of scientific discoveries. Science has turned out to be responsible for our planet's slide toward a global catastrophe, which could happen just in the 21st century, as well as for the growth of material and spiritual wealth of Earth's modern inhabitants. This poses a real threat to the economy, politics, environment, defense capability, and the entire social development. We must save the scientific potential and its ability to render effect on the machinery, technology, education, and other social activities. However, in order to regain its lost prestige in the society, science itself should change significantly by reorienting its development to the goals of the survival and sustainable development of mankind. The exit from the systemic crisis is seen in science's functioning in the framework of the values of the future "sustainable society," the sphere of mind (whose formation was predicted by Vladimir Vernadsky) rather than in the framework of industrial and consumer values ruining both man and nature. Overcoming the contradictions arising in the "man-society-biosphere" system, the creation of positive scientific prerequisites for the survival of our civilization in the 21st century and in the whole third millennium is associated with a new type of science (the noosphere-sustainable science).

Future specialists in the field of social sciences and humanities related to studying man and society, human behavior, should understand not only the essence of the foundations of modern natural sciences but also their possible impact on human activities, which will increase in the future. That is why they need to study integrative and general-scientific leads of creative search, the synthesis of scientific knowledge in the course of solving complex and especially global problems. Scientific knowledge appears not as something far removed from the social sciences and the humanities but as an integral part of the holistic science, significant for it. Until recently, the history of human existence has been virtually alienated from nature and begun to contradict the natural processes and cycles in the biosphere by the very fact of its existence. This "mistake" of the whole historical and socio-humanitarian knowledge, often ignoring the essential links between man and nature, is very costly for the human environment.

The hypothesis of the eternal existence of mankind on Earth, which it could transform with impunity and act in relation to nature so as everything be "for the human benefit," was one of the basic provisions of history and other social sciences, which has been very rarely discussed in science. This hypothesis was considered even axiomatic for quite a long time, until practice did not disprove its anthropocentrism by the quickly impending global environmental crisis. This was the crisis of not only the "antropochauvinistic" vision of history but the idea of progress, which the whole civilizing process was "adjusted" under. Science might actually lose the subject of its research simply because an offensive planetary anthropoecological catastrophe is very likely within next few decades, which could turn to be omnicide, that is, the death of reasonable and many other living creatures. The traditional sociological approach has been proven to be essentially futile, limited by the development and existence of a model of unsustainable development and its spatial and temporal limits.

Our consideration of the concepts of modern natural sciences not just in the "social and natural coordinate system" but in the prospects of transition to sustainable development is a special feature of this book. This transition will require the formation of an integral social and natural system, that is, the "mankind-nature" system which could indefinitely long conserve its interacting components, the civilization, and the biosphere. And from this point of view, it seems very important to emphasize those of the modern natural science concepts which would promote the most efficient movement of mankind to the objectives of sustainable development.

Science, focusing on new civilization goals, should be more engaged in the study of the future, that is, the process of "futurization" of research activities and the use of their results in practice (especially in education) will proceed in prospect. Qualitative changes in science, frequent changes of its paradigms and pictures of the world, pose new challenges for the researchers involved in the methodological and socio-philosophical problems of education and training, since the regularly increasing amount of scientific knowledge is difficult to be confined to the accepted standards of time. It is not always clear which achievements of science and how they should be included into the current educational programs. That is why there exists a complex relationship between education and science. Education depends on science because one should have knowledge in order to distribute and transmit it. But knowledge is obtained by scientists. Therefore, education is in a direct dependence of science. On the other hand, education is primary in relation to science because the knowledge assimilated in the process of education is the basis of new scientific findings.

Designing a unified advanced "science-education" system would require fundamental changes in the role relations between the scholar and the teacher in the sense that the scientist should become a teacher and the teacher should become a scholar. This also means that educational institutions should be increasingly engaged in research, and research organizations should necessarily perform a new, educational function, which has been realized only at the level of individual members of research teams. The production of new scientific knowledge and its transmission while learning will be, in the future, a common scientific and educational system working for sustainable development and functioning on the principles of this civilization model

PART 1

CHAPTER 1

INTERRELATION BETWEEN THE NATURAL-SCIENTIFIC KNOWLEDGE AND THE HUMAN KNOWLEDGE

CONTENTS

1.1 Problem of "Two Cultures" 2
1.2 Relations between Philosophy and Science 4
1.3 Features of Socio-Natural Laws 10
1.4 Integration of Science and Education 12
Keywords 14
References 15

ABSTRACT

The interrelation between the natural-scientific knowledge and the human knowledge is traced. The problem of "two cultures" is characterized. The relationship between science and philosophy is touched upon. The problem of the integration of science and education is posed.

1.1 PROBLEM OF "TWO CULTURES"

Despite the rising status of integrative trends in the natural sciences and in science, in general, the English writer and physicist C. Snow (1905–1980) formulated a thesis of the danger of opposing the natural-scientific culture and the humanitarian culture in the mid-20th century. The idea was that the development of the natural sciences and the humanities would not go to the desired level of integrity. Scientists' opinions on this point diverge. Some of them fix the further "gap" between the natural-scientific knowledge and the humanitarian knowledge, predicting approximation of an era of the "plurality" of cultures; they also predict the increasing trend of differentiation of the knowledge about nature and man. Others note the raising status of the humanitarian disciplines in relation to the natural and technological sciences. There are such who argue that the historically appeared subordinate position of the humanities in relation to the sciences of natural laws is preserved. While more than 30 years ago, during the emergence of debates on the "two cultures" problem, in fact, the absolute dominating status of natural sciences in the hierarchy of modern knowledge was spoken of, by the end of the 20th century, naturalists increasingly turn to the laws traditionally revealed by the human sciences. Moreover, this appeal is associated not only with searching for answers to philosophical, ideological, or social questions but also directly affect the scope of their professional activities: identifying the "limits" of penetration of biological sciences into the "mystery of the living" or analysis of the sociocultural consequences of the modern orientation of physical knowledge. Representatives of the humanities sometimes state (their position is often supported by naturalists feeling "restrictions" of the traditional natural approaches to cognition of nature and man) that only the humanitarian sphere of knowledge, related to the true spiritual values, would lead to cognition of nature and man in all their diversity. Nevertheless, the increasing interconnection and interdependence of natural sciences, technical sciences, and human sciences and art is obvious.

The relationship of the natural, technical, and humanitarian sciences (and art) does not mean the absence of differences between them. The perfection of Planck's or Einstein's theoretical constructions causes a feeling of admiration in a physicist, comparable, say, to the contemplation of the paintings by Titian or Gauguin. Moreover, the new elements introduced into music and painting by, say, Wagner or Cézanne, can be compared with discoveries in the field of natural sciences, which have led to the scientific revolution. If Isaac Newton were not born at his time, the corresponding laws of mechanics would have been discovered, no doubt, sooner or later, by another physicist. At the same time, works of art bear a stark reflection of the personality of their creator (this is typical, but to a lesser extent, for the sphere of science). The music by Mozart or Beethoven, the paintings by Velazquez or Dali, the novels by Theodor Dostoevsky, the prose by Böll are associated with these persons only. Consequently, the "gap," on the one hand, between scientific knowledge and human knowledge, and, on the other hand, between science and culture has a real basis. If the above differences between them were absolutized (e.g., the objectivity of scientific knowledge and the subjectivity of the humanities were exaggerated), by the end of the 20th century, emphasis is increasingly set on such elements which unite them.

The united, or a "third culture," is: (i) the type of sociocultural integrity formed in the process of overcoming the "gap" between different spheres of modern scientific knowledge and art, and (ii) going into such a level of sociocultural development of our civilization where the unity and interrelation of natural sciences, technical sciences, and human sciences would be obvious. The reality of this "third culture," that is, the feasibility of integrative trends in science, is supported by the level of development of modern knowledge. First, the traditional differentiation of scientific knowledge, characteristic of the natural, technical and humanitarian sciences, has prepared the ground for interdisciplinary links in the existing system. Second, the apparatus of modern scientific knowledge is actually adapted to the realization of integrative concepts, which are caused by the internal logic of the formation of science, the universal design, and methods of scientific thinking. Third, the universal (global) problems arising in the framework of civilizations in the late 20th century demand, for their solution, activation of processes associated just with integrative tendencies in the structure of science. Vernadsky's forecasts are coming true, who as far back as in the 1930s noted the reality of wiping off the borders between individual sciences, the suitability for scientists to specialize by problems rather than by sciences. In any case, the traditional disciplinary approach, identifying its specific limitations, is gradually replaced by the problematic approach, where the synthesis of scientific and practical concepts occurs in the context of solving certain tasks (or their system) of social practice. The transition from the disciplinary to problem-disciplinary development of science creates theoretical, methodological, and social prerequisites for constructive realization of integrative trends in the system of modern natural sciences, technical sciences, and human sciences. Going into the level of the "third culture" implies mutual enrichment of the principles and methods of reflection of the objective reality inherent in natural sciences, humanities, and artistic cognition. Conceptual thinking, predominant in science, and creative thinking, which determines the style of art, forms an interconnected unity, expressing the indivisibility and interdependence of science and art. At the heart of the "third culture" is the concept of the "unity of the world," which determines, ultimately, the unity of natural sciences and humanities. The unity of our knowledge of the world based on a unified system of methods is deduced from the unity of the scientific and sociocultural being. The idea of the common character of theoretical and sociocultural foundations of scientific knowledge thus manifests itself. Therefore, within the framework of the "third culture," the unity of scientific knowledge is not achieved by negation of the specifics of its different areas but is expressed in a variety of forms.

1.2 RELATIONS BETWEEN PHILOSOPHY AND SCIENCE

Philosophy has originated as a form of reflection and comprehension of reality. Philosophy and science are historically interrelated and oriented to solving a similar problem, namely, the identification of regularities in nature and the essence of man. Initially, especially within early philosophical systems, before the middle of the 19th century, philosophy existed mainly in the form of "natural philosophy," trying to interpret the "essence" of diverse things and phenomena of reality in their "mental" ("speculative") integrity. At the same time, developing philosophical systems tried not only to formulate the theoretical basis of ideology but also claimed to be the "highest level" of cognition, the role of "the science of sciences." As special sciences developed, philosophy was losing its former importance. The situation in philosophy was somewhat similar to the tragedy of Shakespeare's King Lear having distributed his wealth among his daughters. Special fields of knowledge were detaching themselves from the body of philosophy and tried to forget about the relation with it. "Positivist philosophy" was to relieve the situation. In the framework of it, the thesis of the necessity of converting philosophy to a special particular discipline with a relatively limited subject of research (the language of science, the methodology of scientific cognition, etc.) was developed. Positivists, especially post-positivists, predicted the "death" of philosophy in the traditional sense. The development of science has had a fundamental impact on philosophy, which was to find its place "under the sun." And in this "drama of ideas," philosophy was not lost and preserved its targets and a high status in culture.

The philosophy of science is such an area of scientific knowledge which studies the philosophical and methodological aspects of the development of natural sciences, technical sciences, and human sciences (including social sciences).

The interrelationship of philosophy, natural sciences, and other particular sciences is implemented in several ways. First, the philosophical analysis of different fields of science, interpretation of their achievements and identification of their possible philosophical and methodological consequences; second, the use of the philosophical and methodological apparatus (categories, laws, principles, etc.) for analysis of scientific knowledge, going into a new level of theoretical cognition; and third, the perception of the achievements and results of natural scientific cognition by philosophical knowledge. In several historical periods, the relationship between philosophy and science was posed and resolved in different ways. As will be discussed below, in the ancient period, the notion of the features of nature, arising in a generalized philosophical form, had the character of natural philosophy, with its predominantly speculative interpretation. Philosophy was treated as "the science of sciences," offering its a priori schema of cognition to other sciences.

During the Renaissance, the process of separating sciences from philosophy began, which had an impact on both science and philosophy. On the one hand, the object of philosophy was as though getting narrower, sections becoming independent sciences (mechanics, physics, etc.) were detaching themselves from it. On the other hand, it expanded because sciences needed their philosophical-methodological basis and a fundamental understanding of their results. In the early modern period, the differentiation of natural sciences continued, the process of separation completed, but, at the same time, the relationship between science and philosophy strengthened and went to a more fundamental level. Currently, philosophy gets not only empirical material for its own constructions from the natural sciences but also contributes to overcoming the theoretical contradictions arising in the emerging scientific knowledge.

Every historical period of the relationship between philosophy and science has its own style of scientific thinking, which is a specific system of principles, laws, and categories of the theoretical assimilation of objective reality. Analyzing the scientific achievements of the late 18th–mid-19th centuries (Kant–Laplace's cosmogonic hypothesis, the discovery of the cell, Darwinism etc.), Friedrich Engels revealed the evolution of the style of scientific thinking in his "The Dialectics of Nature." His concept was that the ancient era was characterized by a dialectical way of thinking in its spontaneous-naive form. In the context of the Middle Ages, the scholastic style of scientific thinking predominated, having been developed in the framework of religious discussions. It is underlain by the subordination of the scientific values to the religious ones (science is the "handmaiden of theology"). In the New Age, a metaphysical style of scientific thinking developed. On the one hand, within its framework, the dogmatism of scholasticism was overcome on the basis of the development of natural sciences, and, on the other hand, the capabilities of natural scientific cognition were absolutised. This led to some one-sidedness of the cognitive process, that is, to the exaggerated status of mechanistic laws and to the predominant analytical approach in comparison with the synthetic one in the study of nature. The spontaneously naive form of the dialectical style of scientific thinking allowed one to capture the picture of the phenomena of nature in the whole; however, some real connections and relationships between natural objects and phenomena were sometimes lost. On the other hand, the metaphysical style of scientific thinking puts emphasis on the particular scientific worldview provided by mechanistic natural sciences.

At the turn of the 19–20th centuries, the idea of the dialectical style of scientific thinking was developed by Vladimir Lenin in his "Materialism and Empirio-criticism. Critical Comments on a Reactionary Philosophy" (1908). It was shown that the views of a number of philosophers (e.g., the

German-Swiss philosopher Richard Avenarius [1843–1896]) and naturalists (e.g., the Austrian physicist Ernst Mach [1838–1916]), who advanced the theory of "criticism of empirics" (empirio-criticism), gave evidence of a more in-depth knowledge of nature rather than the "disappearance" of matter. It should be noted, however, that Lenin's criticism of empirio-criticism was mainly ideological. Prominent Russian naturalists have contributed to the philosophical and methodological understanding of science. The mathematician Nikolay Lobachevsky (1792–1856) justified the fundamental philosophical idea of geometric relations being dependent on the nature of material bodies. The physicist Nikolay Umov (1846–1915) considered the relationship of physical knowledge with the whole system of scientific concepts of the natural and social reality, sought to justify the ways of harmony of the "man–biosphere–cosmos." The chemist Dmitry Mendeleyev (1834–1907) was never closed in the circle of his professional interests, analyzing the direction of Russia's development. The plant physiologist Kliment Timiriazev (1843–1920), on the basis of his fundamental work on photosynthesis, was thinking of the positive prospects of civilization. In the human physiology by Ivan Pavlov (1849–1936), a system of theoretical principles was formulated, underlying the study of the human mind. The scientific works by Vladimir Vernadsky (1863–1945) is characterized by profound philosophical orientation.

The Soviet period of the development of the philosophical problems of natural sciences cannot be assessed unambiguously. In the 1930–1950s, the Soviet Marxist philosophers sometimes intervened in studies in particular sciences (with their criticism of "physical idealism," "Weismannism–Morganism," cybernetics, etc. from dogmatic positions), which prevented the rapid adoption of Western scientific achievements and undermined the relations between philosophers and naturalists for decades.

There was an unprecedented "explosion" of interest to the philosophical problems of natural sciences in the USSR in the 1960–1970s. Representatives of the natural sciences (less affected by the dogmatism of the past) came in philosophy. The domestic philosophy of science following the Marxist tradition, within its framework, was able to approach an adequate philosophical and methodological analysis of the contemporary trends of science. These were philosophical questions of natural sciences which became the area where a significant increment of knowledge occurred. First, the relationship between philosophers and naturalists restored and strengthened. Second, the philosophical and methodological problems of the whole system of modern scientific knowledge of nature (the philosophical questions of astronomy, physics, biology, cybernetics, computer science, space exploration, systems theory, etc.) were actively developed, receiving strong positive response worldwide. Third, within the framework of the philosophical questions of natural sciences, problems of the present stage of the science development were studied (the scientific and technological revolution, the "man–biosphere" relation, global issues, informatization, etc.).

In the West, the relationship between philosophy and science was actively developed in the framework of positivism (from Lat. *positivus*) since the middle of the 19th century. *Positivism* is a doctrine which denies, on the one hand, any cognitive value of philosophical studies, and, on the other hand, claims that only particular (empirical) sciences can be a source of real knowledge. The founder of positivism, the French philosopher Auguste Comte (1798–1857), put forward a thesis according to which metaphysics (philosophy) should be eliminated, and science should be limited to the description of specific phenomena. Positivism, having proclaimed the slogan "Science does not need any philosophy," absolutized the value of particular (positive) sciences in the cognition of objective reality. *Neopositivism* is the positivism of the 20th century, in its various forms and directions (logical positivism, scientific empiricism, etc.), continued to study the problems of the relationship between philosophy and science from the standpoint of formal logic, analyzing the structure of languages, and so forth. In the framework of logical positivism, the principle of verifiability (from Lat. *verifacare*—to prove the truth) was formulated. This principle means that the truth of any statement about the world must be ultimately established by comparing it with some sense data. In accordance with this principle, cognition cannot (and must not) go beyond sensory experience. Later, an adjusted version of this principle was formulated, originating from the necessity of partial coherence of sensory data and theoretical positions.

In contrast to logical positivism, the arising critical rationalism tried to identify fundamental mechanisms of links between the theoretical and empirical levels of cognition, to overcome the one-sidedness of positivism. In this context, one of the founders of critical rationalism, the English philosopher Karl Popper (1902–1994) developed the idea of the existence of "three worlds." Namely, the "first world" is the world of physical objects; the "second one" is the world of consciousness states; and the "third world" is the world of objective contents of thinking. In the traditional conception of science (Descartes, Berkeley, Hume, and Kant), the relations of the "second world" with the "first" one were mainly considered. Popper ascribes science to the "third world," comprising a set of scientific problems, disputable situations, hypotheses, rational schemes, and so forth. A tripartite structure of scientific research is provided, namely, scientific problem—hypothesis—experimental testing, which allows one to rationally organize his cognitive process.

Another principle, the principle of falsifiability (from Lat. *falsus* and *facio*), means checking the validity of theoretical propositions (hypotheses, laws, and theories) in the process of their disproof when comparing with data obtained in experiment. The basis of this principle is the formal logical relation, according to which any theoretical statement is deemed rebutted if its negation logically follows from a set of consistent statements of observation. This principle of scientific cognition is one of those which allow one to distinguish between science and pseudoscience.

The English philosopher of Hungarian origin Imre Lakatos (1922–1974) proposed the concept of "research program" as an alternative to Popper's theory of scientific knowledge. In its framework, the theoretical and logical foundations of science are distinguished, including the set of most important ideas, theories, and hypotheses. Three elements of the "research program" are fixed, namely: the "hard core," taken conventionally and therefore "conclusively as any decision to be made in advance"; the "positive heuristics," which dictates the choice of research problems; and the "protective belt of auxiliary hypotheses" to be put forward to justify the research program and to explain scientists' cognitive actions.

Paradigm shifts in science are made on the basis of changes in research programs rather than on the basis of changes in separate theories. The history of science is considered as the replacement of some scientific programs by other (competing) ones. The idea of the "research program" is treated as the original unit of measurement of the dynamics of science and cognitive activity.

In the framework of post-positivism, a "historical trend" in the philosophy of science has been developed. Its adherents (the Polish philosopher Ludwik Fleck [1896–1961], the American philosopher Paul Feyerabend [1924–1994], et al.), adhering to various viewpoints on the problem of the relationship between philosophy and science, however, agree, at least, that the traditional positivist (formal or structural) approach to the dynamics of science requires some corrections. The idea is that an adequate study of a scientific phenomenon implies not only exploring the internal processes in science but also analysis of the impact of some sociocultural factors related to the civilization development conditions on it. In essence, the formerly alternative research programs, "philosophical questions of natural sciences" and "philosophy of science," get substantially converged. The traditional opposition of natural sciences and human science is being overcome, and the formation of a new style of scientific thinking, based on the construction of a complete image of objective reality proceeds, that is, searching for the scientific pattern of the world.

Thus, the end of the 20th and the beginning of the 21st centuries demonstrate the improvement of the status of philosophy in the hierarchy of modern scientific knowledge and the failure of any ideas of its "death." It is philosophy which introduces the necessary humanistic element into the dynamics of natural sciences, not giving science to fall into scientism, allows comprehensive evaluation of the present status of science, and to identify the main directions of its development. Finally, the process of bridging the gap between the "two cultures" goes just under the auspices of philosophy.

1.3 FEATURES OF SOCIO-NATURAL LAWS

The formation of a united culture does not mean purely spiritual union of science and human thinking. This includes the formation of a unified socio-natural system of a global and even cosmic scale. In the socio-natural system, components of both social and natural origin and essence merge; they are in various kinds of links and relations to form some integrity. This socio-natural whole can develop in different directions, either harmonizing its relationships or destroying them. This problem most clearly arose in connection with the formation of the already mentioned disciplines of "synthetic" natural sciences and "mixed" socio-natural directions of scientific research, which include, for example, social ecology and other areas of research related to the interaction between nature and society. Two main approaches can be traced in this kind of research. One of them (which can be called traditional, or sociocentric) focuses on society, the impact of the negative consequences of its interaction with nature, and the activities to be implemented by mankind to harmonize its relationship with the environment. The second approach (unconventional, or naturocentic) focuses on the impact of society upon nature, on its protection from the damaging effects of man. The existence of these two approaches that have different options and variations associated, in particular, with various philosophical interpretations, is due to the fact that the two objects (or systems), society and nature, interact. Depending on which element is given priority, one or the other approach manifests itself, each of which is individually nonsystemic. Obviously, along with these one-sided approaches, a more general one is possible, which would consider the interaction of society (man) and nature as a certain socio-natural system. Within this approach, one cannot consider that the laws which social ecology begins to discover and study are individually either social or natural, and, therefore, a new kind of laws, namely, socio-natural laws, is introduced. It should be noted that the problem of laws, more general than purely social or natural ones, exists not only in the field of social ecology but also in all areas of knowledge closely related to the interaction of society and nature.

The laws of social ecology, in our opinion, should be considered not just socio-natural studied by social and natural sciences only. Machinery acts as a mediating link in the interaction between society and nature. Of course, we can assume that machinery interacts with nature "on the side" of society, and, in its essence, is of social character; therefore, one should not introduce it as a "third member" in the description of the laws of interaction between society and the environment. However, this is incorrect, and the introduction of machinery (production) significantly alters natural processes. The main feature of the processes occurring in the biosphere is due to the fact that in the system of cycles associated with the presence of three main functional elements of the biosphere (producers, consumers, and decomposers), there are the reproductive conditions for the existence of each of them. However, if production (machinery) is added therein as the following functional element, the evolution of this four-element system will aim to "excluding" the human industrial activity from the system or to limit its scope. A kind of rejection reaction proceeds, eliminating such an element from the biosphere organism which is dramatically different from the rest of its elements by functional properties. This implies at least two ways: first, a significant limitation of the devastating impact of production upon the biosphere; second, taking production activities outside the biosphere—into space. In epistemological terms, the noninvolvement of machinery in the "society–nature" system and that of the science studying the interaction between these components is also unacceptable. Ignoring machinery and engineering sciences simplifies the situation only apparently, because the laws of the art are not reduced to social laws, as well as to the laws of nature.

1.4 INTEGRATION OF SCIENCE AND EDUCATION

Any national doctrine of education has the transition to the sustainable development of society as a priority objective, which largely depends on the educational level of its citizens, their knowledge of the legal and ethical norms governing the relations of man to nature and society, and the ability to use this knowledge in their daily and professional activities, their ability to understand the essence of the current socioeconomic transformation, their commitment to the ideals, principles and ethics of sustainable development. The educational level of the population and the guaranteed opportunities of being educated are recognized as the basic human development indicators by the international community. The accumulation and reproduction of scientific knowledge, cultural values, and ethical standards proceed due to education. The basis of the educational process should be the concept of the development of the world as a whole, the idea of the development of humanity as part of the interaction process between society and nature, humanism, respect for the historical and cultural heritage.

The formation of a unified culture will impact on enhancing the relationship between science and education, and the relationship of such areas of the educational process which are still separated. This is a more systematic approach than one currently being implemented in Russia and many other countries. The systemic character of, for example, education for sustainable development is, among other things, in the fact that social-humanitarian, natural-scientific, and technological educations should be united into one. This, in particular, has begun to occur in the form of ecological education. Ecological education already appeared several decades ago, but its demerit was in that it was developed and is still developing as a new, diversified sector of education, still poorly oriented to the whole complex of the socio-natural goals of sustainable development. The future education will be of systemic advanced character, performing the socio-natural function of training the man to early activities for civilization survival in the context of the ongoing global anthropo-ecologic crisis and an exit from it by the transition to sustainable development. We call this new education as the education for sustainable development, or the noospheric education, which will be realized only in the case of free scientific thought "turning" to the sustainable development of the noospheric orientation.

Noospheric transformations in science will affect different groups of sciences differently. The natural sciences will be least affected, although the problems will be posed there in terms of the objectives of the transition to sustainable development. In addition, the natural sciences experience the effect of all other branches of knowledge in the process of the formation of their noospheric status, since the integrative and interdisciplinary relationships will be enhanced. As for the technical sciences, their substantial greening and humanization should occur. As noted, powerful integrative processes aimed at addressing the common scientific interdisciplinary problem of transition to sustainable development will be a characteristic feature of the future of the noospheric science. This will require a fundamental restructuring of the structure and functions of the major scientific organizations and their management. It is important to move the new orientation of science to education in order to create such conditions that would allow the formation of a new system of education in the 21st century, adequate to the new civilization strategy of integrated social and natural orientation. The role of education in society is not only in knowledge transfer from generation to generation but also in preparing the individual for the exit from possible global crises and disasters, which can be overcome by advancing knowledge and actions only. The inclusion of timing functions and orientation to future into education, along with knowledge transfer, would change its very understanding.

The education for sustainable development involves the observation of a number of new educational principles, technologies and innovations, the observation of a number of the following recommendations which should find their educational and methodical support:

1. It is advisable to revise curricula, programs, state educational standards of all disciplines and specialties from the standpoint of the problems of the future, with particular attention given to the ideas of sustainable development.
2. The inclusion of sustainable development ideas into the social and economic life of our country and the world community will require

a radically new public policy and strategies in education, government support of all educational institutions, projects and programs to actively develop and implement a new model of the civilization development.

3. Education in the 21st century should be designed to dramatically change the minds of people for several generations, to form new general civilized values which would largely contradict the stereotypes of people's thinking in the industrial and postindustrial society.
4. The formation of education for sustainable development should be reflected in an appropriate system of indicators and criteria for decision making, in the assessment of skills, knowledge and practices, psychological tests, and so forth. It is also advisable to develop indicators characterizing the degree of facilitating the transition to sustainable development of every educational institution, of various educational systems (higher education, vocational education, additional education, etc.).

Since the transition of every separate country, region, or locality to sustainable development is impossible in principle (the biosphere being a single entity), effective international cooperation, including the field of education, is mandatory. The new educational system should evolve as a system open in time and space, widely using distance learning methods, as part of the globalization of the educational system, aimed at the transition to a new civilization development strategy.

KEYWORDS

- **natural sciences**
- **humanities**
- **"two cultures"**
- **science and philosophy**
- **education and science**

REFERENCES

1. Goldfein, M. D.; Ivanov, A. V.; Malikov, A. N. *Concepts of Modern Natural Sciences. A Lecture Course*; Goldfein M. D., Ed.; RGTEU Press: Moscow, 2009. [Russ].
2. Goldfein, M. D.; Ursul, A. D.; Ivanov, A. V.; Malikov, A. N. *Fundamentals of the Natural-Scientific Picture of the World*; Goldfein, M.D., Ed.; SI RGTEU Press: Saratov, 2011. [Russ].
3. Goldfein, M. D. *Implementation of the interdisciplinary principle in the ecological education of humanitarian students*, Proceedings XXIV Int. scientific-method. Conf Pedagogical management and progressive technologies in education, Penza, 2012. [Russ].
4. Mirinov, V. V. *Modern Philosophical Problems of Natural, Technical and Socio-Humanitarian Sciences: A Textbook for Postgraduates*; Mirinov, V. V., Ed.; Gardariki Press: Moscow, 2006. [Russ].
5. *Philosophy of Modern Natural Sciences: A Textbook for Universities*; Lebedev, S.A., Ed.; FAIR Press: Moscow, 2004. [Russ].
6. Il'insky, I. M. *Between the Future and the Past: Social Philosophy of the Occurring*; MGU Press: Moscow, 2006. [Russ].
7. Kapustin, V. S. *Synergetics of Social Processes: A Textbook*; MEI Press: Moscow, 1999. [Russ].
8. Goldfein, M. D.; Kozhevnikov, N. V.; Ivanov, A. V.; Kozhevnikova, N. I.; Malikov, A. N.; Timush, L. G. *Fundamentals of Ecology, Life Safety and Economico-Legal Regulation of Nature Usage*; RGTEU Press: Moscow, 2006. [Russ].
9. Ursul, A. D. *Education for Sustainable Development: Scientific Basics*; RAGS: Moscow, 2004. [Russ].
10. Ursul, A. D. *Russia's Transition to Sustainable Development. Noospheric Strategy*; Noosphera Press: Moscow, 2004. [Russ].
11. Danilov; Yu, A. *Wonderful World of Science*; Progress-Traditia: Moscow, 2008. [Russ].
12. Il'in, I. V.; Ursul, A. D. *Global Evolutionism*; Moscow University Press: Moscow, 2012. [Russ].
13. Polak, L. S.; Mikhaylov, A. S. *Self-Organization in Non-equilibrium Physicochemical Systems*; Nauka: Moscow, 1983. [Russ].
14. Prigogine, I.; Stengers, I. *Order Out of Chaos. Man's New Dialogue with Nature*; Heinemann: London, 1984.
15. Trubetskov, D. I. *Introduction to Synergetics. Chaos and Structures*; Editorial URSS: Moscow, 2004. [Russ].
16. Trubetskov, D. I.; Mchedlova, E. S.; Krasichkov, L. V. *Introduction into the Theory of Self-Organization of Open Systems*; Fizmatlit: Moscow, 2002. [Russ].
17. Ursul, A. D.; Ursul, T. A. *Universal Evolutionism*; RAGS Press: Moscow, 2007. [Russ].
18. Crutchfield, J. P.; Farmer, J. D.; Packard, N. H.; Shaw, R. S. Chaos. *Sci. Am.* **1986,** *254*(12), 46–57.
19. Ervin László. *The Age of Bifurcation: Understanding the Changing World*; OPA: Amsterdam B.V., 1991.
20. Turner, M. S. Origin of the Universe. *Sci. Am.* **2009,** *301,* 36–43.

21. Ursul, A. D. *Scientific World Pattern of the 21st Century: Dark Matter and Universal Evolution*; Eurasia Safety: Moscow, 2009.
22. Cherepashuk, A. M.; Chernin, A. D. *Universe, Life, Black Holes*; Vek-2: Friazino, 2003.
23. Chernin, A. D. *Cosmology: Big Bang*; Vek-2: Friazino, 2005.
24. Lighthill, J. *The recently recognized failure of predictability in Newtonian dynamics*, Proceedings of the Royal Society, 1986; Vol. A407, pp 35–50

CHAPTER 2

SOME GENERAL TRENDS IN THE DEVELOPMENT OF MODERN NATURAL SCIENCE

CONTENTS

2.1 Nonlinear Science in the Modern Scientific Picture of the World 18
2.2 The Concept of Universal Evolutionism 28
2.3 Fundamental Role and Practical Significance of Information in Nature and Society 37
2.4 The Expanding Universe 48
2.5 Nonlinear Dynamics in Geosystems 53
2.6 Dark Matter (Energy) and Universal Evolution 63
Keywords 82
References 82

ABSTRACT

The trends of development of science at the present stage are characterized. The role of nonlinear science in the modern scientific picture of the world is emphasized. The concept of universal evolutionism is described. The role and importance of information in nature and society are highlighted. The use of nonlinear dynamics in geosystems is shown. The relation between the dark matter (energy) and the universal evolution is demonstrated.

2.1 NONLINEAR SCIENCE IN THE MODERN SCIENTIFIC PICTURE OF THE WORLD

> *Synergetics can be considered as a forum, on which scientists of different disciplines have met to exchange their ideas of how to treat big systems.*
>
> — *Hermann Haken*

An appropriate scientific apparatus is required to describe the global evolutionary processes. Nonlinear science is the core in the formation of a new scientific picture of the world (the terms synergetics, nonlinear dynamics, and the science of complexity are often used as well), with an emphasis on the theory of self-organization.[1]

President of the International Union of Theoretical and Applied Mechanics J. Lighthill wrote the following in an apologetic tone (Lighthill, 1986).

---

[1]For example, Alwyn Scott's book describing the basic model of nonlinear science, their properties and methods of their study is named "Nonlinear Science: Emergence and Dynamics of Coherent Structures" Oxford: University Press, 2003. In fact, explaining the title of his book, the author writes: "It is known that in the case of considering non-linear systems, the whole is more than the simple sum of its parts and is manifested in the formation of new structures that are spatially or temporally coherent." (One of the definitions of coherent structures (CS), say, in turbulent flows, is as follows: CS in turbulent flows are long-lived and ordered large-scale formations on the background of small-scale turbulence for this type of flow. The problem of the emergence and self-maintenance of CS is common for hydrodynamics, plasma physics, ecology, biophysics, etc. (see Scott's book, wherein he classifies ring and spiral waves, nerve impulses, solitons, etc. as CS). "The understanding of this fact was a turning point in the history of science and has led to a change in our ideas of the internal organization of scientific knowledge. These new structures are very large entities having their own unique properties, characterized by certain times of existence and entering into specific interactions with each other. Since, these interactions are also nonlinear, the new dynamics induces other structures on a higher level of description" (Scott, p. 14).

"We are all deeply conscious today that the enthusiasm of our forebears for the marvelous achievements of Newtonian mechanics led them to make generalizations in this area of predictability which, indeed, we may have generally tended to believe before 1960, but which we now recognize were false. We collectively wish to apologize for having misled the general educated public by spreading ideas about the determinism of systems satisfying Newton's laws of motion that, after 1960, were to be proved incorrect ...."

What did Dr. Lighthill apologize for?

There is no need to prove that the main power of science is its ability to predict. For example, eclipses can be predicted for hundreds of years from the laws of gravity. However, weather cannot be accurately predicted for some reason, although the atmospheric currents obey the physical laws as strictly as the planets do. It is impossible to accurately predict the course of a mountain stream or the trajectory of a ball in billiards. What is the matter? Why a clear link between the cause and the effect is lost in these phenomena? In fact, it was believed until recently that the accurate predictability of, say, weather could be achieved: one only needs to collect and process more information.

Laplace's (1749–1827) ideas are also known, who wrote as far back as in 1776 (quoted from Crutchfield, 1987): "*We ought then to regard the present state of the universe as the effect of its anterior state and as the cause of the one which is to follow. Given for one instant an intelligence which could comprehend all the forces by which nature is animated and the respective situation of the beings who compose it—an intelligence sufficiently vast to submit these data to analysis—it would embrace in the same formula the movements of the greatest bodies of the universe and those of the lightest atom; for it, nothing would be uncertain and the future, as the past, would be present to its eyes.. The human mind offers, in the perfection which it has been able to give to astronomy, a feeble idea of this intelligence. Its discoveries in mechanics and geometry, added to that of universal gravity, have enabled it to comprehend in the same analytical expressions the past and future states of the system of the world. People owe to the powerful devices they use, and a small number of relations used in his calculations. However, the ignorance of various causes of certain events, as well as their complexity, combined with the imperfection of analysis prevents us to achieve the same certainty with respect to the vast majority of events. So it is that we owe to the weakness of the human mind one of the most delicate and ingenious of mathematical theories, the science of chance or probability."*

Laplace believed that the laws of nature imply complete predictability and strict determinism, while randomness is caused just by the imperfection of our observations. And it seemed that he was right in principle for more than a hundred years. The application of Laplacian determinism to social phenomena has led to the well-known philosophical conclusions of the absence of people's free will, predestination of their behavior. It was believed that a single human's effort could not have a significant impact on the course of history, that an individual's activity did not affect social processes. Development could only be progressive, without any alternatives. Moreover, it was believed that the past could be of historical interest only, and any return to the past had some new basis. As for alternatives, they might only be occasional deviations from the mainstream, being subject to it.

The picture of the world, drawn by the classical mind, represented it rigidly connected by cause-and-effect relationships, being dependent on neither small effects at the lower levels of existence nor small influences from space. Moreover, causative purposes are linear, and the consequence is, if not identical to the cause, at least proportional to it. The course of history could be calculated from cause chains indefinitely stretched into the past and future. Evolution is retro-cognizable and predictable. The present is determined by the past, and the future is by the present and the past. The "control action → desired result" chain was considered the traditional approach to the management of complex systems with an implication of "the more energy you spend, the more the return will be." Although it was already known that many efforts could come to naught or even did harm if contradicted to the own direction of self-organization of complex systems. The linear mathematical apparatus brought to a high degree of perfection (part of the mathematical culture of, primarily, physicists) was an additional justification for the domination of linear representations.

The term "nonlinearity" itself bears an imprint of the once widespread misconception of the allegedly dominant role of linearity in the world around us: its creators considered linearity as primary, while nonlinearity was seen as something secondary, derived from linearity, and they defined it by negation. The modern physicist, had he occasion to reformulate the definition of such an important thing as nonlinearity, would have done, most likely, differently, considering nonlinearity as more important and widespread of the two opposites, would determine the linearity as "non-nonlinearity."

The mathematical behavior of nonlinear systems is described by nonlinear equations containing the studied values raised to powers above unit (or coefficients depending on these values). According to Ilya Prigogine (1917–2003), one of the founders of the science of the complex, some control parameter changes in the differential equations describing the evolution of a system. At a certain value of this parameter, at least two ways of evolution of the system occur, corresponding to a bifurcation, that is, the transition to a new dynamic mode, to another world (bifurcation from the Latin *bi*—double and *furca*—fork, which means a fork or split in half) (Prigogine, 2002).

The science of the complex behavior of nonlinear systems, their evolution in time and space, vibrations and waves in them, the development of various instabilities and their stabilization, the appearance of chaos and the birth of structures is called *nonlinear science*. The theory of nonlinear oscillations and waves is most important in nonlinear science. Nonlinear effects are described by nonlinear functions, which enable one to describe oscillations, resonance emissions, saturation, and so forth. Linear functions, on the contrary, are "able" of either growing or decreasing: the same increment of a linear function's argument corresponds to the same increment of this function itself. A nonlinear function behaves differently: the same increment of its argument may well correspond to different behaviors of the function. The difference between the two worlds, linear and nonlinear, proceeds from this. What does nonlinearity mean for natural sciences? "*Whatever field of natural sciences the nonlinearity of phenomena would arise in, it is deeply functional. The nonlinearity in physics is an account of all kinds of interactions, feedbacks and fine effects, which escape from the linear theory's coarser network. The nonlinearity in chemistry reflects feedback in the innermost reaction mechanisms. The nonlinearity in biology is full of high evolutionary sense: only strong nonlinearity favors biological systems ... to hear a creeping snake's rustle and not to be blinded by a close flash of lightning. Those biological systems, which were not able to cover the range of vital environmental objects, simply have died out, being unable to withstand the struggle for existence*" (Yu. Danilov, 2008).

The understanding of the world being not linear has led to the destruction of many traditional, immutable ideas. Some aspects of the new view of the world from the viewpoint of nonlinear science are as follows:

1. There are several alternative ways of development for complex systems (after any bifurcation point). If it was believed earlier that there were historical regularities determining the global process as the course of history in only one direction known beforehand, it has become clear now that the historical process includes alternatives and options, presented, for instance, in the form of several political platforms. At the bifurcation points, that is, at certain stages of evolution, a certain predestination of the development of processes occurs. Moreover, the present state of a system is determined by not only its past, its history, but begins to form from the future in accordance with some expected state. Nonlinear science is typically associated with natural sciences, but this is not true. Apparently, it seems equally important to master the concepts of the modern picture of the world for humanitarians, the more so if (s)he is also going to study global processes, and, say, for a physicist.
2. Nevertheless, the discovery of deterministic chaos was the most important discovery in nonlinear science. Beyond the bifurcation point, the system can exhibit chaotic behavior, obeying certain laws. There is order in chaos! The most amazing thing is that, under certain conditions, the motions of very simple systems get not only similar but undistinguishable from random motion. How could this be explained? There are several explanations, the so-called algorithmic approach in the theory of dynamical systems being one of them. To explain this, let us imagine the phase space, the usual space of coordinates, and space of velocities (or pulses) of the system.

If a dynamical system (even very simple) is put in the phase space, its role will consist in transforming the randomness of its initial conditions into the macroscopic randomness of the system's motion. If the so-called local instability exists in the system, when close trajectories diverge, at some stage the movement is determined by the details of the initial conditions and is highly dependent on them. Assume that the phase space is limited. Then, sooner or later, the diverged trajectories will go back to each other. And this will happen many times. As if a kind of mixing of the phase space occurs, manifesting itself in the chaotic motion of the phase trajectories.

The second law of thermodynamics and the laws of biological evolution were discovered almost simultaneously in the mid-19th century.

Rudolf Clausius formulated the essence of equilibrium thermodynamics as: "The energy of the universe is constant. The entropy of the Universe tends to a maximum." The Darwinian basic laws of biological evolution appeared at the same time, being fundamental importance for science and life. A question of the compatibility of classical thermodynamics and the idea of time evolution of the spontaneous formation of more and more complex structures then arose. Indeed, thermodynamics stated that the entropy of any closed system increased with time and reached its maximum value when the system reached the state of thermal equilibrium. In other words, if the system is left to its own resources, it will always seek to achieve the state with a minimum degree of order permitted by the initial conditions. But this contradicts the possibility of the continuing process of structure formation. If biological systems are spoken of, constant exchanging of energy and substance with the environment is required for their life, that is, they are open and do not obey the laws of statistical physics for closed systems. In an open system, there may be an increase in the degree of its ordering, and its entropy reduces, which is quite consistent with the laws of nonequilibrium statistical physics formulated by Boltzmann, Gibbs, and Einstein. But they were not used in biology; moreover, the possibility of explaining biological phenomena from the standpoint of physics was questioned. Probably, these doubts were first dispelled by Erwin Schrödinger (1972). Just in his book "What is life? The Physical Aspect of the Living Cell," it was shown that the main feature of biological objects, that is, the exchange of energy and substance with the surrounding nonequilibrium medium, creates the necessary preconditions for self-organization phenomena. This will be discussed in more detail in Chapter 4 in relation to the consideration of open systems. The openness of a system means that it contains sources and sinks; substance, energy, and information exchange with the environment.

The formation of spatial order from disorder, the appearance of complex 3D structures in a homogeneous medium, and so forth are associated with the phenomenon of self-organization. What is self-organization?

The emergence of the Benard cells in a fluid layer heated from below is a classic example of self-organization. This is a structure of hexagonal prismatic cells arising from the "struggle" of instability and dissipation (friction, viscosity) in the medium. As a result of convection, instability occurs, leading to increased perturbations of the velocity field and temperature within a range of spatial scales, which is only possible in the presence of dissipation. As a result of competition, only a grid of a certain scale survives. Hexagons are formed by synchronization of the phases of the grids with different spatial orientations.

The process of generating electromagnetic radiation in lasers is another example of self-organization. The continuous laser is a highly nonequilibrium open system formed by active atoms (for example, a mixture of helium and neon atoms) interacting with the electromagnetic field of the resonator (an oscillating system consisting, e.g., of two mirrors, one of which is translucent). This system is disturbed from the equilibrium due to the constant energy influx from an external source of incoherent optical pumping. The incoming energy is not stored in the laser system but and continuously leaves it as electromagnetic radiation and heat flux. Upon reaching a certain value of the pump power, the laser radiation becomes coherent, that is, begins to be a single wave train where the phases of the waves are highly correlated at macroscopic distances. This transition to coherent oscillations can be interpreted as self-organization, establishing order in the laser system.

We note in passing that Hermann Haken (b. 1927), who coined the term "synergy" (synergetics), departed from the effects of lasing in his creation of a new scientific lead. Even from a consideration of these two, in essence, classical examples it can be understood that the phenomena of self-organization are associated with the formation of spatial order from disorder, that of complex 3D structures in a homogeneous medium. In the future, we will follow this definition: self-organization is the establishment of spatial structures in a dissipative nonequilibrium medium (which may evolve over time), whose parameters are determined by the properties of the medium and are weakly dependent on the source of disequilibrium (energy, mass, etc.), the initial state of the medium and the boundary conditions.

All complex natural systems, from galaxies to cells, from a cyclone in the atmosphere to a whirlpool in the creek, are open and dissipative systems. Moreover, almost all of them are nonlinear. The global oceans and the Earth's climate system, including the atmosphere, hydrosphere, and cryosphere (ice and snow cover) can serve as examples of such systems. Self-organization processes should go in these systems. The birth of hurricanes and typhoons, oscillations of the boundaries of ice sheets on land and at sea, the formation and decay of fronts in the atmosphere and the ocean, processes in the Earth's mantle are undoubtedly examples of self-organization. Not fewer examples of self-regulation can be found in human society as well.

It should be emphasized that the phenomena of self-organization in nonequilibrium systems are fundamentally different from ordering phenomena (phase transitions) in equilibrium systems. In the latter cases, order increases with decreasing temperature: a fluid is converted into a solid with decreasing temperature, the spins of individual atoms can be arranged to form ordered structures characteristic of ferromagnetic or antiferromagnetic materials, the transition to the state characteristic of the phenomenon of superconductivity can occur in metals. The nature of equilibrium phase transitions and that of self-organization effects in open systems are obviously different. But there exists a profound analogy between these phenomena: in both cases there is a process of restructuring or the emergence of order, if the latter is understood as symmetry breaking. Let us return to the above examples. Indeed, homogeneous liquid is more symmetric than the same liquid after the occurrence of even regular convective flows in Benard's experiments; the crystal formed on its cooling is less symmetrical than the original liquid; a paramagnet with a random orientation of the magnetic moments of individual atoms is more symmetrical that a ferromagnet, where all these moments are aligned in the same direction and there is no spatial isotropy. The homogeneity of the medium, that is, symmetry with respect to translations in space or in time, violates the occurrence of any spatial or temporal structure. Self-organization is closely related to another global natural phenomenon—turbulence.[2] Both effects are observed in strongly nonequilibrium systems of flowing type.

In the Benard problem, at high temperature differences, the system switches into the turbulent convection mode, while the intensity of the external impact is small, and the state of the system (open system) is close to equilibrium. In other words, under increasing intensity of the external influence, the transition occurs from the mode of thermal equilibrium to the turbulent one. The transition to turbulence may occur abruptly or may take a range of values of the parameters characterizing the degree of external influence upon the system under consideration. In the latter case, the transition to turbulence is accomplished by successive complication of regular structures, whose formation is self-organization in nonequilibrium open systems (Trubetskov, 2004).

---

[2]One of the definition of turbulence sounds as "Turbulence is such a state of the medium in which motions (turbulent fluctuations) of different scales are excited, with an energy transfer among them. Under the scale is understood the order of the distances over which the speed of motion changes significantly.

The types of the ordered behavior of strongly nonequilibrium open systems may be different. This may be, for example, dissipative structures, which are formed and stored by energy and substance exchange with the environment in nonequilibrium conditions. Another type of regular behavior is periodic self-excited oscillations or, for distributed systems, their analogues—autowaves.

Studies in recent decades have shown that self-organization, that is, the origin, evolution, and death of macroscopic structures under nonequilibrium conditions, is characteristic of physical, chemical, biological, social, and other systems. It is important that the establishment of general laws of self-organization opens up the prospect of designing complex artificial self-organizing systems. Furthermore, opportunities may appear to control the evolution of processes in already existing natural systems, and the possibility of creating artificial beings with predetermined behavior and properties. This will lead to radical changes in the world.

The concept of self-organization. The processes of self-organization are considered in synergetics as key ones in the development of complex systems. The systems must comply with the following provisions:

1. Self-organization is the process of evolution of a system from disorder to order. Naturally, the entropy of the system in which self-organization occurs should decrease. However, by no means this contradicts the law of entropy increasing in closed systems, that is, the second law of thermodynamics. Self-organization processes occur in open systems. If self-organization occurs in a closed system, it is always possible to outline an open subsystem, in which self-organization proceeds; at the same time, disorder increases in the closed system as a whole.
2. Self-organization proceeds in such systems whose state is essentially distinct from equilibrium at present. Equilibrium disturbance is caused by external influences. In the example with Benard's convection cells, the external influence is shown as heating of the vessel, which results in distinction of the temperatures in separate macroscopic areas of the liquid. In electric generators, an external influence (voltage created by a source) leads to a nonequilibrium distribution of electrons. The state of a system far from equilibrium is unstable with respect to a state near equilibrium, and, owing to this instability, processes leading to structure formation occur.

3. Self-organization is only possible in systems consisting of a large number of particles. In some cases it is obvious; for example, macroscopic spatial structures contain a plenty of atoms and molecules. However, if we consider the self-oscillations of populations, it is possible to state that such self-oscillations are impossible at small number of individuals in the population. The matter is that the emergence of fluctuations (macroscopic inhomogeneities) is only possible in systems with many particles.
4. Self-organization is always related to spontaneous symmetry reduction. A beautiful symmetric snowflake has, nevertheless, lower symmetry than unstructured water vapor. The ideas of such changes in symmetry have gained good development in the modern theory of the microworld, and in the description of phase transitions in physics (e.g., the liquid–crystal transition). In general, the processes of self-organization are in many respects similar to phase transitions, so they are often called as kinetic phase transitions. The difference is in that microstructures (e.g., a crystal lattice) emerge upon phase transitions, while the system remains homogeneous in a macroscopic volume.

2.1.1 PRINCIPAL STATEMENTS OF THE SYNERGETIC CONCEPT OF SELF-ORGANIZATION

1. The objects of study of self-organization are open systems in a nonequilibrium state, which are characterized by intense (flowing) exchange of substance and energy between the subsystems and between the system and its environment. A specific system is immersed into its environment, which is also its substrate.
2. The environment surrounding the system should be considered as a set of its constituent objects being in dynamics. The interactions among the objects under study in the environment are characterized as short-range (contact interactions). Objects can be realized in a physical, biological, geological or other lower-level medium, characterized as gas-like, homogeneous or solid-like. Long-range interactions, field and indirect (informational) ones, are realized in the system.

3. The processes of organization and self-organization are distinguished. They have a common feature, namely, the increase in ordering owing to proceeding of the processes opposite to establishment of thermodynamic equilibrium of independently interacting elements of the environment (and moving away from chaos by other criteria). Organization, unlike self-organization, can be characterized, for example, by the formation of homogeneous stable statistical structures.
4. Self-organization results in the emergence, interaction, inter-assistance (cooperation, coevolution), and even regeneration of some dynamic objects (subsystems) which are more complex, in information sense, than those environmental elements which they arise from. The system and its components are essentially dynamic entities.
5. The directness of self-organization processes is caused by internal properties of the objects (subsystems) in their individual and collective manifestation, and influences from the environment which the system is "immersed" into.
6. The behavior of the elements (subsystems) and of the system as a whole is essentially characterized by spontaneity, that is, behavioral acts are not strictly determined.
7. Self-organization processes occur in the environment along with other processes, in particular, those having an opposite direction; and separate phases of the system's existence can either prevail over the latter ones (progress) or concede them (regress). The system as a whole could have a steady tendency or undergo fluctuations to progressive evolution or to degradation and disintegration. Self-organization can be based on the process of transformation or disintegration of the structure, which has emerged earlier as a result of some reorganization process.

2.2 THE CONCEPT OF UNIVERSAL EVOLUTIONISM

We are on the threshold of a new culture, the synthesis of global spiritual consciousness and global scientific knowledge.

— *Nicolai Moiseyev*

2.2.1 PRINCIPLES OF DEVELOPMENT AND EVOLUTIONARY NATURAL SCIENCE

The notion of "change" is known to describe the state of matter in motion. The notion of "movement" is broader than that of "development." If movement is any process of changes in matter, then development represents a certain type of changing, which has some essential features. Development is understood as the direction of change in the content of a process, phenomenon, or material object. Directedness is the most important and invariant feature of development, which is, in contrast to movement, lies in the fact that development is characterized by a single, internally interconnected line. Development processes have two main directions: either from the simple to the complex, from the lower to the higher, or vice versa—from the more complex to the simpler. Such a direction of development, which is characterized by increasing complexity, organization or transition to the higher, is called progressive development. That type of development which is characterized by the transition from the higher to the lower, is regressive. In some cases, when studying the development of matter in the Universe, evolution and progress are mentioned, but, in general, these notions are not the same. The concept of universal evolutionism uses the notion of evolution in the same sense that the notion of development, that is, as a directional change in the content of the object. The concept of development has integrative character and promotes both the process of synthesis of natural-scientific, humanitarian and technical knowledge and strengthening of the interrelation among the sciences about nature and the man, which leads to the emergence of new generalizing and often universal notions, concepts, and theories on development. The penetration of the idea of development into science was going through overcoming metaphysical ideas, and modern science, primarily its natural and social topics adopted this idea in varying degrees. The modern science is not mere adequate reflection of the phenomena and structures of matter and the laws of functioning—but also that of their changes, development, and reflection of the regularities of evolution of material entities. This situation is typical for individual branches of science in which the idea of development penetrated at different times. If the idea of development, mainly due to the works of Darwin, was adopted by biology as early as in the 19th century, a different picture was observed, say, in astronomy. Undoubtedly, the Kant–Laplace cosmogonic hypothesis (the formation of the planetary system from a cloud of diffuse matter) did introduce these ideas into astronomy before the evolutionary theory did it into biology. However, the concept of progressive development was applied to the solar system only, not to the whole Universe. In biology, however, evolutionary knowledge actually was immediately extended to the entire biosphere. That is why, right up to the 20th century, astronomy presented the Universe as static (in contrast to the modern image of an evolving Universe).

One of the major questions at the focus of the concept of universal evolutionism is about the degree in which the terrestrial (geocentric) and space (cosmic) aspects of evolution are combined in the problem of development. If the ancient thinkers simply combined both terrestrial (including humans) and space issues, then the cosmic and terrestrial problems were developed in different (as if parallel) planes.

The ideas of development could be combined just in the framework of particular sciences, and, especially, in geology, where evolution was considered from geocentric positions (up to the works of Vladimir Vernadsky). This fact points to limited capabilities of analysis of the problem of development in particular sciences, which is due to the lack of the broadest view therein, peculiar to philosophical knowledge only. It is obvious that a prerequisite for combining the terrestrial and cosmic aspects of evolution is a more and more in-depth study of the idea of development by philosophy, which combined and summarized particular scientific and universal (general) development ideas. This synthesis shows that the terrestrial and cosmic development processes are not a kind of unrelated natural processes but aspects of a single entity, the *Universum*, under which the Universe as all real things available to observation in a wide philosophical sense is understood. Therefore, the scientific and philosophical outlook and methodology have appeared that sound basis which has tied the terrestrial and cosmic, particular and general, and so forth development processes into a scientific concept of evolution. Such a synthesis of the development aspect happens now, but at a new qualitative level, caused by the revolutionary advances in astronomy and space exploration.

2.2.2 UNIVERSAL EVOLUTIONISM AS A GENERAL SCIENTIFIC PRINCIPLE

The ideas of evolutionism appeared, in one form or another, in ancient times, when thinkers began to think about the development of the world and humanity's place in it. Evolutionism as a scientific outlook appeared in the 18th century only (Kant, Herder, Buffon, et al.), and it is more typical for the 19th century, especially in Hegel's doctrine of the universal forms of development. The term "evolutionism" appeared due to the Englishman Matthew Hale, who introduced it in relation to the human body in 1677. In a broader biological sense, it was used by Walter Baldwin Spencer (1852), referring to the development of all living matter.

In the second half of the 20th century, problems on a global scale appeared, requiring a more meaningful explanation of the processes occurring in the world. It was necessary to move from the abstract dialectical pattern of development to a universal model, which would proceed from the basic concepts of development in the natural, social and other sciences, and especially—in the interdisciplinary research, without which it would be impossible to form the concept of universal evolutionism in principle.

Currently, general scientific laws and trends of development are sought rather than universally philosophical ones, that is, the identification of trends and patterns of development of all things which are of universal invariant nature and can pretend to explain the appearance of humanity and to predict its further interaction with the nature of the Earth and the cosmos. Universal evolutionism should be distinguished from global evolutionism, whose essence is limited to the planetary-scale evolution and focuses on revealing stable trends, principles, and laws as universals of heterogeneous structures of matter. Universal evolutionism should focus on common invariant trends and regularities of development, which manifest themselves at different levels of matter organization. This concept should be underlain by the idea of the universality of the evolution of various structural levels, identifying the universals of evolving structures. Universals (from Latin *Universalis*) are general concepts with philosophical content. Under evolving structures are understood certain structures or their combinations, which develop in a certain direction and are included by the researcher into the process of evolution.

The idea of universal evolutionism, most common in recent years, is due to synergetics, which has extended many of its principles and laws from inanimate nature to wildlife and human society. This approach to universal evolutionism uses the concepts and methods of synergetics as an interdisciplinary research field to establish laws of self-organizing systems (as the spontaneous emergence of complex ordered structures).

From the standpoint of evolutionary natural sciences, progress in nature can be traced back to the elementary particles. Atoms, as compared with elementary particles, are more complex and organized systems; their properties are not identical to those of the elementary particles. The transition from the elementary particles to atoms, from one qualitative level to another one, from the relatively simple to the complex, is one of the first known steps in the ascending branch of the evolution of inanimate nature. Molecules are the next step in the development of material systems; their different species was significantly more than atoms (approx. several millions). The molecular (chemical) evolution has resulted in the appearance, in particular, of complex organic compounds such as polysaccharides, polynucleic acids, and proteins. A new quality of a complex-ordered, in space and time, system of organic matter directly related to the emergence and development of life has appeared at the top of the chemical evolution. From the primary organisms which might resemble modern viruses, unicellular, then multicellular organisms have evolved—the whole hierarchy of levels of biological evolution, during which microorganisms, plants, and animals emerged. Together, they accounted for 30–100 million different species. And the main line of progress was in the development way of the animal world only, which has led to the emergence of higher mammals—primates (apes and humans).

Further, most intense evolution proceeded on planets, on some of them life (the biosphere) appeared with the organism as its structural unit, then a society (the sociosphere) with the man as its structural unit. Interconnected links, structures (systems) and structural elements, constitute the so-called double-stranded series of progressive evolution (or an evolutionary series).

Consider a range of conditions inside which structures and structural elements of the stages of development can exist in nature. Elementary particles exist within a very wide range of conditions, virtually everywhere, except the physical vacuum (they can be in it—but cannot constitute it). One can talk about a certain probability to meet an elementary particle in a finite space volume of the observable Universe. We denote this probability by *Pe*. Atoms exist within a narrower range of conditions: only on planets, in outer space, and on the surface of some classes of stars. Molecules do exist there as well, but the range of their existence conditions is even narrower than for atoms. This is due to the fact that extension of, for example, the temperature range may lead to dissociation of molecules to atoms (or free radicals or ions), which is due to the nature of the chemical bonds in molecules. Denoting the probability of finding atoms and molecules in a given finite space volume through Pa and P_{mol}, respectively, one can write the following inequality, reflecting the narrowing of the range of existence of the corresponding structural elements:

$$Pe > Pa > P_{mol}$$

At last, the probability of the existence of organisms in the environment, P_{org} is less than P_{mol}. These inequalities characterize the probability of not only structural units but also of their corresponding structures.

With the reduction of the range of existence of the development stages in nature (of course, in the observable Universe), the space volume occupied by structures decreases, that is, $V_M > V_3 > Vn > V_{biosph.}$, where Vi are corresponding volumes of distribution of the metagalactic, stellar, planetary, and biological structures.

Similar inequalities can be written for the mass (Mi) and energy (Ei) characteristics of the structures of stages:

$$M_M > M_3 > Mn > M_{biosph}$$

$$E_M > E_3 > En > E_{biosph}$$

using the well-known Einstein's mass–energy relationship ($E = mc^2$). All these inequalities express the relevant laws (universals) of the evolutionary series. Some relationship exists between these trends in the development of nature.

Information which can be stored, processed, transferred, and so forth is considered a universal criterion to unequivocally determine many characteristics of the development of material objects in nature. When studying material objects, phenomena, and processes, of primary interest is information storage, that is, only that part of information which, according to Léon Nicolas Brillouin, is called as *fixed*. It may not be transferred from one object to another, since the main thing consists in that it acts as a certain characteristic of a rather steady distribution or a variety of elements in the given structure of object. It is meaningful to analyze only finite, limited in space in time systems and at a certain level because without these restrictions, the quantity of information (or the quantity of variety) in any object, owing to the inexhaustibility of matter and its development and the general relation among all phenomena, is infinitely great. The assumptions taken enable one to detect some quantitative information regularities of systems. On the progressive line of the evolutionary series, the diversity, number of types of structural units—different representatives of the levels—increases. Given the variety of species is due to their information content, it can be concluded that the amount of information increases with the transition from the lowest level of the evolutionary series to its highest level. The diversity of species is closely linked to the information content of the structural unit of the step. The greater the amount of information in a structural unit, the larger the number of combinations which its components may provide; hence, the greatest possible types of structural elements of the step are possibly generated. But the *possible* number of species is not yet *real*. It is easy to verify that the probable diversity of species, with a combinatorial point of view, is much higher than the actual number of species diversity in nature. However, in general, an increased amount of information in the structural units of the evolutionary series is associated with increased variety of the development levels. This means that, in the process of development, not only the amount of information in the structural units increases but the quantity of information corresponding to the whole step increases as well. This also implies that the growth in the amount of information characterizes both the two interrelated chains of the development series. The previously established inequality $Pe > Pa > P_{mol} > P_{org}$ can now be compared with the inequality of information content $I_p > I_{mol} > I_{org} > Ie$, where the appropriate amounts of information relate not only to individual units but also to all (mostly composed of them) structures.

Therefore, through the use of an information criterion, new opportunities to determine the extent, rate and direction of the evolution of material systems in nature are revealed. The accumulation of information manifests itself in the following cases. First, at the development of some level (or a stage of the development of matter), within which, that object is higher by development which contains more information. Second, the law of information accumulation is also valid for the main line of progress associated with the transition from one stage (level) of matter to another (higher). If at the same stage of development, one succeeds in comparing any specific structural units by the amount of information, for different stages it is necessary to choose average representatives: the "average" elementary particle, "average" atom, "average" molecule, and "average" organism.

These are some universal patterns of evolution, revealed in a study of series of the basic structures of matter.

2.2.3 *THE CONCEPT OF GLOBAL EVOLUTIONISM*

One of the key ideas of European civilization is the idea of the development of the world. In its simplest and undeveloped forms (preformationism, epigenesis, and the Kantian cosmogony), it began to penetrate the science in the 18th century. But the 19th century can be rightly called the century of evolution. First, in geology, then in biology and sociology, more and more attention was paid to the theoretical modeling of developing objects. In the sciences of the physicochemical cycle, the idea of development had difficulties. Until the second half of the 20th century, the initial abstraction of the closed reversible system with the time factor playing no role was dominated. Even the transition from classical Newtonian physics to the nonclassical (relativistic and quantum) ones has changed nothing in this respect. However, in classical thermodynamics, the notion of entropy and the notion of irreversible time-dependent processes were introduced (the notion of "arrow of time" was introduced in the physical sciences). Classical thermodynamics studied only closed equilibrium systems, while nonequilibrium processes treated as perturbations, minor deviations that can be ignored. Penetration of the evolutionary idea into geology, biology, sociology, and humanities in the 19th century—the first half of the 20th century—occurred quite independently in each of these branches of knowledge, although interrelations among sciences and complex "ideological flows" between scientific directions, of course, took place. The philosophical principle of the development of the world (nature, society, and man) had no expression which could be common, pivotal for all the natural sciences (as well as for the whole science). Every branch of natural sciences had its own (independent) forms of theoretical methodological specification.

By the end of the 20th century, the natural sciences approached the design of a *uniform model of universal evolution*, revealing the general laws of nature to tie the origin of the Universe (cosmogenesis), the emergence of the Solar System (heliogenesis) and Earth (geogenesis), life emergence (abiogenesis and then biogenesis), and, at last, the man and society (anthroposociogenesis). This model is being developed under the concept of global evolutionism. In this concept, the Universe appears as a natural whole developing in time, and the whole history of the Universe, from the Big Bang to the emergence of humanity, is regarded as a single process, in which the process of the evolution of cosmic, chemical, biological, geological–geographical, and social systems is linked successively and genetically.

The concept of global evolutionism highlights an important regularity, namely, the directedness of the development of the whole world to increase its structural organization. The whole history of the Universe, from the instant of singularity till the emergence of man, is presented as a single process of material evolution, self-organization, and self-development of matter. An important role in the concept of universal evolutionism is played by factors of the evolutionary process: the emergence of new structural elements (mutations in the evolution of organisms), systems exchanging structural elements carrying information (genetic drift, organism migration etc.), selection of the most effective formations (natural selection in the evolution of organisms), and so forth. In the process of global evolution, levels of organization of material systems are outlined (see Reimers' system level hierarchy, 1994). Any qualitatively new level of matter organization finally asserts itself when it gets able to absorb the previous experience of the historical development of matter. The principle of global evolutionism requires not only the knowledge of the temporal order of the formation of matter levels, but also deep understanding of the laws and internal logic of the Universe's development as a whole.

In this regard, of great importance is the so-called *anthropic principle*, which determines that the emergence of humanity, the cognizing subject (and, therefore, the preceded organic world) was possible due to the fact that the key properties of our Universe are just what they are. This principle reflects the deep inner unity of the laws of historical evolution of the Universe, the Universum, and the background of the emergence and evolution of the organic world until antroposociogenesis. According to this principle, there is some type of universal systemic linkages which determine the holistic nature of existence and development of the Universe as a certain systematically organized fragment of infinitely diverse material nature. Global evolutionism, on the one hand, gives an idea of the world as integrity, letting consider the general laws of existence in their unity, and on the other hand, directs modern science to identification of specific patterns of evolution of matter at all levels and stages of its self-organization.

2.3 FUNDAMENTAL ROLE AND PRACTICAL SIGNIFICANCE OF INFORMATION IN NATURE AND SOCIETY

We have changed our environment so radically that now have to change ourselves, to live in this new environment.

— *Norbert Wiener*

Currently, information is rightfully considered as critical resources and factors of social development. Just a few decades ago, such parameters as the weight of steel produced, mined coal, electricity, oil, and so forth were called as the characteristic parameters of development. Now, information and means of its obtaining, processing and use, occupy first places among the society development criteria. *What is "information?"* Despite the publication of several works on the problem of information which contained an analysis of this phenomenon from broad philosophical and methodological positions, the "informational" aspect of worldview has not yet become an essential component of thinking. Simplifying and schematizing the situation, we can say that the natural-scientific worldview was based on the "substance-energy" foundation. Such notions as "substance" and "energy" have passed from philosophical categories to indicators of human activity. At certain stages of social development different tasks were posed, and although information is an eternal problem, just in the second half of the 20th century, there appeared circumstances which advanced it as a leader in both thinking and human activities. Under information in a broad sense, it is understood the main part of such an attribute of all matter, as reflection. A peculiarity of this part of the matter is that it can be objectified, transmitted, and can generally participate in various forms of motion, which are realized in nature and society, and in the information and cybernetic machinery (computers). Reflection itself is significantly associated with its material carrier and characterized by spatial, energetical, and other parameters. Only that part can be separated from its material carrier which can be transmitted and translated, generating reflection (image) in another object (the receiver of information). The image formed in the mind of one person under the influence of information transmission by another person, never precisely coincides with the transmitter's image, because each has its own individual differences. The information passed from one person to the other one is that what is common between them. In the course of research of information reflective processes, it has been found that information is only transmitted through the diversity of technical means. The basic unit of information is the difference between two elements (or states); they are "0" and "1" in computers, which can be transmitted through communication channels or stored in a computer memory. This is also the basic unit of measurement of the amount of information, called the bit. The byte containing 8 bits is more commonly used unit of measurement in computer science, with each byte corresponding to one character, that is, a letter, number, or other symbol. Accordingly, 1 kB is 1024 bytes, 1 MB is 1024 kB, and 1 GBis 1024 MB. This representation of information as a variety of zeros and units is caused by the design of the computers, as their electronic elements (optical elements in photonic computers) can be in two states, namely, "on" (one) and "off" (zero). Therefore, the concept of information as the side of reflection, which expresses diversity, is objectively implemented in computers.

There is no consensus in the literature about whether information is the property of all matter or it characterizes only the two higher (biological and social) stages of its development. The first viewpoint seems more promising, the more so that experiments that carry ultralow concentrations of substances in liquids, transmission of genetic traits by abiotic communication channels, the development of synergetics, and other sciences of inorganic nature provide new arguments for the existence of information in inanimate nature. The very development of microprocessor technology, realizing the idea of structural information being in the crystals used (chips), gives evidence of not only the prospect of recognizing information as a property of the higher forms of matter organization, but also the availability of information at the very foundation of the Universe as a side of reflection processes. Regardless of the adherence to a particular point of view on the nature of information, there is a trend of recognition of information as a fundamental general scientific category.

Social information has various kinds, in particular, information in inanimate nature, biological, and social information (to form information reality together) can be resolved. In a broad sense, *social information* is the information on both the social form of movements and its other forms received by the society and used in its various activities. In a narrower sense, it is the information about society and man only. This separation is given by the subject of the information process (i.e., that system which receives and uses it) and by the object (which it displayed). The higher kind is the information that passes through the mind of man, and it is possible to resolve its most important form to express the universal humanistic content, the focus on survival and sustainable development of the entire human civilization. For the actually existing global process to become a problem, a condition is needed that is a characteristic of all global problems, namely, there should be a basic contradiction in the development of a problem which threatens the development of the whole society and the individual person in the form of some critical phenomenon. For example, the environmental crisis, the threat of nuclear catastrophe, and so on. This contradiction in the development of social and informational processes is called "information crisis," which appears as a contradictory unity of the "information explosion" and "information hunger." Information crisis manifests itself in the following three issues. First, the existence of a contradiction between the limited capabilities of the person in the perception and processing of information, and the existence of strong flows and arrays of stored information (the first information barrier); this also applies to any group of people not armed with computers (the second information barrier). The total amount of knowledge was changing very slowly at first, but since 1900 it was doubled every 50 years, since 1950 this doubling occurs every 10 years, and since 1970—every five years. By the end of the 20th century, the information flow had increased by more than a few dozen times, doubling every year since 1990 (and even more in recent years). Second, this is the overproduction of a significant amount of information, which either blocks or substantially complicates the consumer's intake of useful, valuable information. Third, it is violation of the integrity of scientific and social communication, which manifests itself in the preference of corporate, elite and group interests, the hypertrophy of departmental goals and needs, the availability of the phenomenon of secrecy, and so forth, that is, there are certain economic, political, and other social barriers, which do not allow information to fully realize its integrative role in the development of modern society. On the one hand, the information factor has produced one of the most profound changes in its history in the life of human civilization over the past decade, namely, has tied the world to a united information system. But, on the other hand, this unity is formal yet, of superficial, technical character, reflecting the fact that the help of information alone does not solve social problems.

Thus, the universality and necessity of information in society make any social activity an information process and, at the same time, any information process in the community is made an activity having all its components, that is, the subject, object, tools, goals, needs, and so on.

The "resource" approach to information and computer science, which will be discussed below, is also based on the activity representation of information processes, on the understanding of there being needs in information, its use for survival and development. Under *resources* is commonly understood this part of the components of social activity or the environment which is used both to develop social production and to meet the needs of individuals as well as the society as a whole. In this case, resources are the need in information, which allows introducing the notions of *information needs* and *development resources*. As a subject of special information activity, information is valued for its content, that is, for its ideal parameters. In this regard, it represents, unlike material resources, an ideal resource saving other material and energy resources. At the same time, information resources (in the broadest sense) include information means themselves—computers, communication devices, models, algorithms, programs, and so forth without which it would be impossible or inefficient to process and transmit information. Information resources also includes the personnel mastered computer and information competence and culture. Thereby, the information resource of development (and in particular, of social management) has two aspects, namely, the ideal one and the material one, as the subject and as the means of information activities.

As is known, resources are divided into renewable (reproducible) and nonrenewable. By its status, information (especially scientific and technical information (STI)) is closer to renewable resources; but it cannot be fully treated as renewable resources, because not all information has the property of reuse by virtue of its aging and some other qualities (e.g., lack of competitiveness). Fundamental STI can be used and reused, and this allows one to classify it as renewable material resources. A certain part of the applied STI can also be used multiple times. The design and spread of personal computers (PCs) open wider perspectives of multiple and rapid use of ideal information resources produced by all mankind. Information as a resource thus tends becoming the common heritage of mankind and a strategic factor for sustainable development.

At the same time, for the wide dissemination and development of information processes in society, such a property of STI as its novelty is very important, and this makes the process of obtaining information as innovation process in principle. The novelty of STI represents just something qualitatively different in comparison with the existing body of knowledge, which is why information acquires particular significance. New information constantly produced in the field of science is, on the one hand, a reproducible resource in various fields of activity, and on the other hand, a forever changing resource, which contains the source of their self-renewal in the very mechanism of its obtainment. The novelty of STI counteracts its aging, deactualization, obsolescence, which leads to the need to solve problems with the aid of new knowledge. The production of new STI, especially fundamental, is an intrinsic property of such a subsystem of social activities such as science. Such properties of STI as its reproducibility, novelty, reusable, have always existed, and, therefore, STI is potentially an important resource for the development of society. Why exactly recently, these properties of information obtained in the field of science, attracted attention and an accelerated process of the informatization of society has began? One cause is that, in addition to military stimuli, some objective needs of society in the large-scale use of STI have appeared. This phenomenon is associated with the transition of the developed countries (and—later—other ones) to intense (and—in the future—sustainable) development, which requires involving new resources, new qualitative factors of development, STI being the most important of which due to the above circumstances. In recent years, information has gained its economic status as a resource and development potential; it has turned out that it can even turn into a commodity, possessing not only its value, but the exchange value as well, can be effectively used to enhance the economic growth as a qualitative factor of development.

Information as a factor of intensification. Already at the stage of the extensive development of material production, it was observed that the growth of output required more than a linear increase in the amount of information, even due to the increase of the "field of activity," the amount of components of production with information links. This even more applies to intense activity, when the qualitative factors of development are multiplied and the quantitative ones are minimized. Speaking of information as the most important factor in the transition to sustainable development, it should be noted that most of the social information produced is not used for various reasons: it is either excessive, redundant, or rapidly becoming obsolete, and so forth. Therefore, the growth of the information content of the sociosphere unlikely can hardly be represented as a process of continuous accumulation of some unchanging atoms of information. This accumulation is really happening, not just in the cumulative process of generalization during conversion of the stored and newly produced information. Of the most value is new and fundamental scientific information, whose movement to the practical and other spheres of activity represents an intensification process.

It is known that to double the production, information must be increased four times, and now we are talking about the fact that the increase in production should go even faster; there is a transition from the extensive to intensive and biosphere compatible economic development. But the matter is not only in the growth of the production volume, but also in the labor productivity, the efficiency of production. This means that to maintain the quality of management to increase productivity by N times, it is necessary that the labor productivity of the persons involved in management would increase by N^3 times. Thus, the labor productivity of managers, scientists, communicators, and all those involved in the creation and processing of information becomes a deterrent of social development at a certain level of the technical equipment of society. These mathematical relationships illustrate the reasons for the enhanced growth of the information sphere. The need to increase information processing in the management sphere at the transition toward sustainable development is dictated not only by socioeconomic processes but also by socio-environmental ones, which, in principle, cannot be implemented without switching to the computer technology, without achieving the continuous informatization of society and its interaction with nature.

Information industry is an interdisciplinary complex, providing the development, production, treatment, and support of the application of information machinery and information technology. The rapid growth of the production of computers was a response to the increasing information needs of society, related to both the development of material production and the needs of other areas of activity (especially science). The emergence and development of computers is a necessary part of the informatization process of society, but the possibility of their introduction into various social sectors would be unfulfilled if they failed to meet certain (informational) needs of society. Modern material production and other areas of activity are in need of more information services, processing vast amounts of information, as already mentioned. The number of those involved in the information field is growing, and this growth was predominantly extensive, since the labor productivity increased very slowly and lagged far behind compared to the labor productivity in industry. By the early 1980s, the provision of equipment of industrial workers was ten times higher than that in the information sector of social production (in terms of value). Just this latter became a weak link in the further building of the efficiency of social production, and there appeared a need for a substantial increase in the labor productivity in the information sector, where, because of the extensive way of development, more than half of the total employed population in developed countries was accumulated. Introducing the latest computer equipment and communication means was a natural response to the said urgent social need.

Before the advent of computer technology, the scale of the manufacturing process and its characteristics had never been changed by two or three orders of magnitude for a decade in history. For example, aviation and power generation, the most impressive technological characters of the 20th century, rapidly developing, did not reach a hundredth part of the rates of change of the technical specifications which was the norm in computing. At the same time, in computing, the experience gained over the last 20 years is based on a practical basis, amounting less than 0.1% of the real production base, which we have to work today with. Although, by its economic status, information industry is among the infrastructure industries (along with the energy, transport ones, etc.), however, its product (information) gives it a unique specificity. The most important trend in the development of computer technology in the future is the possibility of its intellectualization. Initially, computers (which follow just from their names) were designed merely as powerful calculators to solve calculation problems. However, according to forecasts, the computational functions of computers, data processing are increasingly giving way to performing logical functions, knowledge processing, and this latter function will dominate in the early 21st century. In this regard, information systems have turned out to be highly effective, which were named as expert systems used to analyze the structure of chemical compounds, disease diagnosis, playing chess, and so forth. Expert systems function by designing programs to formalize the rules of inference used by the experts in this or that area (e.g., gross masters in the game of chess). Thus, there appear not merely technical systems operating independently of man but man-machine hybrid information processing systems and the problem of human communication with computers has therefore become highly relevant. Until the fifth generation, the development of computers was associated with the appearance of a large army of mediators between the user and the machine—programmers. Just their performance (of the order of hundreds of transactions per day) was the "bottleneck" in the development of information processing industry.

Now, a new information technology is implemented, which runs on personal computers and allows end users to design software programs on the basis of knowledge bases and other intellectual means without any professional assistance necessary. In the future, such information technology will be created in conditions of which every person working with a computer in interactive mode will use his/her natural language, and a scientist will use his/her special language of science.

Cybernetics and informatics. We next consider the answer to the question: what is the difference between informatics as a science and cybernetics, which had appeared before, in the late 1950s? For the concept of cybernetics proclaimed by Wiener, the idea of control was very substantial, and cybernetics is often defined as the science of the control over complex dynamic systems. This definition minimized the idea of cybernetics, but apparently coincided with some objective tendencies of science development. Now, compared to Wiener's understanding of cybernetics and its subject, a number of scholars insist on the necessity of its limitations by control issues only. It was sufficiently clearly formulated by Nicolai Moiseyev, who stressed that cybernetics is a scientific discipline that deals with general control in various areas of human activity, the natural world and machinery. It is believed that in any case there is no concept of control in computer science, essential for cybernetics, and cybernetics itself exists independently of computers, occupying in relation to it the same place as physical devices in relation to physics. Cybernetics studies the processes of communication, storage and processing of information by computers, but only from the viewpoint of the tasks and functions of control. Informatics explores the processes of transformation and creation of new information from a much broader viewpoint, not limited to control tasks. Based on the above information, it is difficult to draw a clear line between computer science and cybernetics. The idea of control, whose core is information transformation, also exists in computer science as the software control over the computational process. Computer science as science does not focus attention on the problem of control in general, but, mainly dealing with information processing, it is not abstracted from the problem of control, especially in the social sphere. Control is not alien to computer science, as it is internally incorporated in any computer. The core of any computer is a computing complex, the processor that controls the entire set of operations on the basis of a program entered, implementing the entire computing process. A computer is defined by the type of its processor, a central information processing device, managing the other computer devices. Besides, the computer is designed for its use in external (with respect thereto) control processes, such as control over technological processes, or in other information systems and processes where control is also present to a greater or lesser degree. However, it would be wrong to assume that information, the problem of communication, and the information approach itself play a minor role in cybernetics. Obviously, the statement that cybernetics does not use computers would be absurd. Therefore, any one sign is not decisive for the distinction of informatics (computer science) from cybernetics. Does this mean that there are not significant enough differences of computer science from cybernetics? We can assume that cybernetics has transformed into computer science just with time. It seems that there is no reason for both the identification of cybernetics and informatics and their sharp demarcation. Computer science in its present form is not something essentially different from cybernetics, but is one of the modern trends of its development, acting primarily as the science of information processing by computers. It turns out that the range of issues raised by cybernetics and included in the scope of its concepts, does not expand in computer science, and even gets narrower in a certain sense. Thus, we can conclude that the relation between cybernetics and informatics as scientific disciplines is completely determined by the ratio of the general and the particular. But when computer science is thought as a unity of science, technology and production, it goes, in this sense, beyond the scope of cybernetics, representing a broader sphere and system of activity.

Information technologies (IT) have information as a source, which is also the result of their activity. In material production, under technology is commonly understood a set of methods of production, processing, changing the properties, shape, state of matter and energy. IT, dealing with the material carriers of information (substance and energy), is inherently linked with ideal objects (if a relevant part of social information is meant), that is, a "smart" technology. That is why, artificial intelligence is considered the foundation of the new IT.

Automation of such technologies as editorial publishing activity, design works, and so forth leads to the displacement of their paper-based carriers, which allows us to speak of becoming paperless information processing industry. All forms of information motion are amenable to automation, and special hardware is designed for this. Technical means of communication and data transfer are used for information transfer. Obtaining and input of information are provided by sensors, terminals, PC keyboards, and so forth, its storage is done by computer memory, data and knowledge banks and bases; information processing is performed by microprocessors, the output and presentation are by indicators, printers, displays, and so on.

Unlike substance energy technologies (SET), IT deals primarily with a fundamentally different (intangible, ideal) resource and factor. But this does not mean that there is no movement of the substance and energy, belonging to the ideal material carriers of information, during the development of IT. However, the most important thing, for which IT occurred, is the development of intellectual activity, increased creative capabilities of human consciousness. And this has left its mark on social and informational processes, and, in particular, IT, no matter what physical form they are implemented in (pre-paper, paper, or electronic). The idea is that such a general property of matter (attribute) as *interaction* was primarily used in traditional technologies, while *reflection*, closely associated with the previous attribute, is used in IT. From a philosophical viewpoint, it is essential, because, differing and developing, interaction and reflection gradually have led to the formation of such a phenomenon and attribute of the social level of matter, which was given the philosophical status of "ideal." In the activity-based nature of traditional technologies, interaction was lain, whilst for IT it is reflection, or, as said sometimes, "the activities of reflection."

Automatic information processing on a computer is considered to be the new IT (NIT). Now NIT is the usage of the latest achievements of computer science—the latest generation of computers, telecommunication systems, and, primarily, the Internet, artificial intelligence (just the latter has allowed to call NIT as intelligent IT). Among the newest intelligent IT, there are artificial neural network (ANN) (designed like their bioanalogs), evolutionary computation means (based on the formalization and automation of natural evolutionary processes), genetic algorithms to simulate the process of natural selection, and so forth.

The global nature of informatization. Currently, the process of informatization has already covered all countries in the world, which provides the basis for the transformation of informatization into a very significant global trend of development. Of particular importance in this global dimension is merging of computer technology with telecommunications and the design of tools and communication systems, new social and information media, providing a more reliable connection using both traditional and non-traditional means, in particular, providing user access to computer information resources, that is, which recently was named as mediatization. The observed global effect strongly depends not only on the saturation of the world community with computers but also on the connection to the global communications network Internet (this process is increasingly referred to as "connectedness," or "internetization"). Informatization is a global process, not only because all countries and regions can be computerized, but also as a result of the real possibility of transmitting information from any point of the globe to another one, where the users' computers are installed. The process of designing integrated computerized communication systems and their conjugation with the means of mass communication proceeds. The informatization of society includes not only that of individual countries and regions, but that of international links, which will gradually lead to the information unity of the world civilization and to the emergence of a global information civilization. The problem of information and the process of its solution should be rightly attributed to global problems as the problem of informatization has all characteristics and attributes of global problems. The positive and intensive development of the informatization process will not only solve the global problem of informatization, but also contribute to the solution of a number of other global problems, create an intellectual base for the formation of mechanisms of civilization survival.

Survival is a common goal of all mankind, and no survival of any part of the whole civilization of the expense of another one is possible before the threat of global eco-omnicide. Therefore, the leaders of informatization, if guided by universal values and perspectives of joint survival, must not only understand the necessity of creation of the information society on a global scale, but also help to accelerate informatization of the states and regions which are behind the process. At the same time, the global nature of the information society as a stage of the noosphere will not exclude some specificity and diversity of information societies by virtue of historical, economic, cultural, political, ethnic, and other features. The information society will be a society with the maximum degree of diversity, which will allow it to counteract social and natural disturbances, conflicts and disasters. Only such a super-pluralistic society, implementing the strategy of sustainable development, will be able not only to survive, but then to move toward sustainable development.

2.4 THE EXPANDING UNIVERSE

A. Friedmann (1888–1925) was the first who predicted and mathematically described cosmological expansion. The main scientific works of Friedmann deal with fluid mechanics, the theory of gravitation, and theoretical geophysics. By his book "The Hydromechanics of a Compressible Fluid" (1922), he laid the foundations of theoretical meteorology, and in 1922–1923 he proposed nonstationary solutions of Einstein's gravitational equations, having proven the possibility of the existence of a nonstationary expanding Universe (in his papers "On the curvature of space" and "On the possibility of a world with constant negative curvature of space"). The phenomenon of the cosmological expansion is often called the Big Bang.

In 1929, Friedmann's theory was confirmed by Edwin Powell Hubble (1889–1953) who discovered the recession of galaxies.

Hubble's major works are devoted to the study of galaxies. In 1922, he proposed to divide the observed nebula by extragalactic (galaxies) and galactic (gas and dust). In 1924–1926, he discovered constituent stars on the photos of some nearby galaxies, having thereby proven that they are star systems like our own galaxy. In 1929, he discovered the relation between the redshift of galaxies and the distance to them (Hubble's law). In 1935, he discovered asteroid No 1373, which he named "Cincinnati" (1373 Cincinnati). Asteroid No 2069, discovered in 1955 (Hubble 2069), was also named in honor of Hubble, as well as the famous space telescope "Hubble," put into orbit in 1990.

The elementary model of the expanding Universe can be represented as follows. It is based on the experimentally established (by Hubble) recession of galaxies at a speed $\vec{v}$ proportional to the distance $\vec{r}$ to the center of divergence, that is,

$$\vec{v} = H\vec{r} \tag{2.1}$$

where H is the Hubble constant.

Suppose we have a sphere of radius R, the density ρ of substance inside is constant everywhere, and the mass of all substances uniformly filling the ball is M. Assume that the substance of the ball is some gas of particles (no matter what). The main thing is that the pressure inside the ball was negligibly low. Then the only force acting on the particles of the gas will be their mutual attraction, which tends to bring together the particles (ball compression). In this case, the ball is expanding due to the fact that at some "initial instant," all the particles are assigned speeds according to eq 2.1, and they fly out from the center of the ball. Consider a particle of a mass m on the surface of the ball. Its velocity is equal to

$$v_R = HR \tag{2.2}$$

and the kinetic energy is

$$E_{KUH} = \frac{mv_R^2}{2}$$

The attractive force, created by all the other particles, is directed against the movement and equal to

$$F = -\gamma \frac{Mm}{R^2}$$

where γ is the gravitational constant (Newtonian gravitational constant). It tends to stop the motion of the particle and to change extension to compression. For the potential energy of the given particle, we approximately have

$$E_{nom} \approx FR \approx -\gamma \frac{Mm}{R}$$

Then the law of conservation of energy can be written as:

$$\frac{mv^2}{2} - \gamma \frac{Mm}{R} = const = C$$

or

$$v_R^2 = 2\gamma \frac{M}{R} + C_1 \tag{2.3}$$

where $C_1 = 2C/m$. Hence, if $C_1 > 0$ in eq 2.3, the expansion will continue indefinitely; if $C_1 > 0$, then, when $R = \frac{2\gamma M}{C_1}$ is achieved, the expansion will stop and be replaced by compression. The boundary situation corresponds to the equality

$$v_R^2 - 2\gamma \frac{M}{R} = 0 \tag{2.4}$$

Substituting eq 2.2 and the mass of the substance with the density ρ filling the ball, that is, $M = \frac{4}{3}\pi R^3 \rho$, to eq 2.4, we obtain:

$$R^2\left(H^2 - \frac{8\pi}{3}\gamma\rho\right) = 0$$

The substance density in this model of the Universe is

$$\rho = \frac{3H^2}{8\pi\gamma} \tag{2.5}$$

Equation 2.5 coincides exactly with the assessment obtained by Friedman from the general theory of relativity. From eq 2.4, it follows that

$$v_R = \sqrt{2\gamma \frac{M}{R}} \text{ or } \frac{dR}{dt} = \sqrt{2\gamma \frac{M}{R}} \tag{2.6}$$

Separating the variables and integrating the resulting equation from 0 to R and from 0 to t, we find:

$$\frac{2}{3} R^{3/2} = \sqrt{2\gamma M} \cdot t \tag{2.7}$$

where t is the time elapsed since the beginning of the cosmological expansion. Given that $v_R = \sqrt{2\gamma \frac{M}{R}} = HR$, we find $\sqrt{2\gamma M} = HR^{3/2}$. Substituting the last relation into eq 2.7, we obtain the age of the Universe $t = \frac{2}{3H}$.

The Hubble constant is $H = 65 \pm 10 \frac{\text{km}}{\text{s} \cdot \text{Mpc}}$ (Mpc is megaparsec); a parsec is the distance at which the diameter of Earth's orbit is seen at an angle of 1 s. Then $t \approx 10^{10}$ years, which is a reasonable value.

From eq 2.7, we can also estimate the dependence of distances in the world on the age of the Universe. Obviously $R \sim t^{2/3}$.

In the absence of gravity, we would have the inertial scattering of the particles, that is, $R \sim t$. When there is gravity, the expansion is slower than inertia, it decelerates with time.[3]

Nevertheless, the question is natural: "What is expanding?" Friedman's theory describes the Universe as a whole, and therefore its properties (as a whole) appear only in very large scale, available to observation. The basic idea of the general theory of relativity is that the properties of space and time are determined by the distribution and movement of the substance filling this space. The conclusion follows that the space in which the substance expands must itself expand, that is, Friedman's law of the expansion of substance is simultaneously the law of expansion of the world space.

Some researchers believe that all distances in the world are increasing due to the cosmological expansion. It is not so. Nothing on the Earth changes its size, and it itself does not expand on the "cosmological cause." No planets, no stars, no galaxies change their sizes because of this. Galaxies may even get closer to each other at small distances.

Now, most scientists believe that modern cosmology is based on three major experimental discoveries, namely, the recession of galaxies (A. Friedman), cosmic microwave background (CMB), which uniformly fills the entire space of the world (G. Gamow, 1904–1968) and the discovery of the "vacuum of space" (the so-called dark energy or matter) (A. Einstein, 1979–1955). The latter phenomena will be discussed below. Without going into further details regarding the discussions between different researchers on this topic, we consider it appropriate to present the cosmic timeline describing the reliably established events until AD (Tereré, 2009; Table 2.1).

---

[3] Note that the above estimates in the books by A. M. Cherepashchuk and A. D. Chernin are obtained from the approximate equation $\gamma\rho t^3 \sim 1$, which follows simply from dimensional considerations (a single dimensionless combination of γ and ρ can be composed). As indicated in these books, by the 2003 cosmological data, the age of the world is within the range of 13–15 billion years. Thus, our estimate of 10 billion years is very reasonable.

TABLE 2.1 Timeline of the Reliably Established Events of the Birth and Evolution of the Universe

First moments of the Big Bang	
10^{-35} s	Cosmic inflation generates a huge region of space filled with lumpy quark soup (quarks are elementary particles with fractional charge, of which, according to modern concepts, all strongly interacting intranuclear particles are composed)
First moments of the Big Bang	
10^{-30} s	Axions as one of the possible types of dark matter are synthesized. The axion is an ultralight "cold" particle with a mass of about a trillion times less than that of the electron. It can turn into a photon in a very strong magnetic field. The axion is called "cold" because, despite its birth at a very high temperature, it moves slowly. Axions are therefore easily grouped into galaxies
First moments of the Big Bang	
10^{-11} s	Matter prevails over antimatter The transition from the first moments of the Big Bang to the formation of atoms
10^{-10} s	A second possible type of dark matter, the neutralino, is synthesized. The neutralino is the lightest particle of the recently predicted class of massive copies of the known particles. It should have a mass of between 100 and 1000 masses of the proton and can be produced in experiments at the Large Hadron Collider at Conseil Europėen pour la Recherche Nucléaire (CERN) near Geneva
Formation of atoms	
10^{-5} s	Protons and neutrons are formed from quarks
Formation of atoms	
0.01–300 s	The nuclei of helium, lithium, and the heavy isotope of hydrogen (deuterium) are formed from protons and neutrons
Transition from atom formation to the dark era	
380,000 years	Atoms are formed from the nuclei and electrons, releasing the relic (cosmic microwave) radiation
The dark era	
380,000–300 million years	Gravity continues, increasing the density fluctuations in the gas filling the space
Transition from the dark era to the modern era	

300 million years	First stars and galaxies are formed
The modern era	
1 billion years	The limit of our present observations (objects with the highest redshift)
3 billion years	Clusters of galaxies are formed; the maximum frequency of star birth
9 billion years	The birth of the solar system
10 billion years	Dark energy overcomes gravity, and expansion begins to accelerate

2.5 NONLINEAR DYNAMICS IN GEOSYSTEMS

2.5.1 *GENERAL PROVISIONS*

The Earth is a complex open system, in whose history, self-organization processes have occurred for almost 5 billion years and proceed now. The Earth's evolution, according to historical geology, is characterized by irreversible and phasing. In the historical way of the Earth, such "parts" are established by geologists in most detail in the development of the so-called galactic geochronological scale. Stages of quiet development are interrupted by phases of instability, crises, bifurcations, which determined the choice and features of the scenario of the future development of our planet. This is typical for the Earth as a whole, and can be traced separately for geological and biosphere global processes.

The complexity of the historical way of our planet is evident if we consider the coevolution (coadjoint, interrelated development) of the planet's shells (global systems or geospheres), being the largest intraplanetary formations. Patterns of development are largely determined just by the coevolution of the planet's physical shells (the stone substrate, hydrosphere, and atmosphere), living matter (the biosphere) and the man, his rational mind (the noosphere).

In the 1970s, the English engineer and philosopher James Ephraim Lovelock proposed a concept which occupies a prominent place in today's global environmental and earth sciences. Its essence lies in the fact that the Earth as a whole has some properties of a living system (homeostasis, self-regulation, etc.) and this largely determines the nonlinearity of planetary processes. The study of the nonlinearity of geosystems and geoprocesses of various ranks (from local to global ones) is extremely important from both theoretical and practical viewpoints. Moreover, some philosophers consider the notion of nonlinearity as a "fundamental conceptual unit of the new paradigm." In general, an idea of geographical objects as systems possessing a hierarchical structure has been already formed.

Various consequences of the nonlinearity in geosystems are easily detected. For example, the relationship of precipitation (the rate of water deposition on the surface of the catchment area) and drainage (the flow in the mouth of the main river of the catchment basin) is usually nonlinear. Precipitation, falling on the ground, then creates different kinds of moisture reserves and moisture flows, and only part of them gets into the mainstream. Some part of the precipitation initially fills depressions in the relief (open lakes, reservoirs, etc.) and infiltrates the soil to the level of field capacity. Only after this, the process switches to a different mode. This process is highly dependent on the intensity of precipitation. Nonlinearity here can have the critical nature of switching. Some part of the precipitate comes to the surface and subsequently evaporates from the surface. This process is a nonlinear function of the temperature of the evaporating water. A very thin layer of a vapor–air mixture is formed above the water surface, wherein the concentration of water vapor depends on the surface temperature. It follows from Clausius–Clapeyron's equation:

$$\varphi = (\bar{T})\exp\left\{\frac{L}{R}\left(\frac{T}{\bar{T}} - \frac{1}{T}\right)\right\}$$

where φ, T are the water vapor pressure and temperature; L is the latent heat of vaporization; R is the gas constant of water vapor; $\dot{T}$ is some fixed (average annual) temperature. The evaporation rate is usually directly proportional to φ. Therefore, the evaporation rate depends nonlinearly on temperature, which can produce thermal instability of evaporation (a nonlinear effect is again observed). The dynamics of water movement in the mainstream is also nonlinear (as evidenced by the nonlinear hydrodynamic equations describing the motion of water in the mainstream), which generates various effects (which can be explained only by considering the nonlinearity), such as turbulence, the occurrence of benthic and planar landforms of the channel, and so forth.

In the study of the dynamics of geosystems and solving geo-environmental problems, one needs to resolve characteristic times of variation of the studied nonlinear processes. Such processes are important for human life, whose characteristic times fall in the range from few hours (even several tens of minutes) to 100 years. This time interval covers various man-made and natural catastrophic events (avalanches, debris flows, landslides, avalanches of rocks, volcanic eruptions, rapid movement of glaciers, breakthrough of natural dams, earthquakes, etc.); synoptic atmospheric processes with characteristic times within 1–30 days, which determine the state of the weather; seasonal processes (1–12 months); processes associated with interannual variability (longer than 1–10 years); climatic processes (10–100 years or longer). Seasonal processes are related with characteristics which determine crop yields, environmental comfort of living, and so forth. The general condition of flora and fauna, agriculture, and so forth is associated with the interannual (characteristic oscillation periods of 2–3, 5–6, and 11 years are resolved) and secular variability. The secular changes of climate (whose important characteristics currently observed are warming and the increase in atmospheric carbon dioxide) influence the interaction between nature and society. In this connection, there appears a global issue about the viability and sustainability of human development in the further preservation of these trends.

A nonlinear geosystem can have many attractors (domains of attraction), the transition between which is carried out by external influences. In particular, the possibility of the existence of two (or more) stable positions of sea level in the Caspian Sea and other drainage seas and lakes is shown, that is, many of these water bodies can be considered as nonlinear trigger system. This approach allowed for a fresh look at the well-known problem of the Caspian Sea level changes. There are many examples of multistable geosystems, in which there may arise alternative states of the "forest" and "step" tye in the forest-steppe zone, or "wood," "steppe" and "bush" in the southern forest, the occurrence of sharp landscape borders, as well as the phenomenon of "retardation" of the landscape borders on climate change. Triggers have long been found in biogeocenoses. As a result of the development of a nonlinear geosystem, bifurcations inevitably arise. Intransitive properties, revealed in the simulation of climate changes of the Earth are indirect evidence of the emergence of climate bifurcations. Geodynamics and geophysics are the most modern areas of knowledge with regard to the application of nonlinear ideas in the geosciences. In geodynamics, nonlinear concepts are most natural and organic. All elements of these concepts (instability, phase transitions, bifurcation states, the trigger effect, self-organization, etc.) characteristic of the evolution of complex thermodynamically nonequilibrium (i.e., open, dissipative) systems in this case are often present, but not always moved to the fore as backbone ones.

In the 1990s, works by Felix Letnikov were published, where not only the equilibrium thermodynamics but also the self-organization of nonequilibrium natural physicochemical systems with geological consequences was considered on rigorous mathematical level. In particular, it is noted that, for understanding the evolution of geological systems, it is important to determine the degree of their stability in every case, because in a state close to the unstable one, any system is responsive even to slow and random external influences. This area includes all metastable systems, but even among them there is a hierarchy in the degree of response to external noise exposure, which is determined by the binding energy between the components of these systems. For example, the effect of ultrasound on a system of fluids and gases in a stone substrate (the so-called fluid system) is accompanied by its significant changes, while the same effect on a body composed of dense rocks has no effect.

A distinctive feature of an unstable geosystem is its multivariate or *multiplicity*, well displayed in the evolution of ore-forming fluid systems, where the quantitative ratios and compositions of the phases of fluids vary widely, and the factors of self-organization of such systems play a huge role in their subsequent preservation in the accumulation of mineral aggregates. One of the most difficult problems for the synergetics of geological systems is the problem of finding elements of self-organization at different hierarchical levels of evolution, when any subsequent stage may erase information of the previous one. Another problem is the need for an analytical expression of the detected synergetic signs in terms of thermodynamics, mathematical physics, or by designing adequate mathematical or thermal models of the process under study. Establishing, due to seismic data, the tectonic stratification of the lithosphere and the multilayered nature of the mantle convection, along with a variety of nonlinear effects in deep geodynamics, have led to the emergence of a new global-dynamic nonlinear model of Japanese geologists (Maryama et al., 1994, etc.). Synthesizing the tectonic triad: growth tectonics (the tectonics of the Earth's core), plumtectonics (throughout the entire mantle from the core to the

670-km border), and plate tectonics (inherent in the Earth's crust and upper mantle to the 670-km border) is attempted. Around the same period, B. Sokolov's works were published, where oil and gas formation processes are considered as nonlinear and a self-oscillating model of the oil and gas field is offered.

In the first decade of the 21st century, interest in synergetics in the Earth sciences continues to grow. For this new phase, the following main features are characteristic:

1. Attempts of direct cooperation between specialist (geologists, geographers, geophysicists, geoecologists, etc.) with specialists in the field of nonlinear dynamics, the development of special departments in academic institutions and universities.
2. Direct nonlinear simulation of geosystems and geoprocesses at different levels.
3. Increased interest just to the synergetics of global geoprocesses, which can be seen from the number of publications on this topic and specific research activities.

2.5.2 *PHYSICS, SYNERGETICS, AND EARTH SCIENCES*

At the first glance, their relations are harmonious enough, because a powerful interdisciplinary direction, geophysics, has been developing a long time. But on closer examination, in the history of the interaction of these natural sciences, it can be seen that the simplicity of the situation is apparent. Speaking of the unity and diversity of the geological and geophysical approaches in the study of Earth, Samuel Warren Carey concluded that geologists and geophysicists "approach the truth from different sides." Geophysicists consider some aspect of the structure of the Earth, designing a theoretical mathematical model, which limits the number of variables and quantifiable parameters. However, this model may have little (or nothing) in common with the real Earth. Geologists, on the contrary, deal with the real Earth and think more logically. Some authors evaluate the relationship of physics and Earth sciences by characterizing them at some stages as opposition. Scientists did not have accurate methods of measurement of geological time. They could only estimate the relative time, that is, the position of the layers by the animal fossils. And this first confrontation ended by an explosion, a jump. In 1905, the first measurements of the absolute age of rocks by the accumulation of radioactive decay products in them showed that the Earth's age should be measured in billions of years. Geologists, with their archaic method of measuring such as intuition, were then closer to the goal of science (the truth) than physicists with their accurate techniques—but relying on incorrect calculations in this case.

Radioactivity, making it possible to determine the age of the Earth, was useful for another explanation: it explained why the bowels of our planet have a too high temperature. A second confrontation began in 1915, when Alfred Lothar Wegener published his book "The Origin of Continents and Oceans," where the notions of uniformity of the Earth's crust under the oceans and continents were questioned (i.e., the crust under the ocean and under the land is fundamentally different). The possibility of movement of the continents was first suggested as well. The relations between physics and geography have led to the creation of another interdisciplinary field of knowledge, physical geography.

2.5.3 *NONLINEAR EFFECTS IN THE EARTH'S RELIEF*

In geomorphology, there are many forms and structures, whose formation is clearly associated with the manifestation of nonlinear dynamics. Nonlinear processes are likely to contribute to the emergence of trough valleys containing alternating basins and rigels, to relief formation in deserts, and so forth. Different scale manifestations of convective structures on the Earth's surface are examples of the occurrence of spatial rhythmic structures related to dissipative ones. The fractal properties of gully networks and river basins are intensely studied. Another example is the uneven settling of people on the Earth's surface, due to the nature of the relief and getting a mesh, mosaic structure similar to the Benard cells in the convection theory, the emergence and development of a network of settlements, and so forth.

For the relief structures of the Earth in the local and global scale, self-organization is a characteristic associated with the appearance of new macrostructures in geosystems, dying out of mature structures, and so forth.

Examples, in which the role of nonlinearity is decisive, may be a variety of regular structures resulting from self-organization: coastal riptide systems, longitudinal trough valley periodicity, and the regularity in the arrangement of river basins, volcanoes, volcanic islands, and ridge–lake complexes.

Such geoprocesses are known in which self-excited oscillations may occur: wind ripples on the water surface, landslide pulsations on the underwater slopes, vibrations found in models of the atmosphere, oscillations in the interaction of the ocean with the atmosphere, chaotic self-excited oscillations of the hydrosphere and climate, global climatic oscillations in the glaciers-ocean-atmosphere-land system, and the self-oscillating character of moving the lithospheric plates apart. The possibility of self-oscillations in the Earth's bowels has been theoretically established.

V. Chuprynin (2008) emphasizes the obvious geological and geomorphological self-oscillating process, which affects large parts of the Earth. Due to the spatially nonuniform endogenous tectonic processes inside the Earth, the Earth's surface rises (mountain building) in some regions and lowers in the other regions. These processes create horizontal gradients of relief on the surface of the Earth. As a result, the relief turns into a nonequilibrium state. Self-oscillations under the action of slope processes are a widespread process. In general terms, almost same model can be formulated for many slope processes, in particular, avalanches, pulsations of mountain glaciers and landslides on submarine slopes. In addition, a different substance (snow or river-carried solid precipitation) is accumulated in these processes, and nonequilibrium occurs at the expense of other sources of substance rather than at the expense of the processes occurring inside the Earth.

Of interest is the problem of transient processes in the Earth's relief systems. Observations show that, in nature and society, there are two types of changes (over time) of such variables that characterize the dynamics of systems: slow and fast, often abrupt. Moreover, the characteristic rates of these changes (movements) may differ by one to two orders of magnitude or more, although sometimes they may be comparable (but, qualitatively, the characteristics of these movements drastically change). These two types of motion are associated with the alternation of long periods of relatively slow evolution with short eras of rapid changes in the history of different objects.

The transients occurring in geographical systems are rather complex and poorly predictable, since these systems (unlike hydrodynamic ones) typically comprise a plurality of components of different nature. In fact, geography has long implicitly engaged in such processes. For example, the response of geographical objects to various catastrophic events associated with external factors (that of geosystems and ecosystems to anthropogenic impacts, that of the landscape to volcanic eruptions, vegetation, grazing, flooding, or drainage) is studied.

2.5.4 *CRISIS AND CATASTROPHIC EVENTS IN GEOSYSTEMS*

In recent decades, the theory of the concept of critical conditions, or geomorphological thresholds has emerged and being developed in geomorphology. Its essence lies in the fact that, up to a certain instant of time, the state of the relief can be characterized as dynamic (current) equilibrium, but the transition through the critical (threshold) values of some parameters of a geomorphological system is accompanied by changes (the complexity is increased) in the structure of geomorphological systems, often with catastrophic consequences. For example, it has been found that the creep and failure of slopes between the rivers in the temperate zone occur when the atmospheric precipitation rate achieves 55 mm/day, with the duration of this rainfall of 2–3 days. This quantity is proposed as an index of danger.

In order to build a system of natural protected areas in a region, several alternatives are produced in the distribution and characteristics of such territories. Selecting one of these alternatives and its implementation means a bifurcation change of the state of society and nature in the region. In such cases, the society has a certain time interval during which the system is near a bifurcation, and it roughly knows what movement by each of the possible paths may cause in the foreseeable future. Disasters constitute a class of phenomena in which the nonlinearity of natural systems manifests itself in some form. The causes and mechanisms of their occurrence are various. Three types of natural geocatastrophes are counted, namely, *forced geocatastrophes, bifurcation geocatastrophes, and autogeocatastrophes*.

Forced geocatastrophes occur in natural and socio-natural systems under the influence of powerful impulse external factors, such as earthquakes, volcanic eruptions, typhoons (hurricanes), tornadoes, tsunamis, storms, fall of large celestial bodies on the Earth's surface. In view of the high power of such an external influence, the response to them is nonlinear. So, at the fall of a large space body, nonlinear waves of large amplitude appear, the critical tensions are exceeded, gaps appear on the surface, and a piece of land is melted. Another example is the catastrophic consequences of powerful thunderstorms. Such strong impacts on the natural system can cause changes in its internal structure, its transformation into a different system, or just destroying itself.

At a *bifurcation geocatastrophe*, due to the nonlinearity of the system, a catastrophic change of the regime of its functioning arises. In particular, this model describes the mass death of animals and people under drought, rapid destruction of forests at slow climate changes, the transition of the water moving through a channel from laminar to turbulent flow at a certain value of the Reynolds number. An original example of a large man-made disaster is the destruction of Tacoma Narrows Bridge due to the development of self-oscillations in the interaction of the airflow with the construction of the bridge, where the Reynolds number exceeded its critical value.

Autogeocatastrophes are prepared and occur within a system due to its nonlinearity. Stress, energy, and substance (coming from outside) are being accumulated and increased; the negative factors or other characteristics (depending on the type of the system) are increased. This rise occurs up to a certain critical level, after which some nonlinear mechanism is triggered to discharge of the whole accumulated. Some characteristics of the system change stepwise, that is why, a prognosis of autogeocatastrophes in many cases is most probable of all types of accidents considered here. A striking example of this type of disaster is shifts in surging glaciers.

2.5.5 *RHYTHM OF GEOPHYSICAL PROCESSES IN THE HISTORY OF THE EARTH'S EVOLUTION*

Rhythmically constructed strata of rocks, which are often found in the sediments of the Phanerozoic and Precambrian, are often the object of the geologist's work. Usually they are a "stone chronicle" of nonlinear processes in the Earth's history. In this regard, the study of features of rhythm global geoprocesses is of important theoretical and practical significance for the understanding of their mechanisms and forecasting. Many of them are poorly studied. In particular, the nature of the rhythms composed of carbonate rocks remains problematic.

In the study of *the rhythm of sections of the crust,* a large number of various methods can be used, which give the best results in the solution of specific problems at their complexation. Studying cyclites includes: (1) a procedure of testing each element of the cycle and even layer with the aim to estimate changes in the rock properties along the cyclite body; (2) collecting the maximum possible amount of quantitative data, especially over the thickness of layers to detect "embedded" finer cyclicality; and (3) sketching and photographing cyclites. The selected samples are subjected to comprehensive research, including basic petrographic, chemical, and petromagmatic methods. The next step is elucidating the genetic type of the cyclite. Of great importance for the study and interpretation of the nature of rhythms are *mathematical (static) methods for studying cycles*. The computer methods of mathematical statistics, based on decomposition of periodic functions in Fourier series and Walsh functions (these methods belong to the category of linear ones), count the probable relationship of the fluctuations of certain parameters over the section with Milankovitch's astronomical cycles.

The categories of "wave" and "cyclic" geoprocesses are principally distinguished. The *wave* process is a physical process accompanied by alternating deviations of a physical quantity from its conventionally zero (mean) value, where the positive and negative deviations equal in amplitude are equal in energy but opposite in sign. There are many examples in nature where wave processes generate cyclic ones and vice versa. For example, the natural vibrations of the Earth, having a wave nature, can influence the changes in the weak (background) seismicity, causing some cyclicality in its manifestation. At the same time, the cyclical nature of tectonic stress (e.g., caused by the recurrence rates of the plate movements) may cause long tectonic waves (a wave process). A *cycling* process is a physical process accompanied by periodically alternating raising and lowering of a physical quantity of the same sign with respect to its zero (minimum) value (Table 2.2).

"If cyclic processes typically reflect the rise and fall of certain parameters (the number of earthquakes, volcanic eruptions, sunspots, and seismic energy), a wave process, in essence, means the change of sign of the energy released in this process from positive to negative and vice versa. In contrast to cyclic processes, wave processes have their energy of equal magnitude but opposite in sign. For example, these are electromagnetic, acoustic or gravity waves. Therefore, the half-periods of a waves, both

positive and negative, carry equal energy, causes a reaction of the environment to these processes. Meanwhile, in cyclic processes the maximum of the cycle means the maximum amount of energy imparted to the system, and the minimum of the cycle means the minimum energy. In cyclic processes, we deal with processes having only a positive component. Naturally, no negative amount of seismic energy can be released, and no negative number of earthquakes and volcanic eruptions can happen. Therefore, examining and analyzing wave processes as, sometimes, cyclical ones, researchers wrongly interpret the results and lose the physical meaning of the mechanism of this process." (Khain & Khalilov, 2009)

TABLE 2.2 Global Natural Geoprocesses of Wave and Cyclical Nature (Khain & Khalilov, 2009)

Cyclic processes	Wave processes
Solar activity	Electromagnetic, seismic, and acoustic radiations
Oscillations in the level of the seas and oceans	Quadrupole alternating change of the metric parameters of bodies in the field of a passing gravitational wave
Seismic activity	Lunisolar tides
Volcanic activity	Alternating vertical tectonic movements
Periodicity in the tsunami manifestation	Periodic changes in the Earth's radius
Variations in precipitation	Variations of the length of a day
Sedimentation process	Alternating variation of the stressed condition of the crust (compression–tension)

2.6 DARK MATTER (ENERGY) AND UNIVERSAL EVOLUTION

At the end of the 20th century—beginning of the 21st century—several discoveries occurred, which allow for a fresh look at the process of evolution and, in particular, the universal evolution. It is assumed that our Universe is not alone, and no longer represents the Universum containing all things therein. What was earlier called the Universe is now considered one of mini-universes.

Our improved knowledge on the problem of evolution and the fundamentally new vision of universal evolution are largely due to the discovery of new and mysterious substances, "dark energy" and "dark matter." Prior to the discovery of dark energy, already include the so-called dark substance, or, as is more often called, dark mass was classified as dark matter. If this second form of matter occupies about 20% of the mass/energy of the Universe, it turns out that the whole modern science studies only a few percent of the material content of the Universe. Unfortunately, many well-known educational and research publications present the obsolete picture of the world. The modern scientific picture of the world in its most general form should give a systemic and complete image of animate and inanimate nature, man and mankind, obtained on the basis of synthesis and generalization of all scientific knowledge. Moreover, despite the fact that since the second half of the 20th century, the ideas of universal evolutionism have underlain the scientific picture of the world, from time to time, some science unexpectedly makes a contribution which can be very drastic for the understanding of the Universe and man's role in it.

2.6.1 *BASIC FORMS OF EXISTENCE AS WAYS OF MATTER SELF-PRESERVATION*

As was mentioned above, philosophy has always sought to explore the Universe in its entirety and at different levels, to reveal therein the most fundamental and limiting ultimate beginnings and foundations of existence. At the stage of natural philosophy, it sought to penetrate the secrets of nature, to reveal eternal and transient therein, stable and variable, universal and particular. Gradually, with the development of particular sciences and special disciplines, interest in these issues was lost, and the philosophy of science has focused on the philosophical and methodological analysis of particular sciences and topical issues, on the logical and epistemological aspects of the relationship between philosophy and science. The time, when natural philosophy set the tone of cognitive thinking and formed the image of the "ontological framework" of the Universe, is gone; the knowledge flow from particular and other sciences to philosophy has proven to be dominant.

Only 10 years have elapsed since the so-called dark energy was discovered, and searching for observational evidence of its existence actively has begun. Currently, this unusual form of existence of cosmic matter poorly fits in the recently created "substance-evolutionary" scientific picture of the Universe and is little studied due to the very urgent problem of universal evolution in the Universe.

The scientific literature has not yet established common names of the dark forms of matter, and we will try to place them on "conceptual shelves" in order to represent the dark side of the Universe more clearly and understandably for nonspecialists in the field of cosmology. We will analyze, in a discussion and hypothetical form, what major conceptual innovations in considering invisible (in the sense of darkness) forms of matter can be introduced to the concept of the global evolution of matter and into the scientific picture of the world.

Science (though in the form of a discussion) accepts the hypothesis of the existence of a very stable part of the Universe, which will be generally called dark matter consisting of two basic forms, namely, dark energy and dark mass (see Table 2.3). Dark mass is called dark matter in the English-language literature. The notion of matter has extremely common value in Russian scientific philosophy, whereby it is impossible to believe that dark energy is something intangible. It is therefore proposed to consider dark energy and dark mass as forms of matter (as well as substance), which are different in nature.

The definition of the notion of "dark matter" is far ambiguous. Some scientists believe that this notion is poorly applicable, since there already is "dark substance" (dark mass), which, like the term "dark energy," is really dark (emitting no light).

Another name for dark energy is more preferred for them, namely, "the vacuum of space." However, the term "vacuum of space" sounds, to a certain extent, like the term "physical vacuum," which may be fundamentally different from dark energy. According to various estimates, the dark components constitute 96–97% of the whole material content of the Universe. They are invisible in contrast to the luminous (or visible) Universe and are conserved, in some degree, in a weak or even non-evolutionary form amid the distinct evolution of substance. Dark energy opposes gravity and allows, due to the properties of "universal antigravitation," the Universe to expand with acceleration. However, not only the unusual property of antigravity of dark energy attracts attention, but also a kind of "immutability" of the existence of cosmic vacuum, that is, its constant density and negative pressure. It is assumed that the vacuum of space, affecting the expansion of the Universe (antigravity), nevertheless, remains a stable (at least, after the Big Bang), not changing form of matter, which is affected by nothing. In this most common form of existence of matter, according to modern ideas, the property of self-conservation clearly prevails over evolution. Some scientists suggest that the vacuum of space (dark energy) could have undergone several phase transitions before an event occurred which was called the Big Bang. Therefore, a certain transformation of the vacuum could precede the Big Bang, which took several phases. However, after the Big Bang, dark energy might have not changed and will not change.

TABLE 2.3 Fundamental Forms of Matter in the Universe

The substantial (visible) part of the Universe (baryonic matter): 3% of the energy density in the world	*Dark energy* (cosmic vacuum): 67% of the energy density in the world	*Dark mass* (hidden substance): 30% of the energy density in the world
The average mass density (stars, molecular clouds of hydrogen, etc.): 2×10^{-31} g cm^{-3}	The average mass density: 7×10^{-30} g cm^{-3} (the same throughout the Universe)	The average mass density: 2×10^{-30} g cm^{-3}
Characteristics of development	*Characteristics of development*	*Characteristics of development*
1. Evolves (conserved through evolution)	1. Neither changing nor evolving after the Big Bang, uniformly fills the entire Universe	1. Changing but not evolving as a normal substance
2. Obeys the law of universal gravitation	2. Has the property of antigravity, causing the accelerated expansion of the Universe	2. Obeys the law of universal gravitation
3. Has been expanding for about 7 billion years with an acceleration	3. The composition and structure are unknown; assumed as a homogeneous entity	3. It consists of very heavy particles of unknown nature, weakly interacting among themselves and with ordinary substance

Thus, dark matter, that is, that matter which is invisible (it does not radiate), consists of two fundamentally different parts, the antigravitating dark energy with a density of about 67–75% of the energy density of the Universe, and the "hidden substance" whose density is approximately 23–25% of the total energy density of the Universe, exceeding the density of ordinary visible matter (our "luminous" Universe) by 5–6 times. The share of the latter is only 3–4% of the global energy density. Dark mass/

matter, having not yet entirely clear substantial structure and a diffuse form, gathering in the same place, where the usual substance does (and even "attracting" it), in contrast to dark energy, is subject to gravity, and it occurred immediately after the Big Bang and, at first, significantly affected (dominated) the further evolution of matter. Consequently, in the first half of its existence, the Universe expanded with a deceleration. During the second half of that time, the antigravitation forces of dark energy became dominant and the expansion of it was happening with acceleration. Dark mass, undergoing gravitational forces, interacts with neither matter nor radiation, it absorbs nothing and does not shine, but some processes of change are quite clearly seen therein, which are different from the evolution of substance. This is evidenced by the development of gravitational irregularities in dark mass prior to the recombination of electrons. The density of dark mass decreased, so that gravity dominated no longer in a certain period, having yielded the forces of "universal antigravity." Perhaps that "hidden substance" consists of very long-lived components of an enormous density (in the large scale, of neutron stars, cooled white and brown dwarfs, black holes, including the relic ones, and other components having arisen as early as at the Big Bang).

The dark part of the Universe is really the basic component of all the material and energy content of the Universe, in whose foundation self-preservation clearly prevails over evolution, characteristic of the visible Universe. Such a component dominated in the Universe, which neither change nor evolve (dark energy), followed by the slightly changing and almost not evolving part of the Universe (dark mass), and finally, the most studied evolving fragment in the form of ordinary visible substance.

Thus, the modern cosmological (while substantially hypothetical) picture of the world gives us an unknown or little-known forms of self-preservation of matter, which are the cause and source of the existence of stars in the galaxy, galaxy clusters and superclusters, and other forms of baryonic matter (quarks, bosons, and leptons), which composes the "visible" Universe. It is assumed that dark matter and baryonic matter are interconnected, and even the Big Bang can be interpreted as a phase transition of the said part of dark matter into the "baryon" form of existence.

The nature of the property of preservation of large-scale structures in our world is related to the stability of our Universe for many billions of years. This self-preservation as the dominant and fundamental component of the existence of matter implies an asymmetry between the dark and baryonic matter forms. And what is more, this asymmetry must continue to exist, since otherwise the Universe would collapse into a new singularity (which is now excluded in principle by virtue of the nature of the antigravity accelerated recession of the galaxies), or baryonic matter would have long been transformed into radiation. In this connection, it is appropriate to note that, earlier, Friedman theoretically revealed the existence of open and never-collapsing metagalaxy models containing only ordinary substance. It is only important for this type of models that the total density of all kinds of matter is less than a critical density (according to current estimates, the density of our metagalaxy is close to the critical one). In such Friedmann's open model, collapse can be quite avoided without the presence of dark energy.

From the standpoint of modern science, it is important to answer the question: how can matter self-preserve without evolution and, even more, without any changes? It is unlikely to get a full answer to this question, because both thermodynamics and currently popular synergetics do not allow us to build any kind of explanatory model of this self-preservation of matter for nearly 15 billion years. The more so that this is not some small local component of the Universe (such as the central singularity in a black hole) but its overwhelming majority, almost three-quarters of its mass-energy content. If dark energy is conserved over many billions of years, then either some laws of conservation (unknown to us yet) should act inside it, or we must assume there being no motion characteristic of the real matter therein. This means that, according to modern synergistic concepts (applicable to the substantial part of the Universe), an unusual way of matter conservation exists in the vacuum of space. This is still inexplicable from the standpoint of synergetics, which studies the processes of self-organization and self-disorganization, none of which has been found for dark energy as yet. According to modern concepts, dark energy is not a self-organizing substance (although, e.g., the existence of such exotic objects is assumed such as black holes consisting of dark energy—this is an evolving dark energy or a part of it included in some sluggishly proceeding evolutionary process). Thus, we can assume with high probability that there is some sphere of the Universe, and most vast, in the study of which synergetics yet cannot explain its constant or slowly varying existence. The existence of matter there is most associated with rest, self-preservation, and the phenomenal stability of the vacuum of space, pushing away all different forms of the existence of matter from itself.

So, the property of self-preservation of matter and its specific systems is realized in different ways depending on one or another basic form of the existence of matter as its self-preservation. Meanwhile, matter and the forms of its existence have always been associated with change, movement, and development. Actually, this had been considered until recently, when astrophysics and cosmology gave us a surprise of dark matter in its two forms, namely, almost unchanging (dark energy—the vacuum of space) and relatively unchanging (dark mass subjected to gravitational and, possibly, weak interactions only). Just the immutability of dark energy in one form or another gives us no reason to include it in the process of global evolution, although this form of matter is relevant for the evolution of the Universe, affecting it through antigravity. In our view, the global evolution proceeds only in the substantial part of the Universe (where the anthropic cosmological principle holds true and there is such an attribute as information). In other mini-universes and invisible forms of our Universe, such a holistic and temporally continuous process of self-organization as the global evolution has not yet been discovered. In dark energy, despite some "circumstantial evidence," no evolutionary processes known to us have been revealed as yet. As to dark mass, it can, in principle, be considered as a weakly evolving part of the Universe, although its evolution is not similar to the evolution of ordinary substance (due to a different structure and very weak interactions between their constituents and the environment). Dark mass itself is closer to the evolving substantial Universe, because it consists of some of its constituent particles (at the micro-level, these "elementary" particles, perhaps a thousand times heavier than a proton, the nature of them is not established yet) and is mainly subject to the gravitational interaction, that is, one of the four types of fundamental physical interactions. It is possible that dark mass is a kind of "transition state" of matter between the vacuum of space and baryonic matter as the basic forms of self-preservation of matter.

By the "degree of evolutionability," matter exists in several forms only: (i) unchanging (or almost unchanging) in the form of the vacuum of space (which would be prematurely to identify with the physical vacuum probably being another form of matter self-preservation), (ii) in the form of very little changing substantial dark mass, and (iii) the evolving substantial part of the Universe, only where, as we assume, global evolution proceeds (see Table 2.3). With such a "scenario" of the main forms of self-preservation of matter, it would hardly make sense to talk about the fact that the entire Universe as a whole is subject to global evolution: it is only a "privilege" of its substantial fragment. As for the changes taking place in dark mass, they could be called "proto-evolution," as a kind of intermediary phenomenon between the rest of the vacuum of space and the evolution of the substantial fragment of the Universe.

The scientific picture of the world thus can be significantly transformed. If before, perhaps during the most of the last century there was a paradigm change process, the transition from the representations of the static-stationary Universe to the evolutionarily dynamic nonlinear image of the multiverse (which also does not evolve from the viewpoint of of inflationary cosmology), but now peculiar "the negation of the negation" is not excluded. If dark matter really predominates in the Universe, being the basic and most common component of the Universe, this leads, if not now, then in the distant astronomical future, to the scenario offered by a number of scientists, taking into account only the effect of dark energy (the vacuum of space). According to Leonid Leskov, I. V. Arkhangelskaya et al., "as vacuum is unchanged, then the properties of space-time determined by it must be invariant as well. The world where vacuum predominates must obey Euclidean geometry and be unchanging in time in interstellar scale. Consequently, the evolution of the world gradually fades, its spatio-temporal framework, on whose background the cosmological expansion continues, becomes more static."

2.6.2 *DARK MATTER AND GLOBAL EVOLUTION*

The information criterion of development acts during the course of global evolution as continuous self-organization of material systems in the Universe. It was thought previously that this criterion is characteristic of all forms of matter in the Universe. However, in the light of modern cosmological theories and hypotheses, one has to significantly reduce its scope and to limit it mainly to the areas of the existence of baryonic matter, that is, our substantial Universe. In short, the information criterion of evolution is applicable when and where there are differences, heterogeneity associated with information. Information is such a property of matter which expresses such its characteristics as variety, of which William Ross Ashby, one of the founders of cybernetics, wrote. However, in our substantial Universe, variety is in motion and is transmitted from one object to another during interactions. That diversity which is not transmitted and bounded in a certain way, either is part of the structure and content of material objects or constitutes their memory (in information organs specially formed therein). Diversity (or reflected diversity as information), until recently, has not been investigated by physical methods, focusing on the energy aspects of the study. This also applied to astrophysics and cosmology, until unusual cosmic objects got to the scope of their search, where the problem of diversity and heterogeneity was not minor. First of all, it concerns the initial stage of the origin of the Universe, whose size was 20 orders of magnitude smaller than the radius of the atomic nucleus.

The study of other "extreme" states of matter, namely, the black holes and other superdense cosmic objects, also has led to a definite conclusion of the absence of diversity and uniformity in these exotic states of cosmic matter.

Scientists have explored black holes and the problem of collapsing as the compression of a number of space objects into an encapsulated form, including the singularity as a "point" of an almost infinite density. The spatial parameters (volume, area, length, etc.) "roll up," twist into a self-enclosed ring at this point under the influence of colossal gravity, and time "freezes" there. It has been found that there are neither irregularities nor diversity in these singularly encapsulated forms of matter. Under the rule of Bekenstein, collapsing into a black hole is associated with a "smoothing" of inhomogeneities (heterogeneous) and preserving a few mass-energy characteristics only (mass, own angular momentum, and electric charge). In addition, the black hole as one of the exotic forms of matter self-preservation is characterized by the almost disappeared heterogeneities of the matter from which it has been formed. And the new form (black) of matter existence no longer "knows" its past and, in the present, either contains no "past" diversity or it is significantly reduced. As Seth Barnes Nicholson notes, "a huge amount of information is lost forever at the formation of a black hole." This view is debatable, but, in our opinion, has the right to exist since dark mass is not homogeneous: it consists of some particles at the microlevel and macrolevel, which somehow interact (through gravitational and weak interactions) among themselves and with the surrounding space environment. These interactions give rise to some weak and slow changes, which cannot yet be classified as evolutionary processes. However, these "weak" changes in dark mass turn out to be some "transitional" process from the unchanging homogeneous dark energy to the evolving substance of the Universe with a complex and ever-growing variety of material entities. Since some diversity yet exists in dark mass (e.g., gravitational heterogeneities) and there are changes, we can assume that some changes occur there, which could lead to the appearance of protogalaxies. The protoevolutionary changes in dark mass appear to be the necessary "dark material" condition for the further occurrence of evolutionary processes of baryonic matter and accompany them in the future.

The problem of the appearance, disappearance, and transformation of diversity has found its development and generalization of the so-called Penrose's theorem, according to which all things in the Universe, and even itself, can be subjected to collapse. According to this theorem, singularities will inevitably appear at gravitational collapse and, in the case of a catastrophic gravitational compression (the big cosmological collapse), our multifaceted Universe can finish its story and disappear in a homogeneous superdense singularity-2. Penrose's theorem was formulated at the end of the last century, before the discovery of the phenomenon of dark energy, whose powerful antigravity causes the Universe to expand with acceleration and excludes it from collapsing objects, when, as a result of the hypothetical Big Crunch, a singularity-2 can form. Collapse as the transformation of cosmic matter into the microcosmic one and as the process of catastrophic gravitational compression transforms ordinary substance into an encapsulated and extremely compact form with a space-time folded into a microscopic infinitely curved ball ring. Black holes act as one of the basic and fundamental forms of self-preservation of matter in its hidden, "dark" form. The process of formation of black holes as the conversion of visible matter into its invisible form is a degradation process. It is exactly opposite to the global evolution, because it leads to a total loss of the existing diversity at all levels of the existence of matter. During this auto-disorganization process, complex substance converts into a superdense homogeneity and indistinguishability, and all previous diversity converts into a simple and self-identical homogeneous formation.

This "going to the dark" as a kind of "anti-evolution" of the space macro-matter represents not only the downside of global evolution, but also a peculiar form of (regression) development of matter as its survival during transformation into a singular-encapsulated form: the differences disappear but the matter remains—it continues to exist in a new exotic homogeneous superdense form. However, mutual exchange may occur between baryonic matter and dark matter in suspected transformers intermediaries, the so-called gray holes, where one kind of matter turns into another. Black holes as a "capsule" self-perpetuating matter in its superdense form may be converted, through "gray transformation," into white holes (or even into their possible "naked singularity"), where the process of "re-encapsulation" begins, and the matter comes to the path of complexity and growth of the variety of its forms and neoformations. From this viewpoint, the evolution of substance in the Universe as the emergence and growth of its heterogeneity (diversity) explicitly takes place only since the birth of particles, that is, in fact, with the plasma radiation and atomic substance phases of evolution and the splitting of the single fundamental interaction into the known four main types of physical interaction.

Another viewpoint is more speculative because it suggests that heterogeneities still exist in the singular, encapsulated form of matter but have some other (e.g., the virtual quantum or subelemental one) form of existence.

However, in general, heterogeneities, and hence diversity, is an attribute of that form of matter which has adopted the structure of the substantial Universe, which is only a few percents of our mini-Universe. This enables the use of information representations for describing the evolution and self-organization of systems of the baryonic matter in the visible Universe. The information criterion of development formulated at the beginning of the second half of the 20th century is fully applicable to the self-organizing forms of matter only, which contain and vectorily change their variety. Substantially, this criterion displayed long-term processes of the progressive development of material systems, which increased their complexity and the level of evolution, accumulating diversity.

It is possible that the emergence and growth of the amount of information (as the diversity of structures and forms of matter) within only a small part of the Universe is caused by something, some yet unknown law of distribution of the forms of matter in the Universe is valid, similar to the already mentioned negentropic pyramid during the complication of substantial material systems. No coincidence that the appearance and subsequent growth of diversity in the evolutionary processes in the substantial fragment of our Universe is "paid" by greater and greater monotony and motionless in the larger part of it. The less complex turns to be the less variable, while the more complex, increasing its diversity (the information content), increasingly narrows its volume, mass, and total energy. This trend is typical for both ordinary matter and dark matter, wherein the homogeneous and unchanging dark energy exceeds dark matter about three times.

Most of the "dark side" of the Universe is hardly included in the global evolution. But the dark part of our Universe generates some space conditions without which the said evolution likely would not occur, at least in the form in which we know it.

Approximately during the first 300,000 years after the Big Bang, dark mass generated gravitational inhomogeneities of the substance distribution in the Universe, wherein galactic clusters and galaxies were later formed. Without that influence no further global evolution would be, most likely, possible, as the galaxies could not be formed in the absence or insufficient amount of dark mass. In the first minutes and hours after the Big Bang, the distribution of ordinary substance in the Universe was very uniform and remained so until the recombination of protons and electrons at the Universe's age of about 270,000 years. The gravitational condensation of ordinary matter was prevented by the pressure of radiation, which this substance intensively interacted with. Meanwhile, dark mass did not interact with radiation, and nothing prevented the formation of gravitational condensations. Therefore, by the age of about 270,000 years, a definite structure of inhomogeneities consisting exclusively of dark gravitating mass had already formed in the Universe. After the recombination of electrons, ordinary substance just fell into the gravitational potential wells prepared by dark mass before. If the latter would not have time to form "dark" protogalaxies, no galaxies could be formed from ordinary matter in the future, and the said substance would have dissipated over the Universe. Moreover, the modern galaxies and their clusters cannot exist outside of the potential wells formed by dark mass. It is possible that dark mass can interact with ordinary matter not only through gravity. There is an assumption that the interaction of that part of dark matter with ordinary matter is possible through the weak interaction (as in the case of the neutrino).

Could the "dark" part of our Universe be included in the global evolution on the basis of the above? There is still no unequivocal answer to this question, although it can be assumed that it makes sense to limit (to some extent) the global evolution of the Universe by the baryonic forms of matter only. This is due to the fact that the "dark" part of the Universe almost does not evolve in that sense in which modern science and scientific philosophy attach the notions of "evolution" and "development" in the study of the visible part of the Universe. After all, these notions imply that the appropriate forms of matter and their specific substantive entities possess directed changes in the content of material entities, usually irreversible (to keep their entropy). However, although some changes occur in dark mass, this does not allow one to suggest the presence of evolution processes and, therefore, the term "proto-evolution" can be used. Protoevolution is as if not yet evolution in the substantial Universe, but no longer billion years peace of the vacuum of space. Meanwhile, many astrophysicists, cosmologists, and philosophers who study and interpret the phenomenon of dark matter (mostly as dark mass) are inclined to conclude that this part of our Universe evolves. Especially such a conclusion applies to dark energy.

In discussing the problems of the global evolution, no issues related to the role of dark matter in this process have emerged until recently. Basically, what was spoken of was that some of the global characteristics, primarily the basic physical constants corresponding to the four fundamental types of material interactions, their adjustment, and some already known parameters of the Universe (the dimension of time and space, topology, etc.) are such that allow the process of evolution (including the global evolution) on top of which is the man, on whom hopes are now set for its continuation in the cosmological perspective. Meanwhile, the existing dark energy as the vacuum of space with a constant and unchanging energy density has a very significant impact on the process of evolution of the substantial part of the Universe. Dominating in our Universe, dark energy three times exceeds all other forms of cosmic matter together by energy density, inducing strong global antigravity. With the age of the Universe of 6–8 billion years, the era of cosmological expansion began with acceleration due to the fact that the density of dark mass gradually decreased and became lower than the density of vacuum. This antigravity expansion of the Universe was replaced by the cosmological era of gravitational dominance over antigravitation and substantial forms of matter over vacuum (dark energy). The fact that the yet unclear world of dark energy determines the cosmological expansion, which, according to modern ideas, will continue indefinitely, creates confidence in the fact that the Universe is no longer threatened by the Big Crunch, which could lead to a new (second) cosmological singularity.

In theoretical cosmology, other models of the universe have been revealed as well, which are compatible with evolution and global evolution. As noted, there are open and not collapsing patterns/models, which contain no dark energy, that is, the cosmological constant is zero. Closed models of the Universe with dark energy (with a nonzero cosmological constant) are described as well. If all dark energy in such cosmological models would be replaced by dark mass or even ordinary substance, while keeping the average density of matter, the Universe would remain flat and open, it will expand forever, but with power-law deceleration rather than acceleration. However, the reality of the existence of dark energy suggests that the permanent continuation of the super mainstream of global evolution has "required" the real existence of a new form of matter in the Universe rather than theoretical models with no coming collapse. Just this form as dark energy has the property of "universal antigravitation" for the Universe not to go on the regressive degradation path and no universal collapse will be realized. This seems to be one of the main directions of relation of the existence of dark energy and the prospects for the global evolution continuation, particularly in the context of its socio-natural development. Preventing the possible compression of the Universe, the vacuum of space, with its antigravitation, quite realistic warns the universal threat of collapse and is "useful" for further continuation of the global evolution. However, the fact that antigravity will continue to push apart the galaxy faster and faster, will lead to their gradual disappearance. The surrounding space will become increasingly empty (in the substantial sense), transforming the galaxy into an isolated island, independent of the gravity of other space objects.

The problem of dark energy is directly related to the question of evolution (or the lack thereof) and contributed in the cosmological debates for nearly a century ago. The question of the stability of the world was discussed, starting with Einstein, who believed the Universe to be static and unchanging. But in 1917, having applied the general theory of relativity to cosmology, Einstein suddenly discovered that his cosmological model did not confirm the eternity, immutability, and static nature of the Universe. Therefore, to retain the idea of the static and unchanging Universe, he introduced the so-called cosmological constant as one of the fundamental physical constants. The first cosmological model of the world, proposed by Einstein, was a perfectly symmetric model of the Universe in space and time, and this representation had remained up to

Hubble's discovery of space expansion (1929), that is, removal of the galaxies from each other, which dispelled (seemingly forever) the idea of a static, unchanging Universe. However, as noted by Arthur Chernin, "not only the cosmological constant but the original idea of a static Universe have suddenly found a new look and a new life today... But what is striking of all is, perhaps, that the traditional idea of the static world is in remarkable agreement with the phenomenon of the cosmological expansion...." And this is true: the cosmological expansion as a continuation of the Big Bang was caused just by dark energy. And the Big Bang itself could have been caused by this cosmic vacuum (its spontaneous fluctuation), and the very "recession" of the galaxies with acceleration can also be interpreted as a new stage in continuation of the Big Bang. Moreover, the vacuum of space is the same everywhere, it exists around us, and its density and pressure are constant. It has been established that dark matter is not affected by the rest of the material content of the Universe, but it itself, as already noted, affects everything else, both dark mass and the baryonic form of matter. The vacuum of space is not subject to the already known physical laws, in particular, Newton's law, according to which "to every action there is always opposed an equal reaction." The vacuum, as the fundamental part of the Universe, is static and unchanging, and it ultimately determines the properties of space and time. This means, according to Arthur Chernin, that the world, where the vacuum dominates, must be constant in time and homogeneous in space, static, and all its four-dimensional point (event) are indistinguishable. It will be a world without information and a new analogue of thermal death in the "vacuum" version.

Scientists and philosophers as long ago as in the 19th century tried to prove that the Universe is not static, fixed and unchanging, that not only visible mechanical movement of celestial bodies takes place therein, but also evolutionary processes occur, complication proceeds while the transition to a higher structural level and the growth variety of forms and types of matter. This is really what happens if we keep in mind our visible Universe with its baryon form of self-preservation and change of matter, especially on the mainstream of global evolution. But the discovery of dark matter has raised the question of that the vast majority of matter is either unchanging or inactive (dark mass) and somehow (mysteriously) self-perpetuates in some special form unexplainable by science as yet. Even synergetics is applicable to the real part of the Universe only, where self-organization processes and their opposites occur. But how the antigravitating part of the matter was self-perpetuating many billions of years (the most part, probably, even since the Big Bang) is unclear as some laws of conservation acts there, unknown to us. But when it is cleared up, the modern scientific picture of the world will change so much that there will be little left of today's very common and seemingly unshakable ideas. All this suggests that the scientific picture of the world is on the verge of truly revolutionary transformations. Further study of dark matter not only presages a cascade of new scientific discoveries, but also significant philosophical innovations, the growth of the bifurcation state in the entire scientific activity. Will they affect the accepted scientific concepts and philosophical categories, our understanding of their fundamental and primordial nature? After all, they have been formed on the basis of studies of the visible Universe, which changes and evolves. But the "dark part" of the Universe tends to constancy and stability.

Since the days of ancient thinkers, philosophers were tying together matter and movement development. Hegel wrote, "Just as there is no motion without matter, so there is no matter without motion." But now it is argued that, at least in dark energy, that is, three-quarters of our Universe, matter itself exists while no changes, and even more so—no evolutionary processes—are observed. Matter in its dark stable form exists with no evolution, no development. It is unclear how the self-preservation of the three-fourths of matter without evolution happens, in any case, in the visible Universe the self-preservation of matter has occurred and occurs through movement, development, self-organization, and evolution. The recognition of the absolute immutability of three-quarters of the Universe thus casts doubt on the position of the dialectic statement of movement being an attribute of matter.

Thus, the idea of evolution and global evolution in the scientific picture of the Universe is being essentially transformed due to the discovery of the dark sides of the Universe; an entirely new vision of the world is being formed. The type of evolution, which was often called the universal evolution, is not as versatile as it seemed only recently. We have to abandon the term "universal" and give preference to the term "global." Thus, globality is thought but in its "substantial incarnation" only, that is, it is understood that the considered type of evolution covers only the "information" (relatively minor) part of the Universe. It is quite clear that this type of evolution is not typical for the "dark grounds" of the pyramid of matter self-preservation in the Universe, and even more so—for a set of mini-universes similar to it.

2.6.3 *THE PROBLEM OF COGNITION OF DARK MATTER*

The problem of cognition of dark matter forms has its own specifics. Let us take, for example, such a component of dark mass as a *black hole*, which is a closed sphere formed by the collapse of a massive star, and where the substance trapped therein cannot go out even in the form of radiation because of the monstrous compression (gravitational attraction). Therefore, under the black hole is understood a region of "space-time," for which the escape velocity is equal to the speed of light and the gravitational field does not release even photons. The spatial boundary of a black hole was named the event horizon, beyond which it is impossible to obtain any information about the events and states inside the black hole. Any cosmic body, matter and radiation can penetrate, fall into a black hole, but cannot leave it. In short, the cognizing subject is not able to obtain information of the internal state of the black hole if he is outside it. This applies even more to the center of a black hole—a singularity as a superdense state, where, as expected, the known laws of nature do not act. However, cosmic matter that falls into a black hole does not lose all its information, because it can be characterized by means of mass, electric charge, and собственного (its own) angular momentum. In the future, other characteristics and properties of the matter composing a black hole can be found, if it really contains information. Consequently, whatever the black hole has been formed from, the vast majority of the previously existing differences disappear in a superdense (10^8 t cm^{-3}) homogeneous medium inside it.

If the interior of a black hole cannot be perceived by an external observer, the very existence of this latent space object can be found, in particular, due to the enormous gravitational fields representing powerful potential sources of energy, which, in principle, can be released when substance gets to the event horizon of the black hole. At this fall (accretion) of substance on the black hole, a very large amount of energy can be emitted before it crosses the border (the event horizon) of the black hole (the presence of an X-ray halo around it is possible). Of course, if there are no matter and radiation in the vicinity of the black hole, it will remain undetected. If there is some substance in the vicinity of a black hole and it falls into the black hole, the external observer will see the effect of radiation emission (as if particles emerge from this hidden space object). When the substance of stars and gas clouds gets into supermassive black holes, their brightness will be the most powerful in our Universe, since a huge amount of energy is released in this case (by two orders of magnitude higher than in nuclear reactions). Such supermassive black holes with powerful gravity can serve as "dark" (hidden) energy sources in several cosmic processes with enormous energy output and giant luminosity (quasars, the nuclei of active galactics). A black hole with a mass of about 3 million solar masses is apparently at the core of our galaxy.

Does the black hole have any information content beyond its event horizon? According to Igor Gurevich, a black hole does contain information, and its volume is proportional to the square of its mass. And bearing in mind that supermassive black holes exist at the nuclei of almost all galaxies, the amount of information in the Universe is within the range 10^{99}–10^{107} bits. Such a statement makes sense on the assumption of the existence of stable irregularities and "hidden diversity" within black holes. However, it is unclear how to extract this supposed "black information," since no signals emitted from the inner part of such a space object, go outside. It turns out that in the most ("dark") side of the Universe, where there is no evolution, information is either absent or insignificant.

The conclusion of the absence of information in most of the inanimate nature of space has unexpected epistemological implications. That is, our cognition of such forms of matter is either impossible or significantly impeded. However, if there is no information of any material object at all, then it simply must not exist in cognition. Therefore, the presence of knowledge about any material object already indicates that it contains within itself or in its "outer contour," which somehow reaches the cognizing subject. Dark energy is detected by the indication that it affects other forms of matter and some specific material objects. And this means that the vacuum of space as an integral material formation has some minimal amount of information, which somehow "reaches" the cognizing subject. This information may be contained in the "outer loop" of the vacuum of space, that is, as some whole object, where integrity is understood not as some "fencing" from other forms of matter, but as the cumulative impact on them by dark matter.

The problem of cognition of dark matter and other exotic hidden superdense cosmic objects is a special epistemological problem, as the authenticity and reliability of the results of scientific research is judged by indirect, incidental signs of their influence on the normal luminous matter. For example, we do not know how to use synergetics and thermodynamics to study dark energy, because there is not—literally—information enough. However, in cosmological simulations, the analogy between the physics of black holes and thermodynamics is used, as well as between them and the theory of gravity. Our cognition of the vacuum of space is more difficult because it does not interact with anything, including a supervisor, although its impact on the substance in the Universe does not merely exist but is crucial for the future of our universe.

The processes of evolution and related information processes play an important role not only in the universe itself, but in its cognition by the man and mankind in general. The connection between the universe and its properties is reflected in the anthropic cosmological principle, which holds true in the visible universe. This principle speaks of the presence of very fine "tuning" of the fundamental constants of the universe and its global characteristics with the possibility of permanent proceeding of evolutionary processes therein, leading to the formation of complex forms of matter, and, at some point, to the emergence of life and the man who is able to cognize the universe. The basic idea, whatever forms are taken by this principle, is that the man and the substantial universe suggest their mutual existence, the existence of human "needs" the corresponding "space home" in the form of the universe (which now appears as one of the local mini-universes of multiverse). And vice versa: the Universe must be such that man can appear and exist, who can cognize the world corresponding to him. Between man appeared at a certain stage of the evolution of the substantial universe, and all the preceded stages, including the initial stage of the birth of the Universe, there is as if feedback, likely, of the information nature, which follows from the principle of system-temporal integrity. Humanity does not wish to disappear from the face of the earth as well as of the Universe; it tries to extend its existence in the Universe. And, if the spontaneous development of humanity is incompatible with the evolution of nature, then there appears an idea to change the form of development, to fit in the global evolution mainstream. This is the philosophical and ideological imperative of the anthropic principle: man must be present and act in the Universe as long as possible (indefinitely long or desirably always).

KEYWORDS

- **modern natural science**
- **nonlinear science**
- **the scientific picture of the world**
- **universal evolutionism**
- **information**
- **nonlinear dynamics**
- **geosystems**
- **dark matter (energy)**

REFERENCES

1. Alexandrov, S. M. *Nonlinearity of the Relief-Forming Processes and Extreme Situations*; RFBR. IG RAN: Moscow, 1996; p 112. [Russ].
2. Aleshkovsky, I. A.; Ivanov, A. V.; Il'in, I. V.; et al. *Simulation of the Nonlinear Dynamics of Global Processes*; MSU Press: Moscow, 2010; p 412. [Russ].
3. Arnold, V. I. *Catastrophe Theory*; 3rd ed.; Springer-Verlag: Berlin, 1992.
4. Bykov, V. G. *Nonlinear Wave Processes in Geological Media*; Dal'nauka: Vladivostok, 2000; p 190. [Russ].
5. Gabdullin, R. R.; Il'in I. V.; Ivanov, A. V. *Evolution of the Earth and Life*; Moscow, 2005. [Russ].
6. Yegorov, E. N.; Ivanov, A. V.; Koronovsky, A. A.; Trubetskov, D. I.; Khramov, A. E.; Yashkov, I. A. *Synergetics and Geo-Ecology: Experience of Co-Evolution*, Proceedings of the Universities. Applied Nonlinear Dynamics, 2010; Vol. 18, No. 2, pp 145–149. [Russ].
7. Kazantseva, T. T. *On the Linearity and Nonlinearity in Geodynamics in Relation to the Geological Laws of Evolution*, Proceedings of the Department of Earth Sciences, the Academy of Bashkortostan, Ufa, 2002; No. 8, pp 115–123. [Russ].
8. Kapitsa, S. P.; Kurdyumov, B. B.; Malinetsky, G. G. *Synergetics and Forecasts of the Future;* Nauka: Moscow, 1997; p 285. [Russ].
9. Knyazeva, E. N.; Kurdyumov, S. P. *Synergetics: The Non-Linearity of Time and Landscapes of Coevolution*; ComKniga: Moscow, 2007; p 272. [Russ].
10. Kovalenko, V. V. *Нелинейные аспекты частично инфинитного моделирования в эволюционной гидрометеоэкологии* (*Nonlinear Aspects of Partially Infinite Modeling in Evolutionary Hydrometeoecology*); SPb, RSHU Press, 2002; p 158. [Russ].
11. Kotlyakov, V. M.; Lebedeva, I. M. "Penitent" snow and ice, the mechanism of their formation and their indicator value. Izv. USSR Academy of Sciences. *Ser. Geogr.* **1975,** *3,* 26–36. [Russ].

12. *Coevolution of Geospheres: from the Core to the Cosmos*, Proceeding All-Russia Sci. Conf. in memory G.I. Khudyakov, Saratov, Saratov State Tech., University Press, 2012; p 472. [Russ].
13. Kravtsov, Y. A.; Etkin, V. S. *Wind Waves as a Self-Oscillating Process*. USSR Academy of Sciences. *Izv. Atmos. Ocean Phys.* **1983,** *19*(11), 1123–1138. [Russ].
14. Letnikov, F. A. *Synergetics of Geological Systems*; Nauka: Siberian Branch, Novosibirsk, 1992; p 230. [Russ].
15. Naidenov, V. I.; Kozhevnikovf, I. A. *Chaotic Dynamics of the Hydrosphere and Climate*. Dokl. RAS. 2002; Vol. 384, No. 3. pp 385–390. [Russ].
16. Sokolov, S. D. *Bifurcation in Geology, on Nonlinear Geology and Geodynamics*; GEOS: Moscow, 1998; pp 28–36. [Russ].
17. Sorokhtin, O. G. *Evolution and Forecast of Changes in the Global Climate of the Earth*; Institute of Computer Studies: Moscow, 2006; p 88.
18. Haken, H. *The Phenomena of Transition and Transition Processes in Nonlinear Systems, Synergetics*; Mir: Moscow, 1984; pp 7–17.
19. Chuprynin, V. I. *Nonlinear Effects in Geosystems*; Nauka: Moscow, 2008; p 197.
20. Ananiev, B. G. *Systematic Approach in Modern Science*; Progress-Traditia: Moscow 2002.
21. Borinskaya, S. A. *Principles of Evolution in Nature and Society*; Moscow, 2002.
22. Galimov, E. M. *The Phenomenon of Life: Between Equilibrium and Non-Linearity. Origin and Principles of Evolution*; Moscow, 2001.
23. Gurevich I. M. *Laws of informatics as the Foundations of the Structure and Cognition of Complex Systems*; Moscow, 2007.
24. Knyazeva, E. N.; Kurdyumov S. P. *Synergetics: the non-linearity of time and landscapes of coevolution*; Moscow, 2007.
25. Moiseyev, N. N. *Universal Evolutionism*; Moscow, 1991.
26. Moiseyev, N. N. *Universum. Information;* Society, Moscow, 2001.
27. Ursul, A. D.; Ursul, T.A. *Universal Evolutionism (Concepts, Approaches, Principles, Perspectives)*; Moscow, 2007.
28. Ursul, A. D. *Informatization of Society: An Introduction to Social Informatics*; Moscow, 1990.
29. Ursul, A. D. *Scientific World Pattern of the 21st Century: Dark Matter and Universal Evolution*; Eurasia Safety: Moscow, 2009.
30. Cherepashuk, A. M.; Chernin, A. D. *Universe, Life, Black Holes*; Vek-2: Friazino, 2003.
31. Chernavsky, D. S. *Synergetivs and Information*; URSS: Moscow, 2002.
32. Goldfein, M. D.; Ivanov A. V.; Malikov A. N. *Concepts of Modern Natural Sciences. A lecture course*; Goldfein M. D., Ed.; RGTEU Press: Moscow, 2009.
33. Goldfein, M. D.; Ursul, A. D.; Ivanov, A. V.; Malikov, A. N. *Fundamentals of the Natural-Scientific Picture of the World*, Goldfein M. D., Ed.; SI RGTEU Press: Saratov, 2013.

CHAPTER 3

ROLE OF MATHEMATICS IN SOLVING SCIENTIFIC AND PRACTICAL TASKS OF MODERN NATURAL SCIENCES

CONTENTS

3.1 Specificity of Mathematics and Milestones of Its Development 86
3.2 Definition of the Subject of Mathematics 91
3.3 Mathematical Methods and Experiment 92
3.4 Leads, Problems, and Limits of the Mathematization of Science 94
Keywords 97
References 98

One, who wants to solve questions of natural sciences with no mathematics involved, poses an unsolvable problem. Measure what is measurable, and make measurable what is not so.

— *Galileo Galilei*

Any theory of nature will contain only so much real science as it permits the application of mathematics.

— *Immanuel Kant*

ABSTRACT

The role of mathematics in solving the problems of modern natural science is demonstrated. The development of this science is traced. The definition of the subject of mathematics is given. The relationship between mathematical modeling and experiment is shown. The mathematization of modern science is predicted.

3.1 SPECIFICITY OF MATHEMATICS AND MILESTONES OF ITS DEVELOPMENT

The great mathematician of the 20th century Academician Andrei Kolmogorov once noted that there were no penetration limits of the mathematical apparatus into any particular science. "The principal application field of the mathematical method—he emphasized—is unlimited: all kinds of movement can be studied mathematically." From the very instant of its origin, mathematical knowledge showed a tendency to generality. As was already mentioned, Pythagoras and his followers either identified numbers with things themselves or saw the essence and causes of things in numbers, treating arithmetically all beings, including man, the whole cosmos, and so forth. Numbers were mystified, they were given the divine religious sense, and they were elevated to the basic principle of the world's existence. The concept of the Pythagoreans already contained the idea of the universality of mathematics and the future mathematization of scientific knowledge—the idea which could not be properly developed more than two millennia ago, even by the prominent thinkers of that time.

In the modern era, the universality of mathematics occupied the thoughts of René Descartes (see his "Rules for the Direction of the Mind"), who believed that "there must be a general science which explains all the points that can be raised concerning order and measure irrespective of the subject matter, and that this science should be termed mathesis universalis … for it covers everything that entitles these other sciences to be called branches of mathematics." Descartes' "mathesis universalis" was not the contemporary mathematics but a science of the future, an ideal to which mathematical knowledge may tend by developing its universal and general scientific features, by exploring order and measure. The great philosopher and mathematician drew his attention to the possibility of the general scientific nature of mathematical knowledge, raising it above any other science. Subsequently, the idea of the universality of mathematics was again developed by the outstanding mathematician Gottfried Wilhelm Leibniz. The philosopher Immanuel Kant believed that "Any theory of nature will contain only so much real science as it permits the application of mathematics."

However, only in the second half of the 20th century, the advances in applied mathematics and modern computer science allowed us to revive (on a new basis) and to justify the ideas of those thinkers who had spoken about the universal character of mathematics. Now, when the process of mathematization has embraced the natural, technical and social sciences, fundamental, applied, and other sciences, and has shown its efficiency, the general scientific nature of mathematical knowledge manifests itself quite clearly.

Let us consider some main features of mathematics as a general scientific phenomenon and its integration and interaction, in this capacity, with special sciences in the way of knowledge mathematization.

What does mathematics study, what is its subject? Only recently this science was considered as studying the quantitative relations and spatial forms of reality. This concept already isolates mathematics from other sciences studying certain fragments of nature, society, or thinking. Mathematics, in this view, studies certain aspects of reality (which are immanently inherent in all forms of the movement of matter) rather than fragments of it. Therefore, such an interpretation of mathematics as a reflection of certain aspects of reality points to this science as to a sufficiently broad branch of knowledge, essentially distinct from other special sciences.

Some changes occurred in mathematics in the last century, which have expanded the subject and methods of its research and isolated it from particular sciences. Mathematics differs from the particular (especially experimental) sciences, first of all, by it not focusing attention only on the spatial and quantitative relations of the reality. It also investigates, as will be shown below, other aspects of both the external world and our thinking (the ideal world) and creates it as well as reflects it. We have every reason to consider that mathematics is defined not by its subject of research (which is continuously enriched and complicated) but by the method of its approach, representing merely a specific language for logical symbolic data processing in other sciences. It would be strange, however, to count that mathematics has no subject and acts just as a method for other sciences. Mathematics really looks as a method (research means) for external use but inside itself it, nevertheless, is engaged in searching its own subject with the purpose to transform it into a method during mathematization. What proves it is the evolution of mathematical knowledge. The difference of mathematics from natural sciences (as basically empirical knowledge) is also due to that its axioms must be logically consistent, while the requirements of validity, obviousness, visualization, and experimental confirmation are inapplicable. Old mathematical theories are not rejected by new, more general ones (despite changes in the language, research areas and programs, and conceptual priorities), and mathematical knowledge accumulates all the time, that is, develops cumulatively.

At consideration of specific features of modern mathematics and at revealing its subject, method, and prospects, it is necessary to note three features, or three directions of development, which substantially define it as a science. It is, first of all, the modern dominating structural approach, that is, the representation of mathematics as a set of certain mathematical structures, based on the set-theoretic approach, traditional since the time of Georg Ferdinand Ludwig Philipp Cantor. Second, it is the category-theoretic approach developing in last decades, which has declared its claim for the status of the "principles" of modern mathematics. And though it is not recognized so widely as the structural set-theoretic approach but it possesses a powerful ideological supply for mathematical knowledge progress. Third, it is the cybernetization and informatization of modern mathematics, the development of discrete mathematics, and the wide usage of computers (especially for applied purposes). The changes in modern mathematics have essentially affected our understanding of its subject.

This, first of all, concerns the notion of quantitative relations, which the notion of mathematical structure is now included into, in addition to quantities and spatial relations.

According to Nicolas Bourbaki's concept (the 1960s), the definition of the notion of mathematical structure is connected with specifying several relations which its elements are in, and with introducing conditions, that is, such axioms which the introduced relations of the structure should satisfy. Mathematical structures express abstract general steady relations, abstracted from the nature of the initial elements and their interpretations. Nicolas Bourbaki has identified three basic structures. *Algebraic structures* are most often sets in which such a relation between three elements is specified when the third element is unequivocally defined by the two others. Groups, rings, and fields exemplify this type of structures. In *topological structures*, attention is focused on the notions of limit, continuity, and neighborhood, and in *order structures* it is paid to orderliness relations.

The structural approach in mathematics is the further development of the set-theoretic approach to its subject, the successive application of the axiomatic method, and allows uniting all the existing sections of modern mathematics from general positions.

At the same time, the structural set-theoretic representations cannot now pretend to the role of the unique and most general bases of mathematical knowledge, since they could not always be necessary and effective. For example, in relation to the definition of the notion of vector (usually defined as a directed straight-line segment), this approach is rather complicated. The set-theoretic unity of mathematics should not be presented as a benefit in any respect, and the emergence of the category theory is clear evidence of this, because this theory returns to visibility to a certain extent (though this theory should hardly be absolutized, which, as well as the theory of sets, is important for performing a specific role in the development of mathematics). The category-theoretic concept essentially simplifies bulky set-theoretic constructions in mathematics. The term of category in mathematics is adopted from philosophy (Aristotle and Kant), which is due to the aspiration to introduce most general notions into mathematics as well. The mathematical theory of categories can be considered, in a certain relation, as further generalization of the theory of sets, because a variety of traditional mathematical notions, including sets, can be considered as objects of mathematical categories. One can speak about the category of all sets, where the object class will consist of sets, and the class of morphisms will consist of sets of maps. The mathematical theory of categories is, to a certain extent, generalization of the set-theoretic approach but does not replace it completely. The general theory of categories is abstracted from a number of properties inherent in every particular mathematical theory and it is quite clearly revealed by transposition of such theories to the language of the category theory, with the result that not all of their content (including some essential properties) can be expressed on a theoretical categorical language. This fact limits, to some extent, the capabilities of the category theory. The transition from classical mechanics to relativistic mechanics and further to the special theory of relativity can be regarded as such an example in physics. The category theory has arisen on the joint of two most general mathematical theories, as synthesis of the algebraic and topological structures, and the further development of mathematics has shown that this new theory "reflects" something essential to modern mathematics, prepared by its previous development. This means that no appearance of the category theory would be possible without the "set-theoretic stage" of mathematics development, without the elaboration of algebraic theories, topology, functional analysis, mathematical logic, and so forth. In the course of development of these theories and, in general, the set-theoretic thinking in mathematics, it has become clear that most important results in it have been obtained while researching the *relations* between objects rather than these *objects* themselves. In mathematics, the idea of identifying isomorphic objects evolved with the advent of analytic geometry, where a plane could be associated with a set of pairs of numbers. The most significant stimulus for the extensive development of the category theory was the highest stage of the set-theoretic development of mathematics in the framework of Bourbaki's structural-mathematical approach. Perhaps, it is the "structuralization" of mathematics, which has made it possible to clarify the important idea that the morphisms of mathematical structures are more essential than these structures themselves. And if the theory of sets can be figuratively considered as the calculus of objects, then the category theory is the calculus of morphisms.

In every single mathematical theory, morphisms are mapping some objects onto others; they give an "imprint," an image of one object in another one. Due to the fact that the morphisms preserve the structure of their objects, the image of the object is, in turn, an object of this type. This makes it possible to judge on the properties of some object (the original) by the properties of its image, and vice versa. The degree of the image–object identity, the copy–original identity, depends on the properties of the mapping type, that is, the morphism. This reveals such an important fact that many types of morphisms having the necessary properties can be studied without taking into account the internal (mathematical) structure of their objects themselves, but only on the basis of the nature of their relationship with other morphisms.

3.2 DEFINITION OF THE SUBJECT OF MATHEMATICS

The subject of any science, including mathematics, differs from the object of cognition by studying certain features of the latter one. Of the whole real world, mathematical cognition extracts only the aspect of number, shape, space, reflection, and relation—and explores it by specific techniques, creating its own special abstractions to more and more adequately reflect the object of study. Note that mathematical abstractions in the process of their development, the emergence of new levels of concepts, develop qualitatively new properties of mathematical cognition. On the one hand, there is still deeper studying the quantitative relations of reality, the formation of increasingly general and invariant "quantitative" abstractions. On the other hand, these new "quantitative' abstractions reveal more and more profound links with such common elements of quality and content, which are in indissoluble unity with quantitative relationships. However, the abstractness of mathematics from any specific content leads to the fact that it explores not only metric (quantitative) relationships or some abstract symbolic structures and categories but also any content that is subject to logic, deductive analysis. The abstractness from that content is a peculiarity of mathematics, its difference from the natural sciences and other empirical knowledge, which focuses on the substantive laws and tendencies of the object of cognition. This object, just as in philosophy (and science in general) is the whole *Universum* in its material and ideal forms. However, mathematics itself in relation to the *Universum* is merely a special symbolic language and method to study, in axiomatic form, models, operations, mental structures, and procedures, which could give new knowledge about information obtained in other specific sciences. The scope of mathematical creativity is likely designing specific mental structures, a kind of "self-reflection" but not just a reflection of what exists outside of mathematics. And this became apparent, especially after the emergence of the theory of mathematical categories.

The "self-sufficiency" of mathematics, requiring the consistency of its theories only, can be an advantage as well as a disadvantage in the sense that mathematics itself, to be considered a science in the true sense, should go beyond its limits, especially in the sphere of empirical knowledge. In other words, mathematics without the mathematization of other sciences, in principle, cannot exist as a social and cultural phenomenon. This is a kind of the systematic approach to science as the whole phenomenon rather than the sum of its individual disciplines. Mathematics binds them into a coherent and universal whole, demonstrating its generality, universality, and general scientific nature.

It thus becomes clear that mathematics, since its inception, has claimed its universality in the system of scientific knowledge, and this universality and general scientific nature was due to the fact that emphasis in mathematical knowledge was made on the form rather than the content—in both the subject of mathematics and its methods, which allows a much deeper insight into the content because of the immanent existing relation between the form and the content.

Exploring the form itself, in abstraction from its content, using special symbolic deductive systems have shown the possibility of obtaining deeper knowledge in the field of substantive theories. Therefore, the most preferred is the definition of the subject of mathematics given by Alexander Alexandrov a few decades ago: "In general, the subject of mathematics can include any forms and relationships of reality which objectively have such a degree of dependence on their content that they can be abstracted from it completely." We could add Bourbaki's definition of mathematics, which states that, in its axiomatic form, mathematics is a collection of abstract forms—mathematical structures.

3.3 MATHEMATICAL METHODS AND EXPERIMENT

The notions of spatial forms, quantitative relations, mathematical structures, categories, and so forth reflect essential aspects of mathematics as a science having not only its subject but also its method. Mathematics, as a method, acts as some kind of secondary science directed to serving other sciences (e.g., physics) having a specific material object and stimulating mathematical knowledge development. This creates an illusion of mathematics being among sciences, where observation and experiment are crucial. However, mathematics is not based on experimental facts and public practice as experimental natural sciences are; it can create its own virtual-theoretical reality without them, it can precede any experimental theory or cannot find such one for a long time. However, there is one new form of practice where mathematics designs its own experiment. It is a so-called mathematical experiment (or simulation), which has arisen on the way of scientific knowledge informatization and represents one of the forms of mathematics–cybernetics integration.

Mathematical experiment, as Academician Victor Glushkov has noted, “takes an intermediate place between the classical deductive and classical experimental methods of research. Its appearance has led to a new philosophical problem, namely, the necessity of reconsideration of not only the subject of mathematics but also of the method of mathematical research”. Without going into the details of this method, we consider some issues related to the extension of the subject of mathematics. Victor Glushkov notes that mathematicians have long been included a number of theories and disciplines dealing not only with spatial (geometric) but also quantitative characteristics into the composition of their science. This is due to the fact that for the objects studied by mathematical logic, the apparatus of deductive construction is fairly well developed; there are fairly general statements expressed by theorems and lemmas. But a similar apparatus almost has not been developed in the case of the method of mathematical experiment. At the same time, the status quo does not mean that such an apparatus, in principle, cannot be designed. Moreover, the whole experience of the development of science shows that the process of mathematization of knowledge, sooner or later, but inevitably would lead to the design of such an apparatus. Mathematics thus answers to a huge qualitative variety of research objects in various areas of knowledge by adequate capabilities of expanding its apparatus of deductive constructions. An important role is played by computers to automate deductive constructions and to increase the efficiency of scientists’ intellectual work. The method of mathematical experiment, essentially strengthening mathematics’ expansion into other sciences, gives evidence of the essence of the influence of informatics upon the development of mathematics, of the appearance of new essential links between mathematics and cybernetics (computer science) to expand and integrate their original, traditional subjects. Through the use of powerful computers, it has been made possible full mathematical modeling for both internal and external requirements, since no simplification of mathematical models is needed, when used in scientific theories and when processing of empirical data. In particular, the success of cosmonautics (astronautics) in the second half of the 20th century is obliged to the informatization of calculations. With the help of mathematical experiment, much more computational procedures can be carried out using the huge memory and speed of modern computers and an optimal mathematical model can be designed for solving a scientific problem. This especially concerns global problems of mankind, which could not be solved by traditional methods in principle. Mathematical experiment in natural scientific research has given a powerful impact to the development of all the sciences of inanimate and animate nature, raised them onto a qualitatively new level, and promoted the elaboration of deeper theoretical concepts and obtaining new scientific knowledge. At the same time, the informatization of mathematical means has resulted in inclusion of new types of models, sets of programs, and algorithms into the mathematical arsenal as well as in traditional method perfection. Mathematics has been so united with computer science that there was "brain drain" from traditional mathematics to computer science, especially to programming, and this has become the most powerful direction in the mathematization of scientific knowledge.

3.4 LEADS, PROBLEMS, AND LIMITS OF THE MATHEMATIZATION OF SCIENCE

The development of mathematics and the mathematization of other (non-mathematical) sciences are deeply interconnected processes. Mathematics as a system of mathematical structures and categories cannot develop without supplying substantial scientific knowledge with its means and logical-symbolic potential. In turn, the "substantial" sciences (first of all, physics, chemistry, and biology) positively influence mathematics as a science with its own subject and method, contributing to its development.

Going beyond its logical-symbolic autonomy and self-sufficiency, mathematics is often transformed from the subject of the scientist's activity into an efficient method of ordering and accumulating new scientific knowledge. Just the process of applying mathematics in any area of scientific knowledge is the essence of the expanding process of the mathematization of science, which has an integrative nature and generates new knowledge. The mathematization of science, thus, confirms the idea that new knowledge is most often appears in the ways of interaction of scientific disciplines.

Mathematization as scientific knowledge synthesis, having its own content, manifests itself in various forms. First, it is expressed in transferring methods (used at systematization and formalization) into other fields of science. Second, it is shown as quantitative analysis and expression of qualitatively established empirical data with the purpose of their use for the formulation of numerical laws. Third, mathematical models or special mathematical branches of various sciences, for example, mathematical psychology or econometrics, can be its forms.

At the present stage, knowledge mathematization substantially depends on the use of means, and it apparently represents the basic lead of mathematization. It obviously includes, besides mathematics and the discipline to be mathematized, computer science with its methods and technical cognition means. Knowledge mathematization with computers also has different stages and components. The use of computers for faster and exact computing works, for data processing, representing a major lead of mathematization, is its simplest display by means of computers. The genuine knowledge mathematization on the "computerization" way, according to Victor Glushkov, occurs only when computers are applied together with the corresponding sophisticated deductive apparatus and logic calculations, and also for searching for new laws, the design of a special formalized language of science, and construction of more fundamental theories.

Some restrictions of using formal mathematical methods follow from Kurt Gödel's theorems. According to the first Gödel theorem (the theorem of incompleteness), there always exist indemonstrable and, simultaneously, incontestable (in the given system) conclusions in formal systems difficult enough. The second Gödel theorem states the impossibility to prove the consistency of a formal system with the aid of the means of this system. These theorems point to the impossibility of full formalization of rather substantial processes and results of human thinking; they are usually treated as serious restrictions of scientific knowledge mathematization. Generalizing Gödel's theorem and simplifying their proofs, Victor Glushkov has shown that these restrictions are eliminated not only when designing more complex formal systems but also covering most of the content of the fragment of study to be mathematized. If the world is infinite in terms of information, it is impossible to design such a formal theory, which would contain inconceivable, improvable conclusions in principle.

These conclusions are improvable only if we use a finite amount of information already available in abstract thinking, derived from the external world as a result of practice. At outlining the borders of mathematization, it is also necessary to mean that the mathematical language has its natural limits of the knowledge content expression. In general, abstract thinking is itself limited if isolated from the external world; but if this isolation is broken and there appears a link with the external world, these restrictions are removed. Though mathematization is limited historically, its limits are mobile, conditional, and finally caused by the unity of strict and non-strict, formal and substantial methods. In the course of mathematization, the interaction of mathematics with other sciences and that of science as a whole with the external world in the course of practical activity will move the mathematical means usage borders apart more and more. Therefore, knowledge synthesis performed in the course of mathematization generates new opportunities of the application of mathematical means as well as of the enrichment of these means with new features.

The restrictions of mathematics usage in social sciences are due to mathematics, for the present, lacking rather developed means for effective applications in social knowledge. In the development of mathematics, its theories and methods, a special place is hitherto occupied by its closest interaction with the sciences of inanimate nature, despite its principal general scientific character. Nevertheless, the interaction between mathematics and the sciences of inanimate nature and society strengthens at the present stage. Other restrictions are caused by the nature of the social form of movement and its extreme complexity in comparison with natural phenomena. There are much more objective obstacles for the mathematization of social sciences in comparison with natural and technical ones. But this does not imply skepticism about the possibilities of the mathematization of these branches of knowledge. The best hope in terms of the mathematization of social and human knowledge can be associated with the "informatization" of mathematics.

It is also necessary to mean that, as well as any development of knowledge, mathematization is accompanied by certain negative consequences. The language and methods of mathematics are known to be sometimes used for proving false ideas and theories. Mathematical notions, owing to their abstraction from the specific content of the objects under study, may undergo certain manipulations with some purposes different from searching for truth. This was noticed by Goethe, who jokingly wrote that "My friend, the art's both old and new, // It's like this in every age, with two // And one, and one and two, // Scattering error instead of truth." All these discussions of the mathematization limits point to the necessity of examining its efficiency and expediency, which assumes refusal of our simplified representation of it as of a good only, demands control of the process of mathematization to avoid its negative consequences and effects.

As to the positive direction of scientific knowledge mathematization represented itself as an integration general scientific process, it, during its development, generates knowledge synthesis between various sciences, theories, and other forms and means of cognition. The integrative potentialities of mathematics are closely related to the nature of mathematical abstractions, which distract from the specific nature of any object under consideration. Therefore, when a certain mathematical theory (say, statistical information theory) is applied to a specific object of research (e.g., to message transfer process in technical communication channels), it cannot be applied to many other objects (to information transfer in genetic biological systems, between animals communicating among themselves, people, the man and machine, etc.).

The possibilities of mathematical models for investigating objects are not usually restricted by the limits of the initial scope, and their scope of application gradually extends. The translation of some substantial ideas previously related to the mathematical apparatus is also performed through this apparatus which acts like a bridge to facilitate the movement of knowledge from one field of science to another one with new ideas to be generated.

KEYWORDS

- **mathematics**
- **natural science**
- **the development of mathematics**
- **the subject of mathematics**
- **mathematical modeling**
- **experiment**
- **mathematization of science**

REFERENCES

1. Goldfein, M. D.; Ivanov, A. V.; Malikov, A. N. *Concepts of Modern Natural Sciences. A Lecture Course*; Goldfein, M. D., Ed.; RGTEU Press: Moscow, 2010. [Russ].
2. Goldfein, M. D.; Ursul, A. D.; Ivanov, A. V.; Malikov, A. N. *Fundamentals of the Natural-Scientific Picture of the World*; Goldfein, M.D., Ed.; SI RGTEU Press: Saratov, 2011. [Russ].
3. Goldfein, M. D.; Ivanov, A. V. *The Concepts of Modern Natural Science*; Nova Science Publishers: New York, 2013.

CHAPTER 4

THE APPLICATION OF THERMODYNAMICS TO THE OPEN SYSTEMS

CONTENTS

4.1 General Concepts of Classical (Equilibrium) Thermodynamics[1,2] 100
4.2 Concepts of Modern Equilibrium Thermodynamics (Gladyshev's Theory)[3–14] 108
Keywords 115
References 115

Classical thermodynamics is the only physical theory of universal content which I am convinced will never be overthrown, within the framework of applicability of its basic concepts.

—Albert Einstein

ABSTRACT

The application of thermodynamics to open systems is described. The general concepts of classical equilibrium thermodynamics are listed. Gladyshev's theory is described. The physical theory of biological evolution and its physicochemical models are touched upon. Practical aspects of the thermodynamic theory of living beings are provided.

4.1 GENERAL CONCEPTS OF CLASSICAL (EQUILIBRIUM) THERMODYNAMICS[1,2]

The concepts of classical and quantum physics allow one, either exactly or with a certain probability, to predict the state of macrobodies or microparticles. In particular, this concerns mechanical movement regularities which can be described by using spatial–temporal coordinates, the vales of mass, velocity, pulse, wave characteristics, and the knowledge of the fundamental type of interaction. However, there exist some processes whose features can be explained by neither classical physics nor quantum representations. For example, the existence of bodies to them in different aggregation states, the appearance of elastic forces at deformations of systems, possible transformation of some compounds into others, and so forth. As a rule, these and similar processes are accompanied by transition of systems from one state to another with changes in thermal energy. Just such processes and most general thermal properties of macroscopic bodies are studied by the section of physics and chemistry called *thermodynamics*.

Thermodynamics studies a system, that is, a body or a group of bodies capable of changing under the influence of physical or chemical processes. Everything surrounding the system forms the external medium (environment). By a "body," they usually understand any substance having a certain volume and characterized by some physical properties. Thermodynamic systems can be homogeneous and heterogeneous. *Homogeneous* are such systems whose properties in macroscopic relation are identical at all points or change continuously upon transition from one point to another. *Heterogeneous systems* are made up of elements different in their properties and separated with interfaces, on which there is a sharp change of properties. Every system is related with such a notion as the condition of a system, which represents a set of quantitative values of all its thermodynamic properties. At change of at least one of these properties, the system passes from one state to another. Usually, thermodynamic processes proceed at constancy of certain parameters of the system. An *isothermal* process proceeds at a constant temperature; an *isobaric* one proceeds at a constant external pressure; an *isochoric* one is characterized by the constancy of volume; an *isentropic* one features the constancy of the entropy of the system; and an *adiabatic* one goes in the absence of thermal energy exchange between the system and its environment.

Energy is an important notion, which is distinguished according to the types of movement and interaction. In mechanics, this is the energy of forward, rotary, or oscillating motion of bodies; the energy of bodies' location in a force field (the potential energy in the gravitational field, in the field of electric forces, and in the field of magnetic forces); and the energy of elastic deformation (stretching, compression, and shift). In applied physics, it is necessary to distinguish mechanical energy, the potential energy of gravity, the internal energy of heated bodies, acoustic energy, radiant energy, and electric and magnetic energy. In technical and economic classification of energy types, the man-to-nature relation is usually reflected: the energy of solar radiation, hydraulic energy, wind power, the energy of fuel, mechanical energy, the internal energy of steam and compressed gases, chemical energy, electric energy, and muscular energy of the man and animals. In thermodynamics, two main types of energy are distinguished, namely, external and internal. The notion of *external energy* covers the energy of movement of a body (system) as a whole and the energy of the body's location in a field of forces when moving, in which the thermodynamic state of the body does not change. The *internal energy* of a system is the sum of the energies of movement and interaction of all the elements constituting the system, that is, the sum of the energy of forward and rotary motion of the molecules and oscillating motion of the atoms, the energy of molecular attraction and repulsion, intramolecular chemical energy, the intratomic energy of optical levels, and intranuclear energy.

The bases of classical thermodynamics have been formulated in the works by Clausius, van't Hoff, Arrhenius, Gibbs, Helmholtz, Le Châtelier, et al., which describe processes depending on changes of the properties of systems in an equilibrium state. In other words, the thermodynamic description of systems and phenomena is based on the idea of an equilibrium state. Thermodynamics answers the question: Where some process is directed to before equilibrium is reached? Therefore, it does not deal, in an explicit form, with time as a physical parameter and considers no mechanisms of processes. If a system infinitely slowly passes from one state to another through a continuous series of equilibrium states and if the maximum work is thus done, such a process is called *thermodynamically equilibrium process*. Between two next states of equilibrium, the values of any functions of state of the system differ by an infinitesimal value, and it is always possible to return the system to its initial state by infinitesimal changes of the functions of state. The system having made an equilibrium process can return to its initial state, having passed the same equilibrium states in the opposite direction, as in the direct one. Therefore, as a result of a *thermodynamically reversible process,* the system and environment return to their original states, that is, no changes remain in the system and environment. On the other hand, a process after which the system passes through a series of nonequilibrium states is called *thermodynamically nonequilibrium process*. If the system and environment cannot return to their original states, that is, any changes remain in them, such a process is considered *thermodynamically irreversible*.

Equilibrium thermodynamics is, first of all, associated with molecular and kinetic theories about the structure and properties of macrosystems (e.g., the molecular and kinetic theory of gases). The main concept of such theories is that every body (system) (gaseous, liquid, or solid) consists of a large number of molecules being in chaotic movement; the intensity of this movement depends on the temperature of substance, and vice versa. Hence, the structure, properties, and behavior of macrosystems are determined by the movement and interaction of the particles which the bodies consist of. There are quantitative relations describing the molecular and kinetic features of the properties and behavior, in particular, gases, namely, the laws of Boyle–Mariotte, Gay-Lussac, Charles, Clapeyron–Mendeleev, van der Waals, and some others.

At the same time, the scope of thermodynamics is much wider than the molecular and kinetic theory of gases. The matter is that, unlike many provisions of physics and chemistry, thermodynamics does not describe the course of processes in time (i.e., their dynamics) and the structural features of the studied system. In other words, thermodynamics only considers the initial and final states of a system which are characterized by special thermodynamic parameters. Usually, temperature, pressure, and volume and *characteristic functions* such as enthalpy, entropy, internal energy, Helmholtz free energy, and Gibbs thermodynamic potential act as such parameters of state. Just these functions and their values characterize the thermodynamic state of a system and cause the probability of proceeding of this or that process accompanied by changes of thermal energy. The functions of state allow establishing the direction of spontaneous processes and determining the degree of their completeness in real thermodynamic systems. For example, Gibbs thermodynamic potential G can be used for exploring equilibrium (and quasi-equilibrium, as will be shown further) processes and closed (quasi-closed) systems, in which transformations proceed at constant temperature and pressure. Helmholtz free energy F is applicable to studying similar processes and systems at constancy of temperature and volume.

Thermodynamics also features being a phenomenological science. This means that all physical quantities, functions, and laws of thermodynamics are based on experience only, that is, there is no strict theory concerning them. The sense of empirical provisions of thermodynamics can be characterized as follows. (1) Unlike mechanical movement, all spontaneous thermal processes are irreversible (and, first of all, this applies to heat propagation in the environment). (2) According to various forms of movement and various types of energy, there are also various forms of energy exchange. In classical thermodynamics, only two forms of energy transfer are resolved, namely, work and heat. Any kind of energy is an unequivocal function of the state of the system, that is, the change of the energy does not depend on the way of the system's transition from one state to another. At the same time, heat and work are unequal forms of energy transfer as work can be directly used for replenishment of the stock of any kind of energy, while heat can be used for replenishment of the stock of the system's internal energy only. (3) Any physicochemical system left to itself (i.e., in absence of external forces and fields) always aspires to pass to the state of thermodynamic equilibrium; this state is characterized by the uniformity of distribution of temperature, pressure, density, and the concentration of components. (4) The characteristic functions of the state of system are interrelated by certain ratios, for example,

$$\Delta F = \Delta U - T\,\Delta S; \qquad \Delta G = \Delta H - T\,\Delta S,$$

where ΔU, ΔH, and ΔS are changes of the internal energy, enthalpy, and entropy, respectively; T is absolute temperature, which points to their interconditionality and universality relative to the description of both initial and final states of the system.

The physical sense of the concepts of classical (equilibrium) thermodynamics is directly related with its three basic laws which are general laws of nature. *The first law of thermodynamics* in a qualitative sense coincides with one of the fundamental laws of nature, namely, the law of energy conservation. In application to thermodynamics, its formulation is as follows: the quantity of heat given to a system (body) goes to the increase in its internal energy and to delivering work (A) by it (them):

$$\Delta Q = \Delta U - \Delta A$$

The second law of thermodynamics has several formulations. For example, any action associated with energy transformation (i.e., with energy transition from one form to another) cannot occur without its loss as heat dissipated in the environment. More generally, this means that the processes of energy transformation can occur spontaneously, provided that energy passes from its concentrated (ordered) form to a diffused (disordered) form. One more definition of the second law of thermodynamics is directly related with Clausius principle: any process in which there are no changes except heat transfer from a hotter body to a colder body is irreversible, that is, heat cannot spontaneously flow from a colder location to a hotter location. Such energy redistribution in the system is characterized by a quantity which has been named as *entropy*, which, as a function of state of the thermodynamic system (a function having a full differential), was first introduced by Clausius in 1865. Therefore, the more energy dissipates irreversibly as heat, the higher the entropy. Whence it follows that any system whose properties change in time aspires to an equilibrium state at which the entropy of the system takes its maximum value. In this connection, the second law of thermodynamics is often called the law of increasing entropy, and entropy (as a physical quantity or as a physical notion) is considered as a measure of disorder of a physicochemical system. Most often, the system's transition from one state to another state is characterized not by the absolute value of entropy (S) but by its change (ΔS), which is equal to the ratio of the change of the quantity of heat (given to the system or taken away from it) to the absolute temperature of the system: $\Delta S = \Delta Q / T$, J deg^{-1}. It is the so-called thermodynamic entropy. Besides, entropy has a statistical sense since the number of those microstates by which a given macrostate is realized can serve as a quantitative characteristic of the thermal state of the system. In these cases, entropy is calculated by Boltzmann formula (1844–1906):

$$S = k \ln W$$

where W is the number of the microstates to realize this macrostate of the system; k is Boltzmann constant ($1.38 \times \cdot 10^{-23}$ J K^{-1}).

Therefore, upon transition from one macrostate to another, the statistical entropy also increases since such transition is always accompanied by a large number of microstates, and the equilibrium state (to which the system aspires) is characterized by the maximum number of microstates.

In connection with the notion of entropy, the notion of time gets a renewed sense in thermodynamics. As already mentioned, the direction of time is ignored in classical mechanics, and it is possible to determine the state of a mechanical system both in the past and in the future. In thermodynamics, time acts as an irreversible process of entropy increasing in the system, that is, the higher the entropy, the longer time interval has been passed by the system in its development.

It is also necessary to mean that there are four classes of thermodynamic systems in nature:

a) isolated systems (upon transition of such systems from one state to another, no transfer of energy, substance, and information through the system's borders proceeds);
b) adiabatic systems (no heat exchange with the environment only);
c) closed systems (no substance transfer only);
d) open systems (which exchange substance, energy, and information with the environment).

Such a classification of thermodynamic systems, at first sight, leads to a seeming contradiction between the general principle of relativity in nature and the validity of the second law of thermodynamics for whole nature. The matter is that, ignoring exchange processes, the second law of thermodynamics is directly applicable to systems (1) and (2) only, for which entropy is either constant or (which is much more often) increases: $\Delta S \geq 0$. This means that any changes in such systems are characterized by increasing disorder (chaos) until the system reaches thermodynamic equilibrium (or its entropy reaches its maximum value). At the same time, some scientists think that, as the whole Universe as an isolated system obeys the second law of thermodynamics and the entropy of all its parts always increases (due to the infinite accumulation of diffused thermal energy in them), the state of the Universe aspires to chaos and, finally, the so-called thermal death must come. Such a gloomy (but wrong) conclusion is connected with two circumstances: first, the notion of isolated system (as a nature element) is always an abstract notion; second, it was long ago established that entropy does not increase in all systems, particularly in open systems, it may decrease. The latter fact, first of all, concerns biological systems, that is, living organisms that are open nonequilibrium systems. Such systems are characterized by gradients of the concentration of chemicals, temperature, pressure, and other physical and chemical quantities. In this regard, a seeming contradiction appears between the second law of thermodynamics and biological ordering.

The recovery from such a contradiction can be found in the use of the concepts of modern (i.e., nonequilibrium) thermodynamics, which describe the behavior of open systems always exchanging energy, substance, and information with the environment. It should be noted that such exchange processes are characteristic not only of physical or biological systems but also of social-economic, cultural-historical, and humanitarian systems since the processes occurring in them are, as a rule, irreversible. Schematically, the mechanism of interaction of an open system with the environment can be represented as follows. First, such a system at its functioning takes substances, energy, and information from the environment, and then gives it the used ones (naturally, in other types and forms). Second, all these processes are also accompanied by entropy production (i.e., energy losses and the formation of various wastes). Third (and the most important), the formed entropy (characterizing a certain degree of disorder), without collecting in the system, is released into the environment, and the system takes from it new substances, energy, and information necessary for its optimum performance. All this allows drawing a conclusion that the open system, as a result of exchange processes, can be in a nonequilibrium state as well, for whose description, new laws (in particular, nonequilibrium thermodynamics) are necessary.

Briefly, the creation history of the nonequilibrium thermodynamics concepts is represented as follows. In the first half of the 19th century,

Fourier, Stocks, Navier, and Fick derived kinetic equations for the description of nonequilibrium processes such as heat conductivity, viscous liquid flow, and diffusion. Further, methods of the thermodynamics of nonequilibrium (irreversible) processes were used in analysis of the thermoelectric phenomena (W. Thomson) and electrochemical processes (Helmholtz). Onsager's works contributed much to the development of the thermodynamics of irreversible processes, which first introduced the notion of thermodynamic force. Just this quantity characterizes the degree of deviation of the system from its thermodynamic equilibrium and is the cause of irreversible changes in the system. The quantitative characteristics of the corresponding irreversible phenomena are called thermodynamic flows and, according to Onsager's representations, are the derivatives of the parameters of thermodynamic state with respect to time (*thermodynamic rates*). The concepts of the linear thermodynamics of nonequilibrium processes, offered by Onsager, describe the behavior of molecular systems at a phenomenological level and establish a relation between the kinetic factors determining the intensity of cross-processes of transfer of heat, masses, momentum, and chemical reactions. Without pressing in further details, it is possible to draw a conclusion that the main objective of the thermodynamics of irreversible processes consists in formulating a set of equations describing the dependences of thermodynamic parameters on spatial–temporal coordinates on the basis of mechanics and the thermodynamics concepts (supplemented by the linearity and symmetry principles). The modern concepts developing nonequilibrium thermodynamics in its application to biological systems were developed by Th. de Donder, S. R. de Groot, K. G. Denbigh, P. Mazur, P. Glansdorff, and I. Prigogine (it will be discussed in more detail in the chapter concerning the concepts of natural sciences in biology).

Another recovery from the specified contradiction is connected with the development of the concept of thermodynamics based on the possibility to apply classical (i.e., equilibrium) thermodynamics to real (open) systems. The matter is that many facts are now known to allow, in a certain degree, to neglect the irreversibility of some physical processes and chemical transformations (i.e., to study them in a certain approximation). In these cases, it is merely necessary to establish such ranges of changes of the properties of a system, within which it is possible to consider them equilibrium ones. Usually, such states are called quasi-equilibrium, when a process studied by changes of some parameter is steady and has such a time, in comparison with which the times of other processes in the system are either very long or negligible. Such a concept underlies *the law of temporary hierarchies* formulated by Gladyshev, which allows to resolve quasi-closed thermodynamic systems (subsystems) in open biosystems and to investigate their development (ontogenesis) and evolution (philogenesis) by studying changes in the value of Gibbs specific (per unit of volume or mass) functions of formation of a given highest monohierarchical structure from the monohierarchical structures of a lowest level. Proceeding from these considerations, it is possible to make two important conclusions:

In every specific case, depending on the chosen timescale (or process), this state of the system can be considered (with some degree of the accuracy of approximation) as an equilibrium or nonequilibrium one, that is, various ways of description can be applied, depending on the character of the problem to be solved.

The second law of thermodynamics is applicable to real (open) systems as well, provided that the whole system (continuum) "open system–environment" is considered; for such a set, the total entropy will always increase, as for the first component $\Delta S_1 < 0$, for the second component $\Delta S_2 > 0$, and at $|\Delta S_1| \ll |\Delta S_2|$, we have:

$$\Delta S = (\Delta S_1 + \Delta S_2) > 0$$

4.2 CONCEPTS OF MODERN EQUILIBRIUM THERMODYNAMICS (GLADYSHEV'S THEORY)[3–14]

The thermodynamic evolution theory of living beings is based on the evolution concept of biological systems from the viewpoint of one of the physical models of the origin and development of life. In this case, the term "thermodynamics" should be associated with such branches of physics (e.g., physical chemistry, biophysics, and the physics of macromolecular compounds) which study the most complex and general properties of macroscopic systems in thermodynamic equilibrium and transitions between these states. Thermostatics or equilibrium thermodynamics gives exact relations between the measurable properties of systems and answers the question: How deep the process (the evolution of a system) will proceed before equilibrium is reached? Thermodynamic methods are not bound by any assumptions about details of the structure of systems (molecules, cells, organisms, populations, etc.) and, therefore, it, defining the time axis, does not operate with time as a parameter.

At the same time, the researcher may be interested in the changes occurring in a system from the standpoint of the thermodynamics of its structure (rather than the thermodynamics of processes).

If we denote the Gibbs function of our system as G, the Gibbs function of *formation of this system* (structure) can be represented as ΔG. In this case, we can assume that the Gibbs function for simple and complex living systems (as we shall see later) has a real sense at any time of evolution, and its value is determined by characteristic parameters of the system's state. During the evolution of an open heterogeneous biological system (e.g., the biological tissue of a living being), any change or increment (Δ) of the specific value of the Gibbs function G is primarily due to variations of the chemical, molecular, and supramolecular compositions of the system. Calculation of the increment of the Gibbs function of the system for any moment of its evolution (e.g., ontogeny, phylogeny) is done by using thermodynamic methods, which (as already mentioned) do not use time as a parameter. In addition, the specific Gibbs function of formation of structure, measured from an accepted standard, is a measure of the stability of this structure (chemical, supramolecular, etc.), and its increment represents the change in this stability. The values of increments of these functions in an open system can be related to the following two factors. For example, an open supramolecular system can be conveniently treated as a system consisting of two functioning (but not spatially separated) subsystems. One of the subsystems (1) is considered as closed (thermodynamically quasi-closed). Changes in the specific Gibbs function in it are due to rapid processes occurring as if in the system with a constant chemical composition. The other subsystem (2) is open and these variations are due to the varying composition of its chemical substance associated with changes in the nature and amount of supramolecular structures of the open system itself (exchanging substances with the environment). This second subsystem can be considered as kinetically partially closed (kinetically quasi-closed), because supramolecular structures of an increased stability are accumulated therein.

It should also be noted that the term *evolution* is commonly used in its conventional sense in this theory. Biological evolution is regarded as an irreversible (in the evolutionary timescale characteristic of the given object) process of historical changes of living matter. However, in contrast to Charles Darwin's evolutionary theory or the modern synthetic theory of evolution, this phenomenological theory reveals only the thermodynamic direction of evolution and the degree of its course (completeness). The theory of biological evolution, until recently, was not designed to reveal the physical essence of evolution direction. Although Darwin admitted that "the laws of life are part or a result of a certain general law" determining the evolution of matter in general, nevertheless, the development of modern Darwinism is associated with the use of molecular biology data to obtain a better understanding of the mechanisms of genetic variations and selection of methods to manage living natural resources. The thermodynamic evolution theory of living beings on Earth considers a smooth transition of chemical evolution to biological evolution on the basis of a hierarchical arrangement model of animate and inanimate matter. Even though this model, in principle, allows us to study the evolution of populations and human society, only the chemical and supramolecular aspects of evolution are discussed in this theory.

4.2.1 SOME EMPIRICAL RESULTS CONFIRMING THE PHYSICAL THEORY OF BIOLOGICAL EVOLUTION

1. It has been found that the chemical composition of living bodies (e.g., plants and animals) in their ontogeny, phylogeny, and long stages of biomatter evolution changes. In the course of evolution, biosystems (an organelle, a cell, the biological tissue of an organism, the biomass of a population, etc.) are enriched with energy-rich chemical substances displacing water therefrom. These substances are mainly organic compounds—lipids, proteins, polysaccharides, nucleic acids, and so forth. For example, at the moment immediately after conception, the human embryo has at least 95–97% water, while the tissues of an elderly person are water depleted (its amount decreases down to 60–65%). Changes in the composition of tissues in the ontogenesis are most pronounced in young plants and animals. Phylogeny (in terms of changes in the chemical composition) often repeats ontogeny. Analysis of such data shows that, in both ontogeny and phylogeny, the energy consumption of a biosystem grows, its chemical thermal stability decreases, and the thermodynamic stability of the supramolecular structures of body

tissues increases. In practice, when comparing the energy intensities (caloric power) of systems of various compositions, it is useful to take the calorific value of the biomass.

2. The phenomenon of adaptation of the chemical composition of living beings to their environment (the nature of food, temperature, pressure, etc.) is well known. Changes in the composition of fatty acids (fat) in the cells of microorganisms, plants, and animals when the temperature of their environment varies are most indicative. When the body temperature is below the optimum temperature, the content of unsaturated fatty acids increases, while the fat of living beings is enriched with saturated fatty acid residues at elevated temperatures.
3. There are many examples to indicate changes in the melting (denaturation) temperature of various tissues in the ontogeny and phylogeny of organisms.
4. It has been proven that the origin of the optical activity of biological substances has thermodynamic nature.
5. Some evidence points to the possibility of hereditary transmission of certain thermodynamic characteristics.

These and other empirical data suggest that the concept of hierarchical thermodynamics underlies the modern physical evolution model of living beings. One of its most important consequences is the law of nature (apparently general) discovered by Georgi Gladyshev, concerning the existence of unidirectional series of lifetimes of structures of different hierarchies. For example, if we consider a fragment of the hierarchical sequence of biological structures, it turns out that the average lifetime of the cells in the body is much shorter than that of the body itself, which, in turn, is shorter than the longevity of the population, that is, *lower-hierarchy structures in biological systems live much shorter than upper-hierarchy structures*. This means that lower-hierarchy structures (e.g., cells) evolve over times much shorter than the lifetime of the corresponding organism. The biomass of the organism serves as if the environment for the cells, whose parameters and composition are virtually unchanged during short times. The evolution of cells is a nonstationary process proceeding against the background of the incoming (into the cell) flow of chemical substances of an almost constant composition, that is, the cell subsystem has a kind of thermostat, which allows one to speak of the quasi-stationary nature of the cell subsystem by some parameters. For each biosystem (an organelle, a cell, an organ, a body, a species, etc.), its own thermostat (the environment) can be identified, which has its own parameters. The system with its thermostat thus forms a complete thermodynamic system, and, in every case, the state of this system can be considered equilibrium or nonequilibrium with different degrees of approximation.

4.2.2 SOME PHYSICOCHEMICAL MODELS OF BIOLOGICAL EVOLUTION

To identify the driving force at separate stages of bio-evolution, we can consider a more or less simple sequence of the processes of formation and disintegration of a polymer—polyacrylonitrile (PAN). In one of the reactor zones, the monomer (acrylonitrile) vapor is in equilibrium with its aqueous solution where photopolymerization occurs. The resulting polymer is insoluble in water and precipitates as a solid phase. Under the influence of gravity (in a centrifuge), the polymer moves into another zone of the reactor. Under the influence of intense radiation, the polymer depolymerizes and the resulting monomer again goes into the gas phase of the first reactor zone. Then the monomer turns out to be in solution, polymerizes, and so forth. If the depolymerization proceeded completely to form the monomer, the cycle would be stationary and would continue for a long time. However, the polymerization is accompanied by the formation of macromolecules with different molecular weights and a somewhat heterogeneous structure; and the higher-molecular-weight products depolymerize with a lower average rate than the lower-molecular-weight products. This leads to the accumulation of slowly decaying substance in the depolymerization zone. In the thermodynamic relation, this cycle, in the first approximation, simulates the "stages" of ontogeny and biological evolution, the formation of solid polymer simulates the appearance of biomass, and polymer destruction simulates its degradation.

Another analog of the proposed model of biological evolution is an open adsorption (or absorption) system, which is rather slowly fed by a flow of substances of a constant composition subject to phase or chemical transformations in the system.

The evolution of the chemical and supramolecular composition of the substance of a tissue's (biomass') micro-object is similar to the evolution of the composition of substance of a model equilibrium chromatographic column, where chemical reactions proceed (similar to biochemical processes in microdomains of the biomass) and which is in real conditions of the life of organisms (when there are changes in temperature, atmospheric pressure, and other characteristics of the environment). The composition of the chemical substances entering the model column also experiences fluctuations, similar to "composition wobble" of the food consumed by the body. It would be easy to suggest and experimentally prove that supramolecular structures with high thermodynamic stability will be accumulated in this model system over sufficiently long times. These structures should consist of substances and supramolecular structures with high affinity for the sorbent in the column. The local values of the specific Gibbs function tend to a relative minimum. All of these seem real as the biological tissue itself is an assemblage of a large number of model chromatographic columns, where supramolecular equilibria are maintained.

4.2.3 ROLE OF THE THERMODYNAMIC FACTOR DURING GENETIC INFORMATION STORAGE

Biopolymers, primarily polynucleic acids (DNA and RNA), are known to be genetic information carriers. If we proceed from the thermodynamic evolution theory of living beings, we can say with confidence that the role of nucleic acids in the origin and development of life was predetermined. It should be considered that this role could be realized on Earth or other planets only in the presence of liquid water and the atmospheric composition similar to the primary Earth's atmosphere, which promoted photosynthesis. Then, you need to keep in mind that the double helix of the DNA molecule is framed by other chemical components, which are combined into supramolecular structures characterized by a certain thermodynamic stability. This stability is determined not only by the stability of the adenine–thymine and guanine–cytosine pairs but also by the nature of the molecular and ionic environment of the double helix. This explains the increased thermodynamic stability of the DNA double helix, which is estimated by the specific Gibbs function. The melting temperature T_m of the double helix, as that of chromatin, is relatively high and significantly higher than that of many proteins and lipids. This increased absolute value of the negative specific Gibbs function of formation of the double helix provides the fairly high accuracy of replication (self-copying), the long-term stability, and the relatively high resistance to mutations. At the same time, the chemical stability of purine and pyrimidine nitrogenous bases, adenosine phosphoric acids is quite low and, therefore, they have increased chemical energy content. The low chemical stability of the monomer units of polynucleic acids determines important characteristics of their metabolism and contributes to the high exchange rate of these structures in the body. That is why, the "thermodynamic conservatism" of DNA associated with the relatively slow changes in its structure in the constantly reproducing biomass has determined the role of this biopolymer as the genetic information keeper.

4.2.4 SOME APPLIED ASPECTS OF THE THERMODYNAMIC THEORY OF EVOLUTION OF LIVING BEINGS

Without going into further details of this theory, we can formulate a few important practical consequences.

1. Diets that include "evolutionarily young" products of plant and animal origin promote longevity and improve the quality of human life. The indicator of the *evolutionary youth degree* of a natural food product is determined by its chemical composition and supramolecular structure. The chemical composition and molecular structure of a product, in turn, depend on its phylogenetic and ontogenetic ages, as well as on the environment of the organism being the source of this product. The specific Gibbs function of formation of the supramolecular structure of a natural food product is the most important quantitative characteristic of its *gerontologic value*.
2. It is possible to assess (test) the gerontologic effects of drugs on the aging processes of human tissues. Many medicines (vitamins, trace elements, hormone imitators, etc.) help us to maintain a physiologically optimal stability of the supramolecular structures of the biological tissues of the body, which is also estimated by the Gibbs function of the formation of structures.

KEYWORDS

- **classical equilibrium thermodynamics**
- **modern nonequilibrium thermodynamics**
- **Gladyshev's theory**
- **biological evolution**
- **evolution models**
- **the thermodynamic theory of living beings**

REFERENCES

1. Goldfein, M. D.; Ivanov, A. V.; Malikov, A. N. *Concepts of Modern Natural Sciences. A Lecture Course*; Goldfein, M.D., Ed.; RGTEU Press: Moscow, 2009. [Russ].
2. Prigogine, I.; Stengers, I. *Order out of Chaos. Man's New Dialogue with Nature*; Heinemann: London, 1984.
3. Goldfein, M. D.; Ursul, A. D.; Ivanov, A. V.; Malikov, A. N. *Fundamentals of the Natural-Scientific Picture of the World*; Goldfein, M.D., Ed.; Saratov, Russia, 2011.
4. Prigogine, I.; Stengers, I. *The End of Certainty: Time, Chaos and the New Laws of Nature*; Free Press: New York, 1997. [Russ].
5. Haken, H. *Synergetics, an Introduction: Non-equilibrium Phase Transitions and Self-Organization in Physics, Chemistry, and Biology*, 3rd ed.; Springer-Verlag: New York, 1983.
6. Bezruchko, B. P.; Koronovsky, A. A.; Trubetskov, D. I.; Khramov, A. E. *Way to Synergetics: An Excursus in Ten Lectures*; Librocom Book House: Moscow, 2005. [Russ].
7. Gladyshev, G. P. *Thermodynamics and Macrokinetics of Natural Hierarchical Processes*; Nauka: Moscow, 1987. [Russ].
8. Gladyshev, G. P. On the Principle of Substance Stability and Thermodynamic Feedbacks in Hierarchical Systems of the Bioworld. *Bio. Bull. Russ. Acad. Sci.* **2002,** *29*(1), 5–9. [Russ].
9. Gladyshev, G. P. Thermodynamic Self-Organization as a Mechanism of Hierarchical Structures of Biological Matter. *Prog. React. Kinet. Mech.* (An International Review Journal. UK, USA). **2003,** *28,* 157–188.
10. Gladyshev, G. P. Macrothermodynamics of Biological Evolution and Aging of Living Beings. Physicochemical Dietology. Proc. MAN VSh. 2003. No 4 (26), 19–46.
11. Gladyshev, G. P. Macrothermodynamics of Biological Evolution: Aging of Living Beings. *Int. J. Mod. Phys. B.* **2004,** *18*(6), 801–825.
12. Gladyshev, G. P. Macrothermodynamics of Biological Systems and Evolution. *J. Biol. Syst.* **1993,** *1*(2), 115.
13. Gladyshev, G. P. Motive Force of Biological Evolution. *Vestnik RAN.* **1994,** *64*(3), 221–228. [Russ].
14. Goldfein, M. D.; Ursul, A. D.; Ivanov, A. V.; Malikov, A. N. *Basics of the Natural-Scientific Picture of the World*; RGTEU: Saratov, 2011.

CHAPTER 5

GLOBAL PROBLEMS IN THE BIOSPHERIC STUDIES

CONTENTS

5.1 Global Climate Models[1–10] 118
5.2 Biodiversity and the Problem of the Evolution of Species[11–25] 127
5.3 The Biosphere Evolution and the Noosphere[8–10] 145
5.4 Modern Immunochemical and Biosensor Technologies for Analysis of Environmental Ecotoxicants 150
5.5 Biodetoxication of Aromatic Hydrocarbons in Aqueous Media.. 160
5.6 Role of Chemical Substances in Ecology[54,55] 165
5.7 Problem of Microelementoses in the Human Body[54,55] 167
5.8 Globalization Processes in Human Society and their Possible Modeling[56–72] 172
Keywords 187
References 188

Living organisms are a function of the biosphere and closely connected with it, materially and energetically.

— Vladimir Vernadsky

ABSTRACT

The global problems of biospheric studies are listed. Models of global climate changes are characterized. The importance of biodiversity is emphasized; the problem of the evolution of species is highlighted. The concept of the noosphere is described. The importance of immunochemical and biosensor technologies is proven. The detoxification of aromatic hydrocarbons is touched upon. The problem of trace elements in the human body is stated. The globalization of human society is characterized.

5.1 GLOBAL CLIMATE MODELS[1–10]

Earth's climate depends on both cosmic and planetary factors. The parameters of the global climatic system of Earth have been subject to constant changes of cyclical nature with strong nonlinear effects in geological time. Besides, a close relation between climate and global processes in the hydrosphere is apparent, and these processes, in turn, manifested themselves nonlinearly in geological time (Figure 5.1). All the concepts currently known are combined into two groups, namely, *astronomo-physical* and *geologo-geographical*. All of them have planetary action. Each concept corresponds to a certain model of a varying degree of completeness and the applicability of its mathematical apparatus.

The ***astronomo-physical concepts*** (***models***) associate climate variations with fluctuations of the amount and content of solar radiation. They are divided into two subgroups. The first subgroup (Milankovitch's paradigm) is based on the principle of the constancy of solar radiation and the variability of the Earth's surface position relative to sunlight. The second subgroup is associated with changes in the Sun's emissivity.

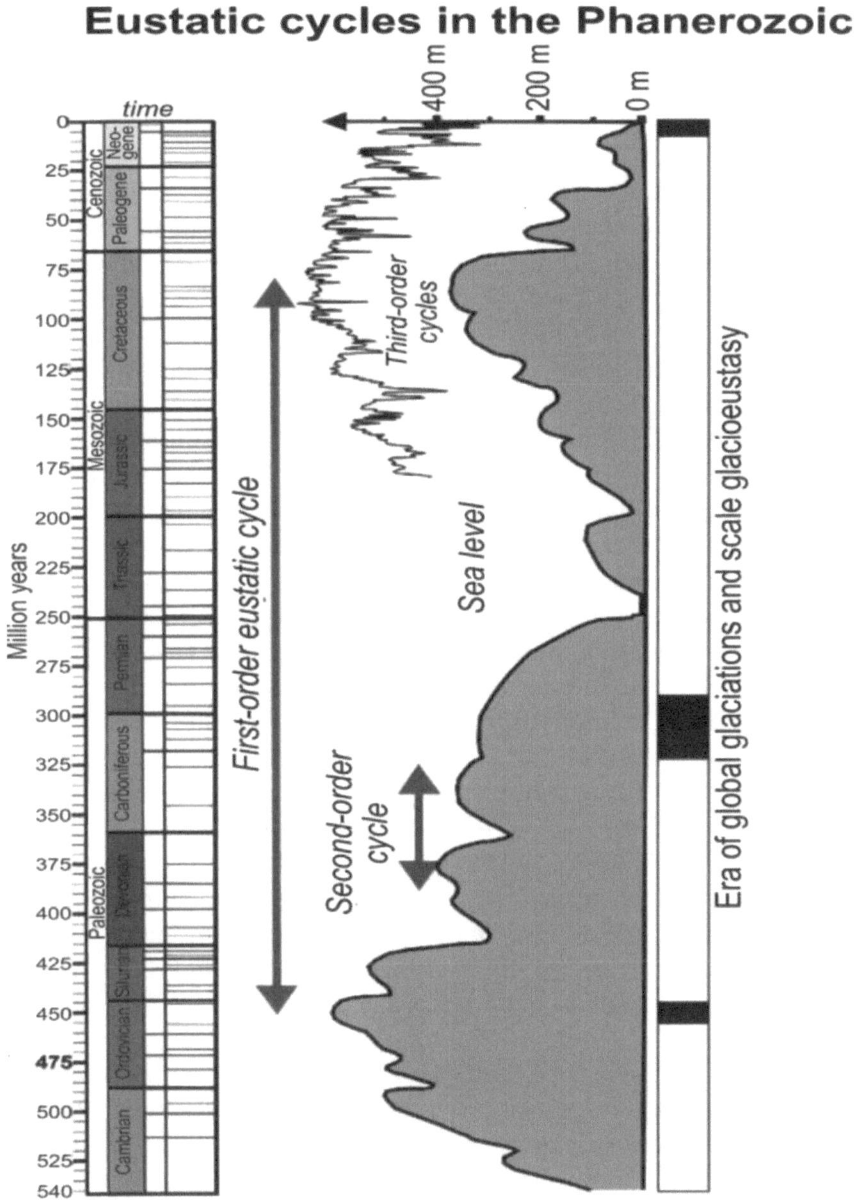

FIGURE 5.1 Global hydrosphere process as changes of the sea level in the history of Earth. (Figure by A. M. Nikishin (MSU), from (Global socionatural ..., 2011)).

5.1.1 *MILANKOVITCH'S ASTRONOMO-CLIMATIC CONCEPT (MODEL)*

The idea of a nonuniform distribution of solar radiation over the Earth's surface due to changes in its orbit elements was first proposed by the English astronomer James Croll in 1875. He suggested a relation between the periodic changes of the polar ice caps in position and volume, due to climate variations, and the variations in Earth's orbit eccentricity and the obliquity of our planet's ecliptic. Milankovitch's works gave fame to this hypothesis, subsequently supported by Wladimir Peter Köppen and Alfred Wegener (Sinitsyn, 1980). The changes in the solar radiation distribution over the surface of our planet were associated with the periodic changes of the following: the tilt of Earth's axis (the ecliptic inclination cycles), the eccentricity of Earth's orbit, and the precession of Earth's rotation axis (the time of perihelion).

The theory of cosmic cyclicity, causing climatic fluctuations, was first proven by M. Milankovitch (1939). The variations in Earth's rotation axis and the elements of Earth's orbit affect the seasonal and latitudinal distribution of solar radiation over the surface of our planet. This distribution is controlled by variations of the following astronomic parameters: the precession, the obliquity of the ecliptic, and the eccentricity of Earth's orbit. The variations in Earth's orbital eccentricity entail changes in the amount of solar energy. Perturbations of the other two parameters lead to latitudinal redistribution of solar radiation. Fluctuations of insolation determining the thermal condition of Earth's surface change are: (1) the intensity of evaporation and condensation of atmospheric moisture; (2) the strength and direction of winds and ocean currents; (3) the duration of the winter and summer; and (4) the position of the upwelling zones. Therefore, there are fluctuations of the warm and cold, and wet and dry climate types (Figure 5.2).

Milankovitch calculated these periodic changes in Earth's orbit elements for several selected latitudes throughout the Quaternary Period. As a result of these calculations, there appeared a curve of oscillations of solar radiation, subsequently compared with the chronology of the Quaternary Period by Frederick Everard Zeuner in 1963. The peaks on Milankovitch's curve corresponding to the periods with cool summers were compared with the glacial epoch, and the peaks corresponding to the periods with warm summers were compared with the interglacial periods.

This correlation was so convincing that many researchers (W. P. Köppen, A. Wegener et al.) accepted Milankovitch's radiation curve as the basis of geological chronology.

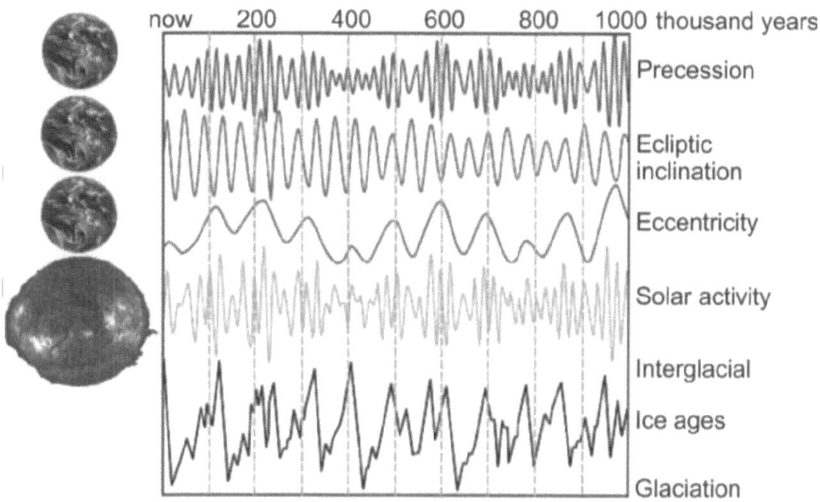

FIGURE 5.2 Milankovitch's astronomo-climatic cycles and solar activity cycles, causing cyclic phase shifts between interglacial periods and glaciation. (From Global socionatural …, 2011).

The rotational speed of Earth's and its orbit trajectory was not permanent in the geological record. In 1989, André Léon Georges Chevalier Berger mathematically calculated the durations of astronomical cycles, taking into account the day duration reduction, the continuously increasing distance between the Earth and the Moon, as well as the changes in the inertia of motion of the celestial bodies in the last 400 million years. It should be noted that, first, the researcher resolved short and long cycles of the ecliptic inclination; and, second, the duration of the inclination ecliptic cycles in the early Silurian approaches the duration of the modern precession cycles.

A number of researchers of the Quaternary Period (astronomers and climatologists) have drawn attention on some demerits of Milankovitch's paradigm. The climatic effects of these changes should have been observed in the Northern and Southern Hemispheres in turn, but the intervals of warming and cooling (including glaciations) were of entire planetary character. Discrepancies between the major events of the Quaternary Period, calculated from Milankovitch's curve, and the data obtained by the tape clays and isotopic techniques were also observed. Solar radiation variations caused by changes in Earth's orbit elements must have taken place in the pre-Quaternary time as well. However, there were no signs of glaciation in both the Paleogene and Neogene (Sinitsyn, 1980). Later, a number of American astronomers found that any changes in solar radiation are small and, therefore, have little climatic significance. The variations in temperature are estimated as approx. 2°, which is not enough to trigger glaciation. In his calculations, Milankovitch ignored the dynamic processes in the atmosphere and lengthening of the duration of the day as a result of the slow Earth rotation under the influence of lunar and solar tides (by 0.0014–0.0024 s per century). For example, the day in the early Paleozoic was 2.5 h shorter than the modern one. Recently, it has been suggested that climatic fluctuations may be caused not so much by Milankovitch's cycles but by the evolution of Earth's climate system.

On the other hand, the results of a study of the Mesozoic-Cenozoic sections of deepwater drilling wells and rhythmically constructed Phanerozoic strata have found a correlation with Milankovitch's astronomical cycles. Nowadays, Milankovitch's model predominates among other astronomical and physical models of climate changes. When it has been ascertained that the time of formation of some selected rhythm types corresponds to the cycle of precession, ecliptics and eccentricity, the climatic or eustatic model (Gabdullin et al., 2005) of the formation of periodites can be transformed into Milankovitch's model. Recently, the "solar Milankovitch cycles" have been suggested, that is, the cycles of precession (95 million years), ecliptic (190 million years), and eccentricity (570 million years) of the Sun's orbital parameters.

Ernst Julius Öpik's hypothesis is based on the assumption that at the development of a star like the Sun, the hydrogen–helium transition inside should create instability and mixing zones, leading to a temporary weakening of solar radiation, that is, cooling. After the restoration of equilibrium inside the Sun, the intensity of solar energy increases (warming). The main drawback of this hypothesis is the absence of any observations and theoretical calculations to prove the very fact and frequency of the Sun's twinkling. The periodicity of solar activity is estimated by different authors as 70 or 200 years, which is reflected in the climatic fluctuations imprinted in the annual rings of modern and fossil trees.

Friedrich Nölke's hypothesis. The essence of this hypothesis is that the solar system periodically meets nebula on its way. During the periods of passing through the darkest regions of such nebulae, some portion of the solar radiation is absorbed without reaching Earth, resulting in a lowered surface temperature and glaciation coming. If the nebula consists of condensations and transparent gaps, the eras of glaciation are separated by interglacial epochs. This hypothesis has a number of questions related mainly to the fact that the effect of solar radiation absorption by interstellar gas is negligible along such a short segment as Earth's orbital radius.

The hypothesis by P. Predtechensky. In 1950, he suggested that solar activity, through condensation processes, affects the basic mechanisms of atmospheric circulation: the west–east transport and meridional exchange. When the solar activity amplifies, the atmospheric circulation increases. The equatorial-tropical zone expands, but the average temperature therein decreases due to increasing cloudiness. The polar regions expand due to advection of the warm air masses of lower latitudes. The temperate zones reduce. The interzonal exchange of air masses weakens, and the equator–pole temperature gradient decreases, especially in the winter. The climate in all circulation zones becomes less continental and uniform over large areas. The extreme types of climates, the desert and arctic ones, disappear. When solar activity weakens, stationary-type processes become dominant. The temperate latitude zones reach their maximum development, while the other zones reduce. Meridional exchange weakens and the west–east transport becomes absolutely predominant. The continental nature of climate increases, and the equator–pole temperature gradient reaches its maximum value. The contrast of the climatic zones gets sharp. Predtechensky classifies the glacial epochs as transient states when the meridional invasion and west–east transport become more or less equivalent, and when the temperature decreases, especially in the summer, with a lot of precipitation. Such conditions favor snow accumulation and the establishment of a glacial epoch. The plurality of glacial epochs is explained by the combined influence of the overlapping cycles of solar activity of various duration and amplitude on circulation processes.

V. Sinitsyn's hypothesis is based on the relationship of long-period climate changes with long-period changes in the geological and biological processes on Earth. There are two types of Earth's development, namely, geocratic and thalassocratic. The geocratic type is characterized by the general buoyancy of the crust masses, the widespread land, the pronounced arid climate, the general weakening of sedimentation, an increased granite formation, a flora crisis, and the almost complete cessation of coal accumulation and bauxite formation. The tallasocatic type is expressed in large submerged areas of the crust, the occurrence of the sea, the moisture-rich climate and dense clouds, increased terrigenous sedimentation, the development of volcanic series and the implantation of intrusions, lush vegetation development, and the intense accumulation of vegetable carbon and coal. There is also a transitional type, corresponding to switch of the two previous types. Contrasting climate, rapid volcanism, dynamic physical and geographical environment and the diversity of landscapes, a clear biogeographic zoning on land and in the sea, increased and complexed sedimentation processes, and flashes of coal accumulation, reef-building, bauxite formation, and halogenesis are typical for it.

Geological and geographical concepts (models) can also be divided into two subgroups: the hypotheses to explain climate variations by changes in the composition and properties of the atmosphere, and the hypotheses to explain climatic fluctuations by changes on Earth's surface.

The role of volcanic phenomena. It should be noted that the significant reduction in solar radiation (by 10–20%) was the result of catastrophic volcanic eruptions: Krakatoa in 1883 (18 km^3 of loose volcanic products was released) and Katmai in 1912 (21 km^3). Modern observations show that big volcanic eruptions were followed by cold years, for example, 1884–1885 (after the 1883 eruption) and 1913 (after the 1912 eruption). However, the intense volcanism of Late Devonian, Early Permian, Late Triassic, and Late Jurassic is not associated with climate cooling. The Katmai eruption occurred 29 years after that of Krakatoa. This periodicity does not meet any known astronomical cycles. It is longer than the Earth cycles around the Sun (11 years), but much shorter than the oscillation cycles of solar activity (400–600 years). This contradiction of modern observations and fossil record cannot confidently assess the role of volcanism as a climatic factor on a global scale. Only local effects on climate can be reliably judged.

The role of the Earth relief. Climate is largely dependent on the relief of Earth. On the other hand, climate forms relief through weathering. This relationship of climate and relief has become the basis for the idea of relief's impact on climate. First, the continent–ocean area ratio affects the general type of climate and atmospheric circulation. Under the predominance of the oceans, whose water is slowly heated up and slowly cools down, the Earth's climate becomes more even and the circulation processes get weaker. The predominance of land over the sea leads to the establishment of a continental climate with a high equator–pole temperature gradient and clear off-season climate contrasts. Atmospheric circulation also enhances. Second, of significance is the geographical location of the major continental blocks. If they are mainly concentrated in high latitudes, the climate is much more continental than in the case of their placement in low latitudes. Third, the average hypsometric level of land is of climatic significance. As it increases, the atmospheric pressure decreases, evaporation increases, and a general decrease in temperature occurs. When the average hypsometric level lowers, the climate of the continents becomes warmer and more humid. Fourth, the role of high-ridge climate divides blocking atmospheric circulation is significant. The Himalayas that separate the monsoon region of Hindustan and the dry region of Tibet are a modern example. Considerations of the relationship between glaciations and the orogenic periods, when the area occupied by land sharply increased with increasing height of the mountains, are of similar meaning. However, the history of geological development and modernity gives many examples disproving the direct link between the paleographic, geographical, and paleoclimatic climatic changes. Glaciations accompanied only the Hercynian and Alpine orogeny phases. There is no evidence of glaciation for all other orogenic phases. The abrupt alternation of glacials and interglacials in the Quaternary Period occurred essentially at constant relief.

The role of the ocean bottom uplifts. The formation of submarine ridges leads to disruption of the circulation of water masses. This entails isolation and a lowered temperature of the polar basins. The relation between the Quaternary glaciation and the raising of the Wyville Thomson Ridge (Atlantic Ocean) can serve as an example of such a hypothesis (Rukhin, 1958). The rising ridge blocked access of the Gulf Stream to the North Polar Basin, which was getting cool and frozen. The cold masses of dry air formed over this basin moved toward the equator and, in contact with the warm Atlantic air masses, contributed to the formation of precipitation (mostly snow). These factors caused the development of glaciation in Europe and North America. In the subsequent lowering of the Wyville Thomson Ridge, the Gulf Stream's warm waters penetrated northward, warming the Arctic Basin. With the warming of the North Basin, the snow and ice covers degraded. Disadvantages of this hypothesis are as follows. First, if the uplifts and subsidences of the Wyville Thomson Ridge would be linked with glaciation, the climate cooling would likely be localized, not appearing on a planetary scale. Second, the uplifts and subsidences of Wyville Thomson Ridge could be limited to the Quaternary Period, but no mechanism of this effect on the pre-Quaternary climate has been detected.

The role of the atmosphere composition. The radiant energy from the Sun is transformed into heat by the atmosphere. The modern atmosphere transmits 48% of the sunlight reaching the planet and holds 93% of its long-wave radiation. The thermal properties of the atmosphere are determined by its composition. The atmosphere composition evolved in the geological history of Earth, and, consequently, the atmospheric properties were changing. Of the multicomponent composition of the modern atmosphere, steam (water vapor), carbon dioxide, and ozone are of most climate significance. Water vapor has permeability for the short-wave solar radiation and shields the long-wave terrestrial radiation. An increased concentration of water vapor entails strengthening of the "greenhouse effect" and global warming. In addition to climate warming, water vapor, condensed in clouds, increases the reflectivity of Earth, causing cooling. The water vapor content is determined by evaporation and moisture factors, depending on the amount of solar radiation reaching the Earth's surface.

With the advent of the first photosynthetic organisms 2.2 billion years ago, the Earth atmosphere became oxygenated and carbon dioxide began to decrease. Modern plants annually absorb about 1/35 of the total amount of CO_2 contained in the atmosphere. In the Paleozoic carbonate sediments, the amount of CO_2 is 15,000 times higher than that contained in the modern atmosphere. The balance of oxygen and carbon dioxide is mainly determined by the vegetable land cover, which, in turn, depends on the climate. The carbon dioxide in the atmosphere is also consumed by weathering processes, peat accumulation, and the formation of carbonate sediments. Carbon dioxide returns through the processes of mineralization of plant and animal remains, respiration, the decomposition of fossil fuels, and the removal from Earth's bowels by volcanoes and thermal springs. The presence of a huge mass of bound CO_2 contained in fossil fuels of the Earth's stratosphere points out that the CO_2 consumption prevailed over its income in the geological past.

The majority of CO_2 is dissolved in the ocean water, where its amount is 50 times higher than in the atmosphere. The CO_2 concentration in the atmosphere is also determined by its exchange in the atmosphere–hydrosphere system. Approximately 200 billion tons of this gas annually flows from the air to water and back. Any changes in this balance and in the oxygen–carbon dioxide balance affect the climate. A general decrease in the CO_2 concentration causes cooling, while increasing entails global warming. Ozone retains the long-wave radiation of Earth and increases the "greenhouse effect" of the atmosphere. Ozone is produced in the ionosphere under the influence of ultraviolet radiation. Strengthening of ultraviolet radiation causes condensation of the ozone layer and global warming. The evolution of the atmosphere composition in the geological record, complex processes of consumption (including biogenic) and income (including volcanic) of CO_2 in the atmosphere, ozone production due to ultraviolet radiation, and water vapor generation associated with the climate-driven processes of evaporation and moistening do not allow to consider the atmosphere as the only climate-forming factor.

In the history of our planet, a number of time intervals can be resolved, corresponding to global climate cooling with the formation of ice sheets, called the *great glaciations* or ice ages (Gabdullin et al., 2005, 2011), namely, Late Archaean glacial (2.8 billion years ago); Early Proterozoic glacial (2 billion years); Middle Riphean glacial (1.2–1.1 billion years) associated with the Grenville orogeny era; Late Riphean glacial I (0.9–0.8 billion years); Late Riphean glacial II (0.8–0.75 billion years); Varangerian (Laplandian) or Early Vendian–Late Riphean glacial (610–570 million years) preceded and accompanied by the Baikal orogeny; Late Ordovician–Early Silurian glacial (Caradoc–Wenlockian) coinciding with the Taconian phase of the Caledonian orogeny; Late Devonian glacial, which coincides with the late Acadian phase of the Caledonian orogeny; Late Paleozoic (Gondwanian) glacial (the Early Carboniferous-Kazan age of the Late Permian); Absheron glacial (Late Pliocene, Neogene); and several Quaternary glaciations: the Oka, Dnieper, Moscow, Early Valdaian (Kalinin), and Late Valdaian (Ostashkov's) glaciations.

5.2 BIODIVERSITY AND THE PROBLEM OF THE EVOLUTION OF SPECIES[11–25]

The emergence and development of living organisms is one of the most interesting global processes occurring on Earth from the beginning of life till the present. Why did some species become extinct while other ones exist today? Why some extinct species existed for a long time while others lived for a very short time (compared to the scale of evolution)? Is there any regularity in the rates of extinction and evolution of biological communities? Whether evolution is possible in other environments? This section tries to find answers to these questions from the perspective of nonlinear dynamics.

Numerous data indicate that since the origin of life (in the first few tens of millions of years), 1–4 billion species have existed on Earth; about 50 million currently exist (the others are extinct). Many works are devoted to the extinction of certain species, or the study of known cases of mass extinction, such as at the end of the Cretaceous Period. Recently, however, works have appeared in which, along with the analysis of statistical features of the historical process of extinction of species, a variety of mathematical models are offered.

It should be noted that the disappearance of taxa from the Earth face, as well as the emergence of new ones, is a natural consequence of the evolution of the organic world—an integral feature of the biosphere development. As a rule, some taxa give rise to other ones (leave descendants) and die out themselves, sooner or later. However, they often disappear completely, leaving no descendants and direct child branches not only of genera and families but also of orders and even classes, whose representatives were previously widely spread and well adapted to the environment. Only such cases are of particular interest.

Extinction is currently treated as a global process going with a more or less constant rate (background extinction), which, however, is broken by relatively short-term (extraordinary) events characterized by extremely intense extinction (mass extinction; Alexeyev, 1998). At present, two fundamentally different definitions of mass extinction are known. One of them is used for statistical analysis, being purely formal. According to this definition, a *mass extinction* means any peak of the level or rate of extinction, separated from the adjacent peak by a minimum. According to the other definition, a *mass extinction* is a “relatively short-time and, in the scale of geological time, synchronous stepwise extinction, during a biotic crisis, of a large number of taxa belonging to different taxonomic and ecological groups; this leads to a temporary reduction in the global taxonomic diversity of the biosphere” (Alexeyev, 1998). Therefore, the events referred to as mass extinction, in turn, are sometimes divided into three groups, namely:

1. extinction events (marked in the history of individual groups, but do not affect changes in the global diversity of the biota or reduce it slightly);
2. mass extinctions-phantoms (events that are considered as mass extinctions from quantitative analysis, but in fact cannot be referred to as such);
3. true mass extinctions (Alexeyev, 1995).

In addition, there are two basic concepts relating to the causes of the extinction of species. The traditional viewpoint shared by most paleontologists is that extinction is a result of some external influence on the ecosystem from the environment (Benton, 1995; Hoffman et al., 1991). There are serious arguments in support of this view, for example, sea-level changes (Hallam, 1989) or the Earth's collision with a large space body at the end of the Cretaceous Period (Alvarez, 1987). Of course, such explanations are not universal, but almost all alternative hypotheses, in essence, also explain cases of mass extinction by external influences, although the range of external factors, which could, in the opinion of various researchers, lead to a mass extinction is quite wide: from climate changes (Stanley, 1984) and the oxygen content in the ocean (Wilde et al., 1984) to rather exotic one, such as volcanism (Courtillot et al., 1988), tidal waves (Bourgeois et al., 1988), changes in Earth's magnetic field (Loper et al., 1988), and even a supernova explosion (Ellis et al., 1995). Let us briefly consider some of the known concepts, some of which being close to the global climatic models.

1. The *galactic mass extinction hypothesis* is related to the rotation of the solar system around the galactic center. Changes in the cosmic ray flux, the intensity of the magnetic and gravitational fields, meeting with dust clouds could thus have an impact on the organic world.
2. *Fluctuations in solar activity* should have led to changes in the level of cosmic radiation and magnetic field strength, the efficiency of the ozone screen, and ultraviolet flux, accompanied by climatic fluctuations, to entail, in turn, the extinction of some organisms.
3. *The impact hypothesis* is that there may be periodic collision of Earth with large space objects, such as asteroids or comets. As direct causes, abrupt climate changes are accepted in this scenario:

the continuous opacity of the atmosphere to sunlight due to the high content of fine particles arisen as a result of the impact, the global temperature drop due to the decrease in the intensity of solar light, storm fires, acid rains as a result of the formation of enormous quantities of nitrogen oxides during the explosion of an asteroid, and so forth.

4. *The explosion of a supernova.* In accordance with this hypothesis, a supernova explosion occurred in the vicinity of Earth and, therefore, a sharp increase in the flux of cosmic rays, first of all, their neutron component, led to the death of organisms as a result of mutation growth.
5. *The volcanism hypothesis.* The release of enormous amounts of carbon dioxide and nitrogen and sulfur oxides during volcanic eruptions leads to abrupt climate changes and the global extinction of organisms.
6. *Sea-level fluctuations* could lead to drying of shallow seas and continental shelves, and, in this connection, the disappearance of the diverse benthic biota peculiar to them.
7. *Changes in the atmosphere composition* lead to sharp fluctuations in the oxygen concentration in the atmosphere, which is accompanied by the selective extinction of the most sensitive groups of animals.

At the same time, many researchers believe that the global process of species extinction is due to solely biological causes and that it is a natural and integral part of the dynamics of ecosystems, occurring independently of external influences from the environment. As causes for the extinction of species, models are proposed based on interspecific interactions, such as the introduction of a new competitor in the previously stable biological system or overly active actions of predators (Smith, 1989). However, such cases of extinction (if any) affect a very small number of species, and, consequently, are virtually indistinguishable against the common background.

Recently, such issues are discussed as the relative frequency of major or minor cases of species extinction. The results obtained are compared with the known data, and a conclusion is made on the basis of the suitability (or vice versa, fallibility) of the model. Some models are based on the biological causes of extinction of certain species, while others basically have the assumption that external factors are the cause of species extinction; there is a group of models that take both biological and environmental factors into account. As the research in this field is relatively recent, it is quite natural that there are not yet established regulations, and a number of models based on various mechanisms of species dying out demonstrate good agreement with real data.

The main source of data relating to the evolution of species is the so-called fossil record, which is defined as a collection of material documents of the organic world of the past, preserved in the sedimentary shell of Earth. The sedimentary shell, formed by successive strata of sedimentary rocks, is a pretty good tool for recording, with some degree of accuracy, all the events taking place in the biosphere. The modern methods of stratification and geochronology allow more or less accurate dating of the events recorded in the sediments, allowing further analysis. Naturally, such a "recording mechanism" is imperfect. The loss of information starts immediately after the death of organisms and only a small part of them, especially having mineral skeletons, gets into the sediment. In the sediments, at the stage of their becoming a rock, a further loss of information occurs, which continues later, when the rocks are destroyed by erosion, sink to greater depths, and so forth. As a result, only small portions of the initial amount of information about the ancient biota may be preserved. Another limitation is due to the fact that modern science is able to extract and "decode" only a part of the information preserved in the fossil record, although these capabilities are continuously expanding.

The fossil record can be divided into separate components: the potential one is the remains located in rocks, which can ever be studied; the unpublished one is the part of the collected materials and collections which is either under processing or has never been treated, or processed but unpublished for various reasons (the death of the investigator, lack of funding, the loss during the war, etc.); and the published one is the information known to the scientific community through publications. The incompleteness of the published data is obvious, but just this component of the fossil record is the source of all available scientific information on the biota in the past. However, the published fossil record is not uniform: at the moment, there are two, largely independent, parts of it. One of them, recorded in marine rocks, has formed the basis for the development of chronostratigraphic scale and provides the most complete and reliable information of historical changes in the biota, but only of its marine component. The remains of freshwater and terrestrial organisms infrequently get into marine rocks, and their fraction is unlikely to exceed 1%. The second part was preserved in continental rocks and is much less representative than the marine one. The recording capacity of marine sediments, as a whole, is much higher than that of continental ones. Accordingly, the fossil record of continental organisms poorly correlates with the marine one. Combining these two parts into one record is not only difficult in some cases but also pointless. Therefore, these two parts of the record can be regarded as relatively independent, which should be taken into account when analyzing published data.

We note in passing that there is no complete information on the number of the described taxa of fossils. Currently, the two databases are most widely used: one compiled by Sepkoski (1993) and the other by Benton (1993). The Sepkoski database contains information on both biological genera (approx. 40,000) and families (about 5000); a vast majority of these are marine invertebrates. The Benton database, by contrast, includes information on both marine and terrestrial fauna and contains data about 7000 biological families. According to some estimates, to date, the total number of genera of fossil plants and animals described in the scientific literature is 50,000–70,000 units, and that of fossil species is 350,000 units (Alexeyev, 1998). However, these numbers do not refer to real taxa, but only to their names. In reality, these figures are estimated to be 20–30% lower due to the fact that the same taxa may appear in the literature under different names, and controversy in the taxonomy continues. If one tries to assess the fraction of the described taxa of the total ones whenever existed on Earth, it can be concluded that today science has information of 5–20% extinct genera, about 50% of extinct families, and 20–30% of extinct species. This number of taxa is enough to reveal main regularities of changes in the diversity of the Phanerozoic biota (the period from the Cambrian Period to the present time).

It is natural that researchers are interested in a classification with the maximum possible degree of detail, for example, at the level of separate species. However, the accuracy with which we can estimate the time of occurrence and death of an individual species depends on the number of the known fossils belonging to this species. In the case of scarce remains, the probability of error in dating is high. As a rule, it turns out that the time of extinction is underestimated in such cases, that is, dated from an earlier period than actually happened (the so-called Signor–Lipps effect; Signor et al., 1982). The opposite effect, sometimes called the Lipps–Signor effect, lies in the fact that for a species, poorly represented by fossils, the time of its origin dates back to a later period. At the same time, research at the level of higher taxa will be more objective.

The time which dates back to the occurrence or death of a particular species is referred to the nearest geological age, which is the interval of geologic time usually determined on the basis of relative geochronology (Table 5.1). This method of dating is common but, unfortunately, raises a number of problems. For example, the boundaries of the conventional geological ages are not known with sufficient accuracy, and are so far disputed (like the ages themselves). Another problem stems from the fact that geological centuries and eras have different durations. For example, the average duration of the Phanerozoic geological century is about 7.3 million years; however, the duration of each age ranges from 1 million to 20 million years.

In order to quantify the global processes of changes in the biological diversity, several indicators exist. The initial data for all calculations are as follows:

1. the total number of taxa that existed for any length of the geologic timescale (the so-called taxonomic diversity);
2. the number of taxa newly appeared during its length;
3. the number of endangered taxa.

Other quantities can be derived from these data. Naturally, the number of extinct taxa within any geological century is not sufficient, since geologic centuries (as mentioned above) have different lengths. To reduce this effect, it has become common to study not only the total number of taxa (species, genera, and families) extinct during some age but also the rate of extinction (the number of taxa died within a unit of time).

The above is illustrated in Figure 5.3. Figure 5.3a shows the number of families of known marine organisms extinct in several Phanerozoic geological periods: each point indicates the number of families died within one geological age. Figure 5.3b presents the same data "normalized" in view of the fact that the geological centuries have different durations: shown is the number of families died within a unit of time (chosen to be equal to 1 million years). One can see a difference between these figures, namely, Figure 5.3b clearly shows that the intensity of extinction decreases over time, which is unnoticeable in Figure 5.3a. However, such an indicator as the rate of extinction contains errors. It is clear that if you want to determine the rate of extinction of species, it is necessary to divide the number

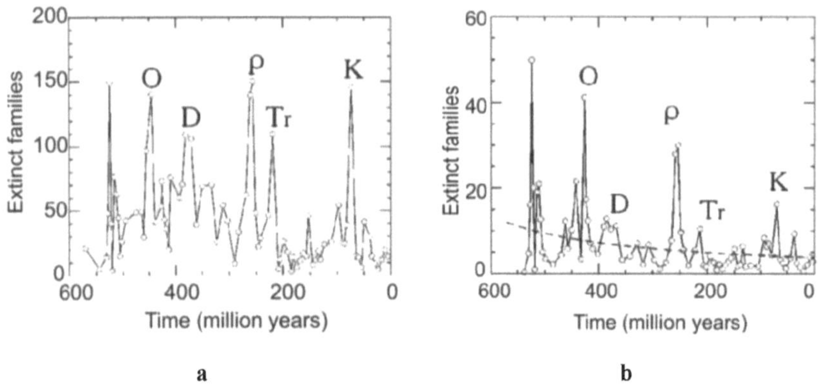

FIGURE 5.3 (a) The number of families of known marine organisms extinct in the Phanerozoic geological periods as a function of time. The number of families died within each geologic century is shown. The figure is based on the data from Sepkosk (1993) and taken from Newman et al. (2003): O, Ordovician; D, Devonian; P, Permian; Tr, Triassic; K, Cretaceous. (b) The same data normalized to the duration of the geological ages: the number of families died within a unit of time (1 million years) is shown. It is clearly seen that the intensity of extinction decreases with time. The dotted line represents the average number of extinct species as depends on time.

of extinct species by the duration of the relevant geological period. However, if extinction occurred nonuniformly, and there were moments when a larger number of species died out (many researchers believe that this is the case), then the results will not be accurate, reflecting a too averaged pattern. For example, the extinction of a sufficiently large number of species within a short geological century would give an anomalously high extinction rate, irrespective of whether this rate really increased or not, compared with the adjacent, more prolonged centuries. Finally, large errors are associated with inaccurate and contradictory geologic timescales. The duration of most units of these scales has been established with some error (which is usually indicated). This means that the actual duration is within a certain range and it is initially necessary to calculate the rate of extinction for the extreme limits of duration. In this case, an interval will be obtained which the true value of this indicator may be within. In order to avoid such errors, such indicators are introduced into consideration as the level of extinction, the background level of extinction, and the extinction intensity.

By the *level of extinction* is understood the amount of endangered taxa, expressed as the percentage of the total number of those existed within a given time period. This indicator indirectly takes into account the duration of the subdivisions of the geologic timescale because the total diversity and the number of endangered taxa will be higher over a longer period. And it introduces no distortions associated with significant discrepancies of the existing geologic timescales when rate indicators are determined.

The *background level of extinction* is the average value of the level of extinction calculated for some few (5–10) nearest periods of time, which can be regarded as "calm." This indicator is needed in order to assess the degree of extinction at some turn, because the background level of extinction will be specific for every group, every taxonomic level, and every time point.

TABLE 5.1 Geochronologic Scale

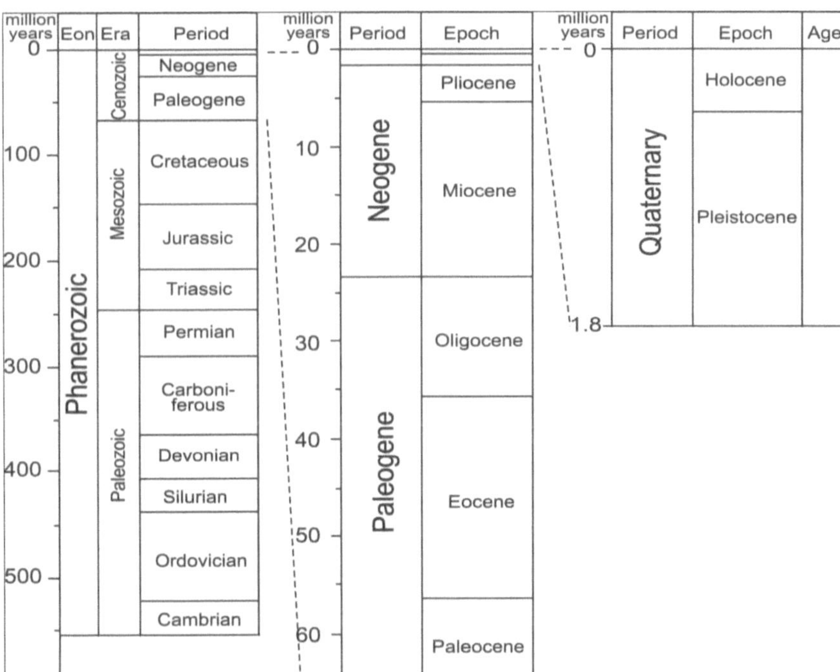

The *intensity of extinction* is the ratio of the extinction level at some critical turn to the background level of extinction. Only this indicator allows one to objectively compare the degree of extinction of various systematic and ecological groups and various taxonomic categories, although it is not free from drawbacks. For example, in the case of an exceptionally high background level of extinction, even the complete disappearance of some group will be expressed only by a small increase in the intensity of extinction and vice versa; if the background extinction rate is zero, any extinction of even one taxon will give an infinite value of the intensity of extinction.

The usage of geological ages as the timeline creates other difficulties as well. For example, as mentioned above, several specimens of the same species, whose fossils belonged to different geological ages, were sometimes assigned different names (Raup et al., 1988). Therefore, due to the boundaries of geological ages, "additional" extinctions appeared—some species could be thought as extinct, although it survived the change of times, but is known under a different name in the following age.

There are other difficulties in studying the birth and death of species. For example, a clear peak of speciation may be a result of the hard work of a researcher (or a research group) to study a specific geological period. The occurrence of a large number of species during a short period (compared to other periods) can be explained only by this period being studied significantly better than the others due to enthusiastic researchers' hard work. A similar cause of distortions and biases of paleontological data is entering new records into the database, which was the result of excavations at a new and fossil-rich site, which may enhance the unevenness of existing knowledge.

Thus, to obtain an objective picture, various quantitative indicators characterizing the global process of evolution should reasonably complement each other. Little representation of fossils after the mass extinction of species is another source of distortions. As already mentioned, many major cases of extinction of species (genera, families) are believed to be caused by changes in the environment. These changes may have violated the conditions under which the organisms become fossils and could be preserved. Consequently, the species that survived these mass extinction events "good enough" are poorly represented in fossils and, therefore, could also be classified among the extinct species.

Given all of the above circumstances, a reasonable question may arise: How reasonably could someone speak of any data and results, if so many inaccuracies and errors are involved? Nevertheless, many researchers in their works have undertaken a number of efforts to neutralize these errors, and the fact that many conclusions made by different researchers independently of each other are the same, let us speak of the reliability of these results.

In recent years, it has been assumed that the distribution of extinctions obeys a power law, at least for significant extinctions of species. In other words, the probability p that some portion s of the existing (at some point) species, genus, or family would die within a certain time interval (period) obeys a power law:

$$p(s) \sim s^{-a}$$

The method of graphical representation plays an important role in analyzing data. If you plot a power-law dependence on a graph where the value of s is plotted along the horizontal axis (x-axis) and p along the vertical axis (y-axis), you will get a curve decreasing rapidly enough as s increases. An exponentially decreasing dependence will qualitatively have a similar behavior (Figure 5.4a). When working with real data, it is almost impossible to visually distinguish (by estimation) one curve from the other one. In this case, it is necessary to approximate the analyzed data, that is, to plot the corresponding analytical curve (e.g., an exponential or power one) on the same graph and to select the values of the parameters, using appropriate mathematical methods (e.g., least squares) to fit the existing data.

At the same time, there is a simple way to determine the nature of the dependence analyzed. If you plot a power-law dependence on a log-log scale (by laying off log s and log $p(s)$ on the axes rather than s and $p(s)$), then, taking the logarithm of eq 5.1, its graph looks like a straight line.

$$\log p(s) \sim -\alpha \log s \qquad (5.2)$$

The exponential dependence plotted on a log-log scale is presented as a curve (Fig. 5.4b). From this figure, first, the qualitative difference in the behavior of these curves is seen, and, second, it can be concluded that if the data are arranged along a straight line when plotting them on a logarithmic scale, the corresponding relationship is a power function. At the same time, looking at the solid curve in Figure 5.4b, one cannot decide whether it is exponential or not, as many other dependences can be represented by curved lines on a log-log scale. Therefore, the log-log scale provides the definitive identification of only one class of dependences:

the power one. To unequivocally conclude that the analyzed data obey an exponential law, it is necessary to plot them on a logarithmic scale, where the abscissa axis is linear and the ordinate axis is logarithmic. In this case, only the exponential dependence is characterized by a straight line (see Fig. 5.4c).

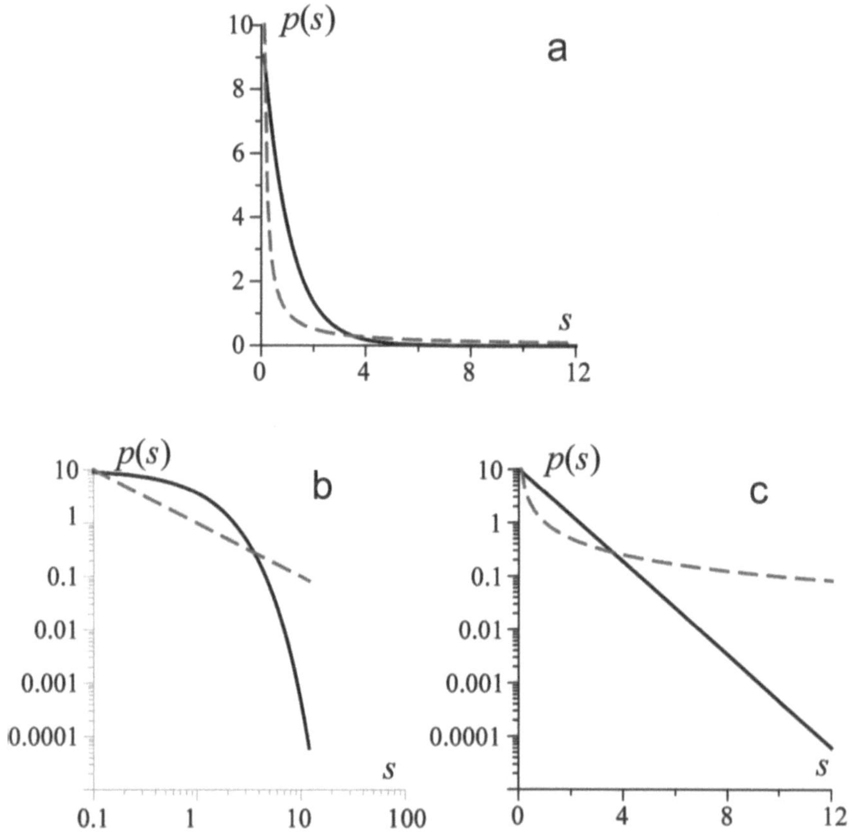

FIGURE 5.4 Power (dotted line) and exponential (solid line) dependences shown on the same graph: (a) the axes on a linear scale, (b) a logarithmic scale (the ordinate axis is shown on a logarithmic scale, while the horizontal axis is a linear one), and (c) a log-log scale.

Figure 5.5 shows the same data as in Figure 5.3 replotted on a logarithmic scale. The data seem to fit a straight line (corresponding to a power law), but probable errors (shown in the figure as well) are too high and the number of points is small in order to speak of the power-law distribution with confidence. The exponential nature of the distribution of these points is also possible. A possible way to overcome the lack of data is to rank the relevant geological ages: the age within which the largest number of taxa became extinct is assigned the first rank, the second one (by the number of extinct taxa) is assigned rank 2, and so on. Then it becomes possible to plot the number of extinct taxa within a geological age as a function of its rank. It has been shown (Zipf, 1949) that if a distribution satisfies a power law, its ranked distribution will also satisfy this law. Moreover, this ranked distribution is more convenient in the sense that there is no reduction of the number of data as in the construction of the corresponding histogram.

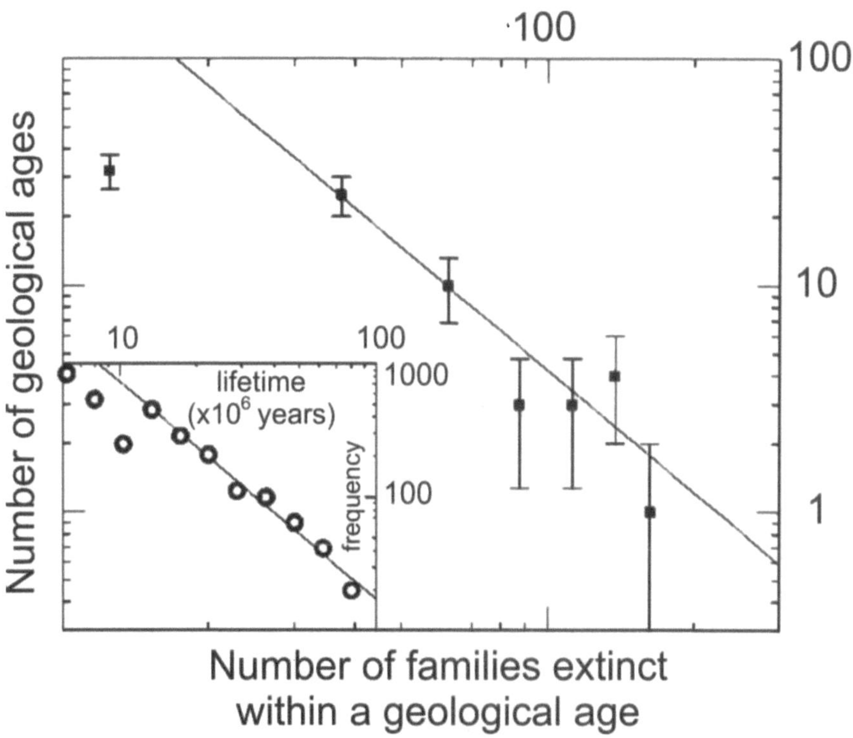

FIGURE 5.5 Data from Figure 5.3 replotted on a log-log scale. Straight line corresponds to a power law (eq 5.2). Inset shows the histogram of the duration of the existence of biological genera (Newman, 2000).

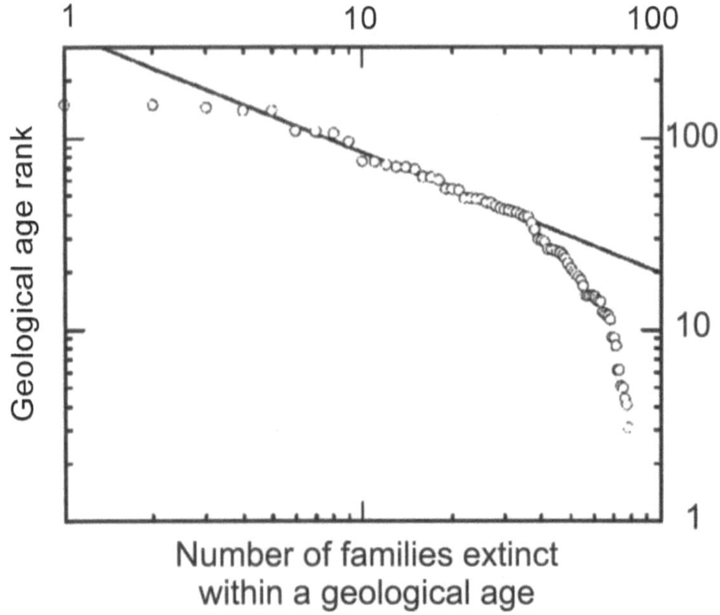

FIGURE 5.6 Rank dependence of the number of extinct families of marine animals within a geological age. Straight line corresponds to a power law (eq 5.2). The dependence is shown on a log-log scale (Newman, 2003).

Figure 5.6 shows the corresponding ranked distribution of the number of extinct families of marine animals within a geological Phanerozoic age, depending on the rank of the age. It is indeed evident that such a ranked distribution is more illustrative than that in Figure 5.5. It can be seen that the distribution points are in good agreement with a power law before those ages when more than 40 families became extinct, followed by a deviation, which can be associated with either inaccurate data or the fact that this distribution corresponds to, for example, an exponential (or any other) law rather than a power one. A deeper analysis (Newman, 1996) also did not make the final clarity on this issue, but most researchers are inclined to believe that the distribution of the number of extinct species obeys a power law rather than an exponential one. Moreover, the coefficient α in eq 5.1 was estimated with an appropriate accuracy, its value is 2.0 ± 0.2.

The mass extinction of species over the past 250 million years being repeated at intervals of approx. 26 million years is another feature of the evolutionary process which attracted attention in 1984 (Raup et al., 1984; see Fig. 5.7 and Raup et al., 1986; Raup et al., 1988; Sepkoski, 1990). A number of scientists are inclined to explain this phenomenon by means of astronomical events. For example, according to one version (Leonovich, 1994), the Sun has an undiscovered twin star rotating along a very elongated orbit with a period of just 26 million years. At the moments of its closest approach to the Sun, serious changes occur on Earth, causing massive loss of existing organisms. There are other possible explanations for this phenomenon; for example, this feature is due to errors associated with poor classification (Patterson et al., 1987; Patterson et al., 1989) and inaccurate dating lines in the existing geologic timescale.

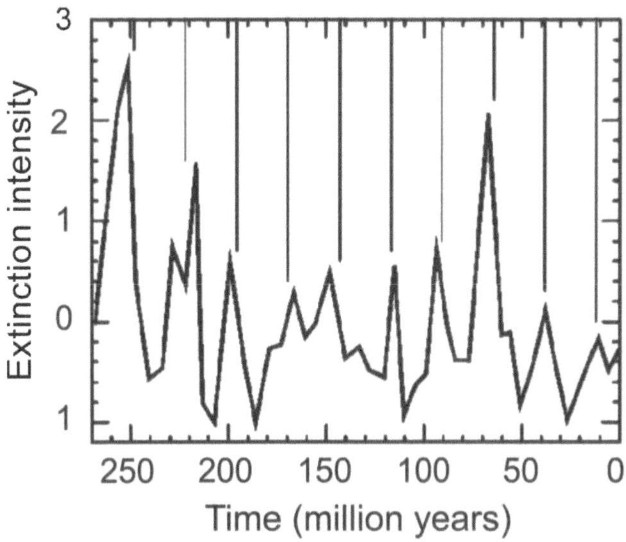

FIGURE 5.7 The number of genera of marine invertebrate animals extinct in the last 270 million years.

Studying the reverse (and also global) process of the emergence of new species is directly associated with studying species extinction. This question has been studied in less detail than the extinction of species. The most obvious fact is that a splash of new species directly follows a mass extinction (Fig. 5.8).

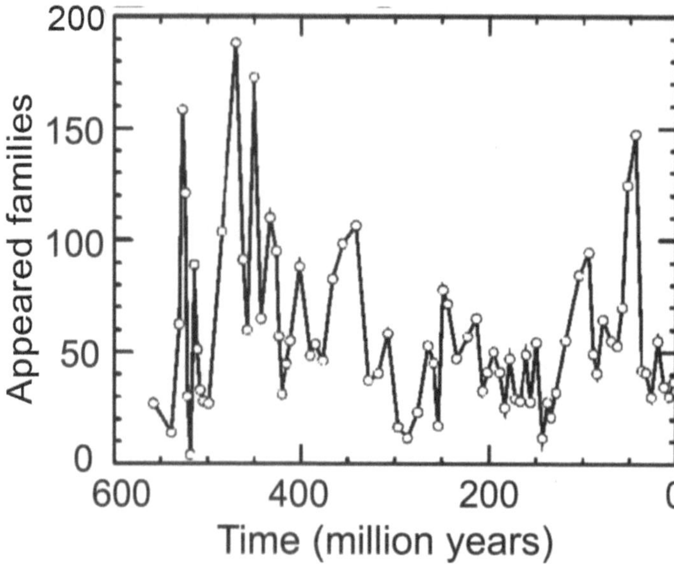

FIGURE 5.8 The number of families of known marine animals appeared in each Phanerozoic geological age as a function of time. The figure is based on the data from Sepkoski (1993) and taken from Newman et al. (2003).

Comparing Figure 5.8 with Figure 5.1 shows that the peaks of the emergence of new families correspond to all extinction peaks, although, of course, the correspondence between these curves is rough. The most typical explanation for this lies in the fact that for new species, it would be easier to take the existing empty ecological niches resulting from mass extinctions, and, accordingly, a peak of the extinction of species should be followed by a splash of new ones, occupying the vacant ecological niches. After all the ecological niches are refilled, the formation of new species decelerates. On this basis, it can be concluded that there is a certain level of saturation of the species which can exist in any ecosystem, which remains constant, except during the periods just after mass extinctions. However, this viewpoint is currently controversial. On the one hand, the modern environmental data support this hypothesis (Rosenzweig, 1995), and on the other hand, the species diversity increases on a longer timescale (Fig. 5.9), as the animals discover the opportunity of using new habitats.

In this figure, the vertical axis represents the deviation from the average extinction rate. The vertical bars indicate periods of 26 million years through which peaks extinctions are repeated (Sepkoski, 1990).

From a comparison of the curve corresponding to the total number of known biological families with the dashed line corresponding to an exponential increase, it can be concluded that the number of families increases exponentially, although logistic growth is also possible.

Figure 5.9 also shows that the increase in biodiversity is interrupted by mass extinctions followed by the reduced number of families.

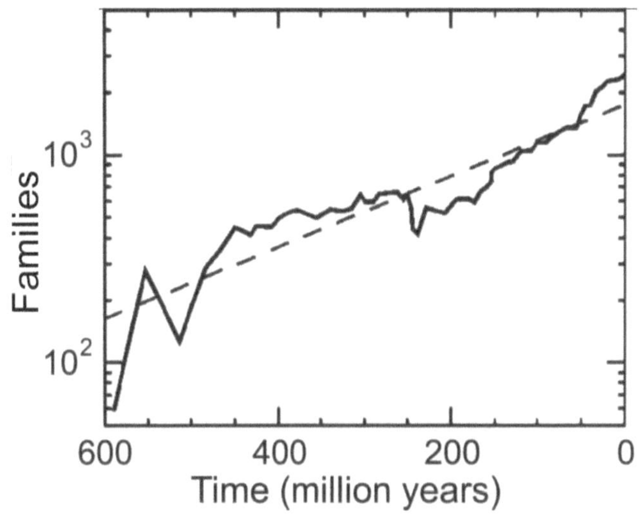

FIGURE 5.9 The total number of known families as a function of time (Phanerozoic). The vertical axis is a logarithmic scale and the dotted line represents an exponential growth (Newman et al., 2003).

Another model was developed, according to which, the variety of marine organisms experiences more or less long periods of growth and saturation, described by separate logistic equations. These periods are interrupted by short-term mass extinctions.

Each segment of the growing diversity is determined by the initial level of diversity immediately after mass extinction, the initial growth rate, and the subsequent levels of equilibrium. The more intense the mass extinction, the higher the level of subsequent equilibrium diversity. At the same time, as noted above, the phenomenon of growth of the biological variety can also be due to the fact that the more recent fossils tend to be better preserved and are easier to find; hence, the increase in the diversity of biological species may be illusory. In any case, there is no reason today to conclude that there is some upper limit of species diversity on Earth.

Another characteristic feature of the evolutionary process is the distribution of the lifetimes of species (genera, families), similar to that shown in Figure 5.10 (see also the inset in Fig. 5.5). At first glance, it seems that the distribution of the lifetimes of biological genera obeys the exponential law. Nevertheless, a number of researchers are inclined to conclude that this distribution is still described by the power law.

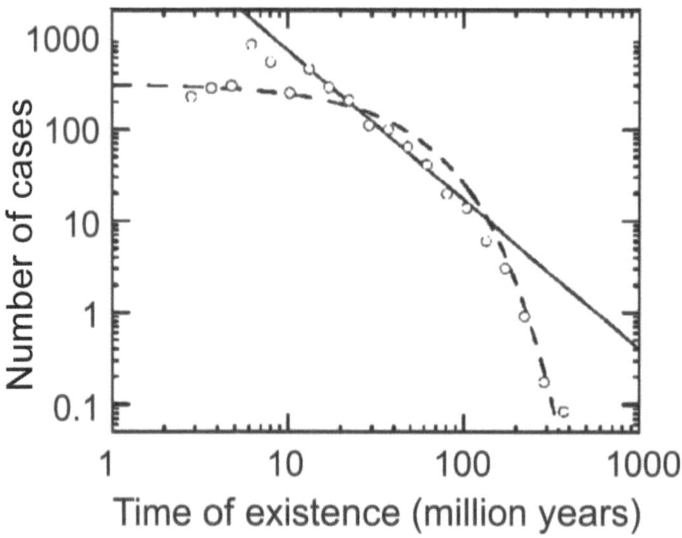

FIGURE 5.10 Frequency distribution of the lifetime of biological genera of marine animals. The continuous line corresponds to the power-law distribution and the dashed line to exponential on a log-log scale (From Newman et al., 1999).

From Figure 5.10, it can be seen that in the range of duration of 10–100 million years, a good correspondence of the data to a power law is observed, while for the lifetimes shorter than 10 and longer than 100 million years, there is a significant deviation of the data from the expected straight line. Nevertheless, researchers believe that the relevant data on biological genera with lifetimes shorter than 10 million years and longer than 100 million years are just not adequately represented in the relevant databases due to systematic errors. First, the biological genera, occurring and dying out within one geological age, in accordance with the common classification, have a zero duration and, accordingly, are not presented in the distribution. Therefore, the biological genera with their lifetimes shorter than the average duration of one geological age (approx. 7 million years) will be underrepresented and the plotted points will accordingly lie below the line corresponding to the power law. Second, as mentioned above, the same species may have different names in different geological ages, especially if this species existed before and after the main, most important time limits. Therefore, the species with their durations long enough are not adequately represented in the distribution too. This effect is shown on durations of more than 100 million years. Consequently, the most reliable data of the biological genera are those whose duration ranges from 10 to 100 million years, which are distributed as shown in Figure 5.10, in the best agreement with the power law (the straight solid line in the figure).

5.3 THE BIOSPHERE EVOLUTION AND THE NOOSPHERE[8–10]

The concept of the biosphere and the noosphere by Vladimir Vernadsky (1863–1945) plays a crucial role in the modern scientific pattern of the world. It treats life as a holistic physical, geochemical, and biological evolutionary process included into cosmic evolution as a separate component. It is the realization of this integrity which largely determines the strategy for the further development of mankind.

Since the second half of the 20th century, particular scientific concepts, losing their autonomy, were becoming more and more fragments of a holistic pattern of the world. They are connected to blocks of the whole picture, which characterize the inanimate nature, the organic world, and the social life—realizing the ideas of universal evolutionism in its own area each. The modern scientific picture of the world is based on the unity and diversity of various sciences. It does not seek to unify all fields of knowledge; there is no desire to reduce everything to the principles of one science only. The idea of the biosphere and noosphere is an important concept in the world picture.

The biosphere (from the Greek *bios*—life and *sphaire*—sphere) is one of the shells (spheres) of Earth, whose composition and energy are determined by the past or present activities of living matter. In 1875, the Austrian geologist Eduard Suess introduced the notion of the Earth crust shells: the water shell was named the hydrosphere, the solid one is the lithosphere, and the area of the Earth crust enveloped by life is the biosphere. Even before, the gas shell of Earth was called the atmosphere. However, the term "biosphere" became widespread half a century later in the works of Vladimir Vernadsky, who referred the whole outer region of our planet Earth as the "biosphere," in which not only life exists, but which, in varying degrees, is modified or formed by life. In 1968, Nikolay Timofeyev-Resovsky in his paper "The biosphere and mankind," clearly outlined the importance of fear and anxiety for the fate of mankind: "Earth's biosphere produces the entire human environment. And our careless attitude toward it, undermining its proper functioning would not only undermine the food resources of people and a number of industrial raw material need for people, but also undermine the gas and aquatic environment of people. Eventually, people without or with the malfunctioning biosphere generally cannot exist in the world. This shows that this is indeed the number one problem and an urgent problem" (Timofeyev-Resovsky, 1996).

What is the biosphere from the standpoint of modern natural science? It includes the troposphere, the hydrosphere, and the upper part of the lithosphere, which are interconnected by complex biogeochemical cycles of substance migration and energy flows. The active part of the biosphere is the totality of all ecosystems (biogeocenoses). All ecological niches suitable for life are occupied in the biosphere, which arose about 4 billion years ago, simultaneously with the appearance of life on Earth in the form of primitive protobiogeocenoses in the primary oceans. Four hundred and fifty million years ago, living organisms began to occupy the land where their evolution accelerated,[1] and, as a result, the ratio of the number of animal and plant species in the oceans and on land is about 1:5.

The main factors of the evolution of the biosphere are geological, cosmic (abiotic); variability, that is, mutations, heredity, the struggle for existence, natural selection (biotic); and anthropogenic, owing to which the biosphere gradually acquires features of the noosphere.

The biosphere is mosaic by structure and composition, which manifests itself in the geochemical and geophysical variety of Earth's face (mountains, canyons, valleys, lakes, oceans, etc.) and in the uneven distribution of living matter over the territory of our planet in the past era and now. Living matter—the totality of all living organisms—forms a special state of space–time processes, determining the evolutionary capabilities of the biosphere. Living matter, trapping, accumulating, and transforming the space and solar energy, acts as a transducer of our planet, its major

[1]Probably, this happened because the environmental conditions on land turned out to be harder than in the ocean.

geological force. This force creates a "face" of the planet, determines its beauty and harmony, individuality, expressing in the very organization of the biosphere, in an amazing variety of the forms of plant and animal life, in the form and colors of landscapes.

In the teaching of the biosphere, there is an important aspect related to the study of the global effects of mankind on the structure and function of the biosphere (mining, the production of biosphere-new substances, environmental pollution, landscapes transformation, man's leaving into space, etc.) and having generated the global problem "the biosphere and mankind."

What is the biosphere in terms of nonlinear science? No doubt, this is a huge nonlinear system, which, even without active external influences, is capable of substantial restructuring. The biosphere as a complex self-developing system has a certain set of mechanisms, some of which play a role of positive feedback (they are responsible for the development of the system, the growth of its complexity, and the diversity of the constituent elements), while others play a role of negative feedback (they are responsible for the existing quasi-equilibrium and for the stability of the components). Great attention is being paid to the study of negative feedback mechanisms, which is natural, because people live in certain environments, having been adapted to them.

Several works of N. Moiseyev are devoted to the analysis of the state of the theory of the biosphere. In one of them, he wrote: "But the study of individual mechanisms, even in combination, is not sufficient to construct a theory of the biosphere. And without such a theory it would be very difficult and dangerous to talk about mankind's strategies with the biosphere... The theory of development of the biosphere cannot be considered complete, as much is not studied yet: the state of its bifurcation states, the conditions of transitions from one state to another, and the structure of attractors, i.e. the neighborhoods of more or less stable states...." Because of the complexity of the set of equations describing the functioning of the biosphere, computer simulation is the only effective way to analyze its behavior. This is the only way because people cannot venture any real experiments with the biosphere. Such a computer model is known as "Gaia." It combines the models of atmospheric and oceanic circulation with the model of the carbon cycle and the energetic biosphere, including solar flux, the formation of clouds, snowfall, and so on. The "nuclear winter" and "nuclear night" hypothesis was confirmed within the framework of this model. The result was tragic: "The Earth after such a powerful impact would cease to be similar to the Earth as we know it in the Quaternary... And this New Earth could not serve as the Ecumene of mankind, the Earth would remain very depleted and, most importantly, without people" (Moiseyev, 1998). Another important conclusion was made as well: the transition to a new qualitative state and the loss of stability threshold may occur under the influence of small but permanent disturbances.

These conclusions are related to the ideas of synergy. The biosphere may apparently have several quite different quasi-stationary regimes. In other words, it has a number of different attractors. And it is possible that the process of evolution of the biota, which has led to the emergence of *Homo sapiens*, could only proceed in the vicinity of one of these attractors. Going to the neighborhood of another attractor would eliminate the possibility of reasonable life on our planet. Therefore, the theory of the biosphere should not be just a collection of the studied mechanisms of functioning of individual elements of the biotic and abiotic components of the biosphere, whose interaction can obey Le Chatelier's principle (which, of course, is absolutely necessary). In order to enable further development of our civilization, we have to study the dynamics of the biosphere as a nonlinear system to study the structure of its attractors and the boundaries between the regions of their attraction.

A new fundamental scientific discipline so appears—the study of the biosphere as a dynamic system. And it is of absolutely applied character, since it becomes the scientific basis for mankind's fateful decisions. The transition of the biosphere from one state to another does not necessarily imply instantaneous overloading, as while nuclear explosions and subsequent fires. A disaster could sneak up quietly as well. And the strategy of human development should not only be consistent with the development of the biosphere, but should be such that the development of the biosphere would go in a channel needed for mankind! In other words, the provision of coevolution of man and the biosphere (or, equivalently, to implement the strategy of sustainable development) requires the development of a special synthetic scientific discipline.

Vladimir Vernadsky as early as in his younger years was pondering over questions about the place and role of the mind as a planetary cosmic force. The idea of strengthening the role of the mind in the development of society is directly related to the notion of "noosphere."

The idea of the "kingdom of reason" was expressed in the enlightenment, when no global economic problems were thought of yet. But this "educational mind" remained unclaimed, and the spontaneous march of capitalism on our planet has led, in fact, to the formation of industrial and consumer (rather than noospheric) values.

The noosphere is currently defined as follows. The *noosphere* (from the Greek *noos*—mind and sphere, the sphere of mind) is a hypothetical future state of society and its interaction with nature, in which mind will occupy a priority status.

Vladimir Vernadsky was one of the first to involve time into analysis of historical time. The novelty of his thought is just in this. In his opinion, the biosphere developing into the noosphere is a natural phenomenon, a deeper and more powerful in its basis than human history. Vernadsky believed that the formation of the noosphere is the mainstream of mankind development and that it will appear spontaneously, like all stages of human evolution.

However, when the global problems were revealed and exacerbated, which threaten the destruction of civilization, it became clear that no spontaneous formation of the noosphere is possible, and that its approach is real only through the social and technological design of the future with the help of the human mind, and, above all, the noospherically oriented science. In the process of real progress on the path to the noosphere, science will, as predicted by Vernadsky, play a dominant role, as well as the intellectually spiritual and morally cultural activity forms associated with it.

Among the main features of the current stage of the noosphere doctrine, the following ones can be listed:

1. The biosphere does not need to be forcibly converted into the noosphere. It must be saved by reducing the anthropogenic pressure by almost an order of magnitude and regulation of the environment.
2. For the noosphere formation, it is necessary to implement the principle of continuity of generations for the survival and continued development of civilization. The essence of this principle is in the equality of opportunities of all people generations on our planet (including the future ones) to satisfy their vital needs.
3. It is necessary to include globalization processes, global problems into the study of the noosphere, including even space activities and large-scale development of the Universe.

4. There must be a new understanding of security issues in connection with the transition to a sustainable development model.
5. The formation of the noosphere should be of fundamentally planetary character, until the formation of the cosmosphere.
6. It is assumed that the creation of the sphere of mind will be accompanied by increased just rational and moral-humanistic values and the formation of the noospheric culture, including the ecological culture and the biospheric ethics.
7. The design of advancing mechanisms of activity is mandatory, including intellectual and communication ones (the noospheric intelligence), management ones (proactive decision-making and proactive management), new education models (noosphere-advanced education), and the noosphere-advanced science.

It should be noted that science itself changes correspondingly to the process of forming planetary mind. The combination of depth and comprehensibility of research reinforces therein. Vladimir Vernadsky, marking a new quality of the 20th century science, wrote: "We continue to be specialized by problems rather by sciences. This allows us, on the one hand, to go very deep into the phenomenon under study, and, on the other hand, to expand its reach from all viewpoints" (Vernadsky, 1998).

5.4 MODERN IMMUNOCHEMICAL AND BIOSENSOR TECHNOLOGIES FOR ANALYSIS OF ENVIRONMENTAL ECOTOXICANTS

Now the increase of the number and type of analyzed objects in modern ecological and medicobiological monitoring requires introduction of relatively inexpensive rapid analytical techniques of environmental pollutants into the practical work of laboratories. Among the basic pollutants are polychlorinated biphenyls, polycyclic aromatic hydrocarbons (PAHs), pesticides, mycotoxins, and heavy metals. The traditional analytical techniques to indicate and identify ecotoxicants (gas chromatography, mass spectrometry, atomic absorption spectrometry, etc.) are labor-consuming and expensive at realization of constant monitoring of the environment and biological media for human health estimation. For example, the cost of one analysis to identify persistent organic compounds reaches a hundred dollars. In this connection, studies to develop alternative, relatively inexpensive immunochemical and biosensor techniques for detection of trace ecotoxicant amounts are urgent.[26,27] In this section, a brief review of modern achievements and basic directions of research on immunochemical and biosensor technologies for analysis of basic pollutants in environmental matrices and biological liquids is presented.

The development and introduction of immunochemical techniques into laboratory practice for the testing of trace amounts of foreign substances (xenobiotics) in various environmental objects, including biological liquids, is a priority lead in studying the status of the environment and human health estimation.

Among immunological techniques, the key role is played by such technologies of immunochemistry that gained a wide circulation in clinical chemistry and diagnostics as heterogeneous (ELISA—enzyme-linked immunosorbent assay) and homogeneous (EMIT—enzyme-multiplied immunoassay technique) immunoenzyme analyses. These methodologies can be adapted for indication and identification of toxic compounds in the environment and biological liquids at diagnostics of occupational diseases of the man. The data presented (Table 5.2) on application of immunochemical techniques in ecoanalytical studies give evidence of the possibility to optimize immunochemical test systems for the screening of xenobiotics in the environment.

TABLE 5.2 Applications of Immunochemical Analysis for Ecotoxicant Detection in Environmental Matrices and Biological Liquids[28–32]

Ecotoxicant	Matrix analyzed
Polychlorinated biphenyls (PCBs)	Soil
Polycyclic aromatic hydrocarbons (PAHs)	Soil, water
Pesticides and their metabolites	Soil, water, air, foodstuff, saliva, blood plasma, urine
Mycotoxicants	Vegetables, grain, fruits

Immunochemical analysis is based on the ability of antibodies to specifically bind to molecules of the detected compound, metal at proceeding of the antibody–antigen reaction.[28–30] At getting polyclonal and monoclonal antibodies specific to xenobiotic present in an analyzed matrix, haptens conjugated with the protein carrier are used. The analogs of revealed low-molecular-weight organic compounds chemically modified by introducing functionally active chemical groups or heavy metals as chelate complexes act as haptens. The methods of obtaining specific antibodies, haptens, and their conjugated protein complexes are systematized in a number of publications and include the stages of immediate synthesis of a hapten, obtaining a conjugate, immunization, isolation, and purification of antibodies. Indication and quantitative determination of analytes in environmental objects are carried out by means of registration of the intensity of antigen–antibody reaction on the basis of usage of labels connected with antibodies or haptens as enzymes and fluorescent or luminescent reagents.

The technique of solid-phase immunoenzymatic analysis (IEA) of pesticides, in which immunoreagents are immobilized on the surface of the solid phase, is a priority lead of immunochemical technologies of screening of health-dangerous compounds. During IEA implementation, the unreacted molecules are separated from immune complexes, and the content of the analyte is evaluated by measurement of the catalytic activity of the enzyme label. The formation of immune complexes occurs during competition of the ecotoxicant molecules in the sample with an analog of the pesticide chemically conjugated with the protein carrier. Two formats of competitive solid-phase IEA are applied. Enzyme-labeled (horseradish peroxidase and laccase) antibodies immobilized on the solid phase and pesticide–protein conjugates are applied in a direct format, and labeled pesticide and immobilized antibodies in an indirect one. The schemes of heterogeneous solid-phase IEA with test systems for herbicides 2,4-dichlorophenoxyacetic acid (2,4-D) and 2,4,5-trichlorophenoxyacetic acid (2,4,5-T) have been optimized. In the ELISA variant, optimization with polyelectrolytes has allowed the preliminary incubation period of the reaction mixture to be considerably reduced, the limiting detected concentration of the herbicide simazine being 0.03 ng/mL. The productivity of ELISA at analysis of foreign low-molecular-weight organic compounds such as pesticides reaches ~100 tests per hour.

Another variant (homogeneous IEA, EMIT) is underlain by the ability of modulation of the catalytic activity of xenobiotic (antigen)–enzyme conjugates at formation of complexes with antibodies. These variants of immunochemical analysis are implemented with no stages of washing off and immobilization of immunoglobulines. For example, an original approach was the application of the bacillary α-amylase as an enzyme marker for detection of insecticides of the pyretroid row (3-phenoxybenzoic acid, permetrinic acid).[33] The formation of the products of hydrolysis of the enzymatic reaction after addition of the substrate such as starch, measured spectrophotometrically, depends on the concentration of the analyte. The limiting concentration of pyretroids revealed by this technique was 3 ng/mL.

Homogeneous polarization fluoroimmunoassay (PFIA) is a promising lead of scientific and practical ecoanalytical research, allowing indication of s-triazines in environmental objects by measurement of the degree of polarization of fluorescence (mP) of labeled structural analogs of the given analytes in the composition of the fluorescent complex (marker antigen–antibody). The change of the polarization intensity of fluorescence of the marker molecules present in the complex occurs proportionally to the quantity of the pollutant in the samples. The technology of analysis with application of labeled pesticide 2,4-D-NH2F is implemented in a variant of fluorescent spectroscopy (PRES/phase-resolved fluoroimmunoassay (PRFIA)) as well. A methodology of biomonitoring alachlor, atrazine, and their metabolites in urine is developed on the basis of microbeads conjugated with analogs (fluorescence microbead immunosorbent assays (FMIAs)) of the compounds detected in samples and fluorescent-labeled antibodies. As a whole, the basic characteristics (Table 5.3) of the most widespread techniques of immunochemical analysis speak for the efficiency of revealing low-molecular-weight persistent organic substances in environmental matrices and biological liquids.

It is necessary to note certain demerits of immunoanalytical systems in detection of pollutants shown in the duration of incubation (ELISA) of reaction mixtures, possible cross-reactions in the presence of the metabolites of the initial compound in samples, and inhibition of catalytic activity (EMIT) of the label enzyme by some components of a sample. To reduce the probability of cross-reactions with components of sample and to increase the sensitivity of immunochemical indication (ELISA), tested compounds can be previously taken from their matrices (foodstuff, soil, etc.) by means of supercritical fluid extraction (SFE) and solid-phase extraction (SP-C18).

TABLE 5.3 Basic Characteristics of Immunochemical Techniques (ELISA, EMIT) Applied to Detect Ecotoxicants[34,35]

Technique	Analyte	Detection limit (ng/mL)	Way of detection
Heterogeneous			
ELISA	2,4-D, 2,4,5-T, simazine, atrazine benzo[a] pyrene 2,4-D	0.1–0.02, 0.05–0.1, 1.0, 0.005	Spectrophotometry Chemiluminescence
Homogeneous			
EMIT	3-phenoxybenzoic acid, permetrine	2–3	Spectrophotometry
PFIA	Atrazine	0.1	Polarized fluorescence
PRFIA	2,4-D	2.2	Spectrophotometry

Immunoaffinity chromatography (IAC), a most powerful technique for separation and purification of analytes, is another important direction of development of environmental monitoring techniques. It can concentrate insignificant concentrations of organic compounds present in environmental objects. It offers to use sorbents with immobilized antibodies with a high affinity to detected xenobiotics as the stationary phase of chromatographic precolumns. At modification of the stationary phase, bacterial proteins (proteins A and G) are used, which are fixed on the chemically activated surface of the sorbent. Previously, with the purpose to activate the sorbent, carbonyldiimidazole, glutaraldehyde, and N-hydroxysuccinimide were applied. Direct immobilization of antibodies on the sorbent is achieved by activation of antibodies by fixing reagents such as carbodiimides, glutaraldehyde, and so forth. The possibility of preparation of immunoaffinity sorbents, including the biotin–avidin complex, is shown. The substances identified in the samples are separated by high-efficiency liquid chromatography (HELC) or in the system HELC-MC. Components can be eluted by changing the ionic strength of the mobile phase, and with the application of chaotropic buffers as mobile phases. Several hybrid techniques of monitoring (Table 5.4) of environmental and biological liquids on the basis of immunoaffinity chromatography and various HELC detection systems are used for screening of micotoxinsmycotoxins, pesticides, and other xenobiotics.[30] Among the immunochemical ways of detection of trace organic compounds, the "flowing" systems of analysis (flow-injection analysis (FIA), flow-injection liposome immunoanalysis (FILIA)) stand out. So, the automated "flowing" immunoanalysis (FILIA) with an electrochemical detector includes immunoreaction column with glass beads bearing antibodies to the analyte on their surface. In this reaction column, introduced liposomes with an encapsulated electroactive marker (NaF) and an analog of the analyte competitively interact with the analyte of the sample at interaction with antibodies. Subsequent lysing of the fixed liposomes by an agent such as Triton X-100 releases the electrochemically detected marker—ferrocyanide (NaF). The flowing immunoanalysis (FILIA), rather simple in operation, ensures realization of monitoring of traces of toxic organic compounds such as herbicides in food products.

TABLE 5.4. Hybrid Techniques of Monitoring of the Environmental and Biological Liquids Based on Immunoaffinity Chromatography and HELC Detection Systems[36]

Ecotoxicant	Matrix	Type of immunoaffinity column	Detection system
Aflatoxin	Milk, grain	Affinity column, Aflatest-P	HELC/fluorescence
Atrazine	Water	Nucleosil-DB	HELC/spectrophotometry
Chlorotoluron	Urine, plasma	Aldehyde-activated silica gel	HELC/spectrophotometry
Carbofuran	Water	Aldehyde-activated silica gel	HELC/mass spectrometry

The trends in the design of rapid monitoring systems for qualitative detection of xenobiotics are reflected in optimization of the membrane variants of immunoanalysis on nitrocellulose strips with visual detection. The technique of liposome immunoaggregation (LIA) with encapsulated fluorescent dye sulforhodamine B and hapten as 2-chlorobiphenyl dipalmitoylphosphatidylethanolamine, contained in the bilayer of vesicles, includes a competitive reaction of these liposomes with the analyte at interaction with the specific antibodies immobilized on the surface of the nitrocellulose membrane. The minimal detected concentration of polychlorinated biphenyls (PCBs) is 2.6 ng/mL.

Considering heavy metals representing a special danger to human health, it is necessary to emphasize that about 20 known metals (Cd/II, Pb/II, Ca/II, Hg/II, Co/II, Ni/II, Zn/II, Cu/II, Mn/II, Fe/III, Au/III, etc.)

are classified as persistent environmental toxins. Quite often with changes of meteorological factors, soil acidity or reservoir acidity, the toxicity of the contained metals may grow. In this connection, periodic monitoring of such objects is required with analysis of many samples. Analysis by means of traditional techniques of heavy metal indication (absorption spectroscopy, X-ray fluorescent spectroscopy, etc.) is expensive and of low productivity. For example, immunoanalysis allows lowering the cost of analysis by more than 50% and raises its productivity. At hybridoma's synthesis of monoclonal antibodies with a high affinity to heavy metals (Table 5.5), bifunctional chelate complexes of analogs of metal ions, covalently bonded with molecules of protein such as bull plasma albumin, are applied. Metal–chelate complexes (Cd(II), Hg(II), Pb(II), Ca(II), Co(II), Ni(II), Zn(II)) were formed on the basis of ethylenediaminetetraacetic acid (EDTA), diethylenetriamine pentaacetic acid (DTPA), trans-1,2-cyclohexyldiethylenetriamine pentaacetic acid (CHXDTPA). The synthesized monoclonal antibodies were used in a competitive format of ELISA. After competitive interaction, the excess of antibodies were removed by washing off with subsequent introduction of peroxidase-labeled anti-specific antibodies (secondary antibodies), whose excess was also washed off. Spectrometric registration (450–650 nm) of the intensity of the reaction was made on a Microplate Reader immunological photometer (Molecular Devices, Sunnyvale, CA, USA). The affinity studies with estimation of the equilibrium dissociation constant (K_d) of the monoclonal antibodies to metal–chelate complexes of various structures have shown that the affinity of antibodies to Pb(II)–CHXDTPA is 10,000 times higher than that at linkage of immunoglobulines with Pb(II)–DTPA. The limiting concentration of Pb(II) detected by the competitor was 0.4 ng/mL (1 ppb).

In the whole, the above data on pollutant screening by immunochemical methods give evidence of the principal possibility to manufacture commercial test systems for mass screening of pollutants in the environmental objects. Further improvement of the existing technologies of immunoanalysis with expansion of the range of synthesized haptens, specific antibodies, and labels will allow carrying out high-grade ecoanalytical control in an automated mode.

The concept of biosensors is underlain by the biochemical recognition principle, which is realized in these devices on the basis of some mechanisms of functioning of biosensor components, such as biocatalytic (enzymes) and bioaffinity (antibodies, receptors, and nucleic acids).[38,39]

TABLE 5.5 Affinity of Hybridoma-Synthesized Monoclonal Antibodies to Metal–Chelate Complexes at Detection of Heavy Metals by a Concurrent Version of ELISA[37]

Metal–chelate complex	Equilibrium dissociation constant (K_d)	Minimal concentrations detected
Ni(II)–DTPA	2.7×10^{-7}	50 ppb
Co(II)–DTPA	5.2×10^{-8}	
Pb(II)–DTPA	1.0×10^{-5}	2 ppm
Cd(II)–EDTA	2.1×10^{-8}	7 ppb
Hg(II)–EDTA	2.6×10^{-8}	2 ppb
Pb(II)–CHXDTPA	8.4×10^{-9}	1 ppb

Biosensors are analytical devices in which a biological component is connected with a transducer. In sensor devices, these components are connected with a transducer element. Biosensors can be classified by the type of their transducer as acoustic (QCM—quartz crystal microbalance, SAW—surface acoustic wave, and STW—surface transverse wave), electrochemical (potentiometry, amperometry, and conductometry), optical (absorption, fluorescence, luminescence, FOBs—fiber optic biosensors, EW—evanescent wave biosensors, ATR—attenuated total reflection, and TIRF—total internal reflection fluorescence), and opticoelectronic (electrochemiluminescence, SPR—surface plasmon resonance, and LAPS—light-addressable potentiometric sensor). Commercial versions of biosensors—RAPTOR portable, FAST 2000 (Research International Inc., USA), and Biolyzer (Environmental Technologies Group, Inc., USA)—are now designed. The possibility to detect a wide spectrum of ecotoxicants (polychlorinated biphenyls, pesticides, insecticides, phenols, etc.) in the environmental objects and biological liquids by similar devices is shown. The biosensors on the basis of enzymes function by transformation of the initial pollutant into a detectable product (potentiometry, amperometry) or inhibition (LAPS, SPR) of the enzymatic activity by the analyzed ecotoxicant. For example, an electrode with the enzyme tyrosinase is effective as a detecting one in HELC at analysis of phenols. The limiting sensitivity of identification of phenol, p-cresol, p-methoxyphenol, and p-chlorophenol in soil and industrial sewage samples was 2–30 mg/L. With the purpose of detection of phosphoro-organic and carbamate pesticides, acetylcholinesterase labeled with fluorescein isothiocyanate (FITC) was immobilized on a quartz optical fiber. The enzyme activity was determined by the quantity of the hydrolyzed substrate (paraoxon, etc.) during registration of changes of the intensity of pH-dependent fluorescence FITC. An acetyl cholinesterase enzyme sensor detects pollutants (carbamates, phosphoro-organic compounds), which are the inhibitors of this enzyme, at the level of 1–10 ng/L. Biosensor identification of low-molecular-weight organic pollutants is possible on the basis of potentiometric pH electrode, amperometric one at measurement of the concentration of the formed hydrogen peroxide; it is acoustic in the case of a piezoelectric crystal carrying cholinesterase as a transducer. A sensor device with the immobilized enzyme urease on the surface of a conductometric transducer detected the concentration of heavy metal ions in the range of 1–100 mM. The mechanism of functioning is underlain by the ability of heavy metal ions (Ag(I), Hg(II), and Pb(II)) to inhibit the activity of urease, which enables determination of the concentration of toxic components in a sample by the quantity of decomposed urea.

The biosensors containing antibodies (a direct format) or an antigen-structural analog of xenobiotic (an indirect format) as a recognizing component are classified as immunosensors.[15] On the basis of the type of the transducer, the immunosensors are subdivided into piezoelectric (bulk acoustic (BA) and SAW), electrochemical (potentiometry, amperometry, and conductometry), optical (absorption, fluorescence, and luminescence), and opticoelectronic (SPR, LAPS, and electrochemoluminescence).

The sensor devices as opticofiber FOBS are designed by the immobilization of antibodies to PCBs, cyclodiene insecticides, and herbicides on the top of a quartz optical conductor. During detection of toxicants in samples, for example, those labeled with a fluorescent label, the analogs of the analytes compete for linkage to the active centers of immunoglobulines. The indirect format of the immunosensor is realized by immobilization of a triazine pesticide on the surface of an optical fiber with the help of glutaraldehyde. In this case, s-triazine is detected with the application of labeled antibodies competing with the required organic compounds.

In turn, functioning of opticoelectronic biosensors with SPR is based on the phenomenon of field attenuation. In such devices, a laser beam passing through a prism is incident on a surface covered with a gold substrate. The beam is reflected from the surface in all directions, except for some critical angle.[40] In this direction, the laser beam excites a plasma wave on the surface, which causes reduction of the reflected light intensity. The critical angle depends on the refractivity of the layer carrying antibodies and adjacent to the transducer's surface. Estimating changes of the critical angle during interaction of antibodies with toxicant molecules, the concentration of the analyte is tested. The designed commercial sensor device Biacore (Biacore Inc., Sweden) detects atrazine in amounts about 0.1 ng/mL. Electrochemical gauges are developed on the basis of electroactive conducting polymers with immobilized anti-PCBs (antibodies). Such an active electrode in the conditions of a pulsing electromagnetic field registers the reaction of binding of the given antibodies with the analytes. The efficiency of arochlorine detection by such devices was shown at monitoring of groundwaters.

PAHs exhibit carcinogenic properties in the metabolism in the human organism. The DNA structure is damaged at biotransformation of PAHs to form DNA adducts. In other cases of pollutant influence, the nucleotide composition of DNA may change, which also results in the occurrence of carcinogenic effects. For detection of such substances of the environment which are capable of chemically inducing DNA damages, electrochemical affinity sensors on the basis of nucleic acids are suitable.

Further intensification of research in the field of biosensor technologies will promote wide introduction of productive microanalytical systems into the daily practice of ecological laboratories.

Comparative analysis of the basic leads of research and obtained results in the field of design and applications of immunochemical and biosensor techniques for analysis of environmental ecotoxicants speaks for their doubtless efficiency in indication and identification of foreign chemical compounds. In combination with traditional techniques of ecoanalytical control, rapid immunochemical and biosensor technologies of screening of foreign chemical compounds will allow considerably lowering the cost of analysis and expanding the range of tested environmental objects at monitoring. The manufacture of inexpensive, commercial test systems on the basis of immunochemical and biosensor technologies for ecological and medicobiological monitoring seems a priority scientific and practical lead in the field of environment protection and human health. Thus, further development and introduction of modern biotechnological techniques into the scientific and practical work of laboratories for control of pollution of working zones, industrial premises, and occupied places will optimize ecological diagnostics and will promote effective rehabilitation of the environment.

5.5 BIODETOXICATION OF AROMATIC HYDROCARBONS IN AQUEOUS MEDIA

The high degree of involvement of the coastal zones of seas and rivers into human economic activity leads to the accumulation of various pollutants in water and precipitations, including aromatic hydrocarbons, both monocyclic and polycyclic, which possess a whole complex of properties dangerous to living organisms. Higher aquatic plants constitute one of the main components of the self-cleaning system of reservoirs due to their ability to absorb, accumulate, and metabolize many pollutants. Transformation of xenobiotics in plants is accompanied by oxidizing degradation processes generally catalyzed by enzymes such as peroxidases and phenoloxidases.

Peroxidases form a widespread group of the enzymes for substratum oxidation in the presence of H_2O_2. Due to their wide functional variety, they take part in many physiological and detoxication processes. Vegetative peroxidases are capable of oxidizing monocyclic and polycyclic aromatic compounds, such as phenol, hydroxytoluene, benzo[a]pyrene, dimethylaniline, and so forth.[41]

Phenoloxidases also contribute significantly to the aromatic hydrocarbon degradation process. This group of the copper-containing enzymes, being present in plants in both active and latent states, includes phenoloxidases of the tyrosinase and laccase type. A wide number of aromatic substances of phenolic nature can serve as a substratum for these enzymes.[42] As the presence of pollutants in water affects biochemical processes in the cells of aquatic plants and, first of all, their enzymatic apparatus, the purpose of our research was studying the influence of some aromatic xenobiotics on the activity of the endoenzymes in *Elodea canadensis*.

E. canadensis cultivated under room conditions was the object of our study. To prepare an extract (a rough enzymatic preparation), 0.5 g of *Elodea* sprouts was crushed and homogenized in a cooled ceramic mortar with 2 mL of 50 mM acetate buffer (pH 5.0) or 50 mM sodium phosphate buffer (pH 6.5) or 50 mM Tris-HCl buffer (pH 9.0). The homogenate was poured into a measuring flask and the volume was adjusted to 25 mL with the same buffer. Whole cells and their fragments were sedimented by centrifugation at 5000 rpm for 10 min. A xenobiotic was added to the extract to a final concentration of 0.2 mM and incubated for 3 h. Monophenols (phenol and toluene) and PAHs (naphthalene, phenanthrene, and fluorene) were used as xenobiotics of the aromatic series. PAHs were previously dissolved in 100 μL of chloroform, phenol in 100 μL of H_2O, and toluene was added without dilution. To analyze xenobiotics in the extract after incubation, they were extracted by 10 mL of chloroform with subsequent application of gas chromatography: a Shimadzu 2010 chromatograph (Japan), an Equity-1 column (Supelco, USA), a flame ionization detector, and helium as the carrier gas.

The enzymatic activity of the *Elodea* extract was evaluated spectrophotometrically on an SF-26 in quartz cuvettes with an optical path length of 1 cm. The amount of an enzyme catalyzing transformation of 1 μmol of the substratum or formation of 1 μmol of the product per minute was taken as a unit of activity and expressed as units per extract milliliter. Specific activity was expressed as μmol/min/mg of protein. The activity of enzymes was measured in the beginning of every experiment and after 3 h of their incubation with the corresponding xenobiotic or without it (a reference). The activity of tyrosinase was revealed by the formation of the oxidation product from L-dihydroxyphenylalanine (L-DOPA) at 475 nm.[43] The activity of laccase was estimated by the formation of the oxidation products from syringaldazine at 525 nm,[44] the diammonium salt of 2,2'-azine-bis-3-ethylbenzothiazoline-6-sulfonic acid (ABTS),[45] and pyrocatechin at 410 nm.[46] The activity of peroxidase was determined at 436 nm by the formation of the oxidation product of ABTS in the presence of H_2O_2 and calculated as a difference between the activities of enzyme in the presence and absence of H_2O_2. Protein concentration was determined by Bradford's technique.[47]

All experiments were carried out in three replicas. Data were processed statistically with Microsoft Excel Office XP, and also by a standard method.

As was noted above, two groups of enzymes can be involved in aromatic xenobiotic neutralization in the vegetative cell, namely, phenoloxidases (of the laccase and tyrosinase type) and peroxidases.[48]

Syringaldazine is a test substratum for laccase activity evaluation.[4] Only trace activity of the extract to this substance not exceeding 0.1 units/mL was revealed during our experiment. When using ABTS (which is also a laccase substratum), no laccase activity was revealed. At the same time, the activity to pyrocatechin (15.4 units/mL) was revealed in the extract—rather high in comparison with ABTS and syringaldazine. Comparing the obtained results with the literature data,[49] it is possible to hypothesize that pyrocatechin oxidation revealed by us may be a result of some reaction catalyzed by an oxidase of non-laccase nature.

L-DOPA is a test substratum for tyrosinase activity estimation. The presence of this enzyme in the material under study with an activity reaching 28 units/mL was found. In the vegetative extract, the activity of one more enzyme (peroxidase) was observed, exceeding the tyrosinase one by three times (86 units/mL).

Therefore, the activity of three enzymes (oxidase, tyrosinase, and peroxidase) was detected in the *Elodea* extract, which essentially depended on pH and the nature of the xenobiotic added.

The optimum of oxidase and peroxidase activity is known to lie within an acidic range,[50] while that of tyrosinase is in a neutral and alkaline one.[51] In this connection, buffer solutions with pH values 5.0, 6.5, and 9.0 were chosen for further research. Their usage to prepare extracts has shown that when pH increases, the amount of the extracted protein increases from 1.75 (pH 5.0) up to 14.1 mg/mL (pH 9.0; Fig. 5.11).

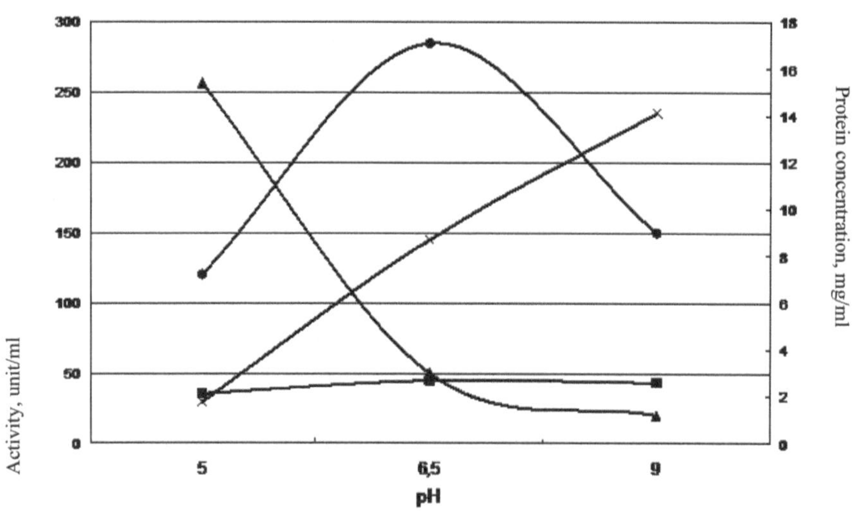

FIGURE 5.11. Dependence of the activity of tyrosinase (■), peroxidase (●), oxidase (▲), and protein concentration (♦) on the pH of the extracting buffer.

At the same time, the highest activity of peroxidase (285 units/mL) and tyrosinase (44.9 units/mL) was revealed at a pH value 6.5 of the extracting buffer, whereas the maximum activity of oxidase was found at pH

5.0. This feature can be a result of both better extraction of the enzymes and their higher stability at acidic pH values. Incubation of the prepared extract at room temperature for 3 h has shown that the detected enzymes keep their activity irrespective of the extracting buffer pH.

During incubation of the *E. canadensis* extract with monoaromatic substances and PAH, the content of xenobiotics (in 3 h) was monitored, and their decrease in the reaction mixture, depending on pH and the chemical nature of the xenobiotic, was thus found. According to the data published earlier,[52] the decrease in phenol increased at pH shifting toward alkalinity and was maximal (53.2%) at pH 9.0. The toluene decrease was also maximum (61%) at an alkaline pH value. According to our data, in the case of PAH, the content of naphthalene did not change in all variants, the maximum decrease in the content of phenanthrene and fluorene (by 23%) was observed at pH values 5.0 and 6.5, respectively, and no decrease of all PAHs at pH 9.0 occurred. The obtained results (Table 5.6) and their comparison with the literature data allow us to assume that the phenol and toluene decrease may be a result of some reaction catalyzed by tyrosinase, since these substances are substrata for this enzyme,[47] and the PAH decrease may be associated with the action of peroxidase and/or oxidase, which are known[53] to catalyze oxidation of similar compounds.

TABLE 5.6 Decrease of Xenobiotics under the Influence of the *E. canadensis* Extract (% of the reference)

Xenobiotic	pH		
	5.0	**6.5**	**9.0**
Phenol	19.8*	22.2*	53.2*
Toluene	57.1*	10.1*	61.2*
Naphthalene	0	0	0
Phenanthrene	23	0	0
Fluorene	12	23	0

*The error of all results did not exceed 10%.

Table 5.7 presents the results of our study of the influence of some xenobiotics on the enzyme activity when an Elodea extract was incubated in buffers with various pH values. It was revealed that, irrespective of pH, the activity of tyrosinase increased by 1.3–4 times depending on the xenobiotic used. Naphthalene and toluene were exceptions, in the presence of which there was an insignificant decrease in the activity of this enzyme.

TABLE 5.7 Influence of Aromatic Xenobiotics on the Activity of the Endocellular Enzymes of *Elodea canadensis*, units/mL

pH	Phenol		Toluene		Naphthalene		Phenanthrene		Fluorene	
	Initial	In 3 h	Initial	In 3 h	Initial	In 3 h	Initial	In 3 h	Initial	In 3 h
					Tyrosinase					
5.0	35.1	67.3	28.1	113.5	103.5	1100.0	65.4	84.6	98.4	1107.6
6.5	44.9	126.2	128.4	107.3	77.8	990.5	75.1	119.5	59.5	1110.3
9.0	43.2	73.5	53.8	93.5	70.3	665.7	72.2	110.0	59.2	999.5
					Oxidase					
5.0	7.1	15.6	18.1	28.1	43.2	116.6	26.6	9.0	32.8	113.8
6.5	30.8	25.2	22.8	12.8	21.4	222.3	25.6	17.1	17.6	5.52
9.0	29.0	17.6	11.8	12.3	16.1	55.2	16.2	12.4	19.0	33.8
					Peroxidase					
5.0	86.0	102.0	166.3	157.1	99.6	998.2	136.2	103.2	139.1	1129.0
6.5	209.3	285.2	243.7	227.9	100.3	1111.8	180.6	120.4	141.9	1149.1
9.0	202.1	150.5	301.0	236.5	80.3	1103.2	147.7	176.3	202.1	1137.6

NOTE: The error of all results did not exceed 10%.

When studying the activity of oxidase, the inhibiting action of xenobiotics at pH values 6.5 and 9.0 was revealed, except for naphthalene and toluene which rendered weak impact only. At the same time, the influence of the substances studied was ambiguous at pH 5.0: monophenols led to an activity increase, whereas PAH led to its decrease.

The peroxidase of *E. canadensis*, unlike tyrosinase, has appeared more subject to the inhibiting action of xenobiotics. For example, all the examined substances reduced its activity at pH 5.0. Phenol was an exception, in the presence of which a slight increase in the activity of this enzyme was noted, which was observed at pH 6.5 as well. At this pH value, toluene and phenanthrene inhibited the activity of peroxidase, while naphthalene and fluorene almost did not influence it. At pH 9.0, monophenols and fluorene inhibited the activity of this enzyme, whereas naphthalene and phenanthrene raised it.

Thus, in the *E. canadensis* extract, the activities of three enzymes (tyrosinase, peroxidase, and oxidase) were revealed. They changed in the presence of a number of aromatic xenobiotics and were pH dependent. In the whole, tyrosinase has appeared more stable against the action of the studied aromatic compounds, whereas the activities of peroxidase and oxidase were inhibited by their presence.

The obtained results are important for our understanding of the adaptation processes occurring in the cells of aquatic plants in response to the presence of aromatic hydrocarbons which may get into reservoirs with oil products and drains of industrial enterprises and, possessing certain solubility in water, may get into plants. The revealed features of the biochemical activity of *Elodea*, depending on influencing factors, can be used to develop ecologically significant biotechnologies.

5.6 ROLE OF CHEMICAL SUBSTANCES IN ECOLOGY[54,55]

In alive nature, many of the known chemical compounds have various specific functions. One of the main functions is realization of certain relations between alive beings and the env*ironment.* Such substances according to the ecological terminology are usually named *chemomediators.* However, depending on the participants of interrelations, chemomediators have special names additionally. For example, in the establishment of relations between individuals of the same species, the basic role is played by chemomediators named *pheromones* (from the Greek *phero*—I carry, hormone—an activator). At interrelations between species and populations, these *are allomones* (from the Greek *allos*—another, different). On the other hand, at interaction of alive beings with the environment in response to the influence of abiotic ecological factors (temperature, light, humidity, acidity, etc.), specific substances can be formed in organisms such as:

- *endometabolites* (from the Greek *endo*—inside, *metabole*—change), which, remaining in the organism, would soften the influence of this or that factor (e.g., the so-called cryoprotectants and antifreezes in wintering animals);
- *exometabolites (*from the Greek *exi*—outside), which are released into the environment and may even change its chemical properties.

Therefore, the first two types of compounds (pheromones and allomones) participate in interrelations of alive beings, the third and fourth types (endometabolites and exometabolites) participate in interrelations of organisms with the environment. At the same time, despite essentially various roles of the said chemomediators in ecology, all of them can perform common functions with the following classification:

- *Protective* function, that is, the defense from a potential predator or parasite (or, at least, scaring it away or reducing its nutrition activity). In particular, such a function is performed by many poisonous substances in plants (codeine, caffeine, quinine, strychnine, etc.; these are mainly heterocyclic, nitrogen-containing compounds). In animals, this function is performed by special releases and toxins. Besides, some hydrobionts (especially deep living) use the phenomenon of bioluminescence, that is, rather weak luminescence of alive organisms, whose basis is chemical reactions of oxidation of some substances named luciferins.
- *Offensive* function, that is, an active influence on an organism. Examples are: (i) the use of chemical substances (as means of attacking a victim) by inedible mushrooms and various pathogenic bacteria; (ii) the influence of neurotoxins of predatory wasps on the impellent function of their prey or that of neurotoxins of snakes, which can result even in death of the object of attack; and (iii) the influence of bioluminescence.
- *Restraint* of competitors (in struggle for food resources, for territory, for female/male, etc.). For example, some higher plants with the help of excreted chemomediators (some organic acids, quinones, phenols, etc.) can suppress the growth of other plants—competitors—or even cause their death.
- *Attractive* function (from the Latin *attrachere*—to attract): attraction of victims (individuals of the other sex, pollinators, seed distributors, etc.); chemosignals of attractive character raise the food, impellent, and reproductive activity (sexual pheromones, flower aromas, etc.).
- *Regulation* of interactions (the behavior of individuals, their reproductive activity, etc.) inside any social group (family, colony, population).

- *Supplying* function, that is, supply of organisms by the substances necessary for the synthesis of chemomediators in them.
- *Formation of the environment*, that is, change of the chemical properties of the environment by the influence of compounds excreted by alive beings on it; first of all, this applies to water and soil ecosystems.
- *Indicator* function. It chemically marks territories, which allows alive beings moving at significant distances to be guided in space.
- *Warning* function. Basically, these are alarm pheromones and allomones to warn alive beings about danger.
- *Adaptive* function. This function helps to adapt to the influence of abiotic ecological factors. These chemomediators are, first, the compounds allowing the inhabitants of salted reservoirs, hot springs, and so forth to exist; second, antifreezing proteins in some animals to prevent crystallization of their intracellular water (even at cooling of the body of such an animal down to −10 to −15°C; third, the so-called cryoprotectants, which hinder sharp reduction (collapse) of the cell's volume at freezing (in particular, glucose in fish and frogs).

In conclusion of our brief consideration of the role of chemical substances in ecology, it is necessary to say that the occurrence and development of rather complex systems of chemocommunications in alive nature contribute to evolution of the organic world. In this respect, for example, of interest is the phenomenon of biochemical convergence, when during evolution various (frequently in no relationship) alive beings get or release identical chemomediators.

5.7 PROBLEM OF MICROELEMENTOSES IN THE HUMAN BODY[54,55]

Long ago, it was revealed that many diseases of alive beings, including the man, are frequently due to either lack or excess of the content of certain macro- or microelements in the organism. Examples are relations between iron deficiency and anemia, between iodine deficiency and malfunctioning of endocrine glands (the thyroid gland), and so forth. In this connection, in recent years, a new scientific and practical lead in medicine and biology connected to proceeding of negative (frequently irreversible) processes in the human organism caused by either lack or excess (i.e., misbalance) of vitally necessary elements has gained development. Such processes named microelementoses have a certain classification.

1. *Natural (endogenic).* They may be inherent (due to the mother's microelementosis) and hereditary (caused by a pathology of chromosomes or genes).
2. *Natural (exogenic).* They are caused by the geochemical conditions of a certain region of the Earth (in essence, by the geochemical heterogeneity of the biosphere), that is, by microelement misbalance.
3. *Technogenic.* They are classified into three types:
    - direct industrial (connected directly with that or other manufacture);
    - indirect industrial (diseases caused by the action of microelements of neighboring manufactures);
    - transaggressive (due to carrying of microelements or their compounds at significant distances).
4. *Iatrogenic.* They cause diseases, mainly in connection with intense treatment of illnesses with various preparations (in essence, secondary microelementoses).

As to the concept of microelement, it means a chemical element which is present in the organism in rather insignificant quantities (hundredth and thousandth fractions of percent of the weight unit, that is, milligram).

The basic vital elements (included in the composition of the body and participating in biochemical reactions) are oxygen, hydrogen, carbon, nitrogen, phosphorus, sulfur, potassium, calcium, sodium, iron, copper, zinc, cobalt, nickel, manganese, silicon, magnesium, selenium, and chlorine. They make about 99% of the elementary composition of the human body. It should be always meant that diagnostics of microelementoses and their influence on the state of human health depend on interrelations and interactions of chemical elements with each other, and on the knowledge of the so-called target organs, that is, those organs which this or that element affects to the greatest degree.

Let us report the data concerning the features of some chemical elements at functioning of the human body.

Calcium. The daily need is 1000–1500 mg. It is important in functioning of the muscular tissue, heart, nervous system, skin, and, especially, osseous tissue. The calcium content may influence the amount of other elements; in particular, its surplus frequently results in zinc and phosphorus deficiency in different organs, and hinders lead accumulation in the osseous tissue. Basic causes of deficiency are poor nutrition; diseases of the thyroid gland; a surplus of phosphorus, lead, zinc, cobalt, magnesium, iron, potassium, or sodium; soft water; diseases of kidneys; vitamin D deficiency; pregnancy; and taking of laxative and diuretic preparations. Organs and consequences of the deficiency in them: osseous tissue (osteoporosis), muscular tissue (pain in muscles, spasms), kidneys (stone formation), thyroid gland (malfunctioning), immune system (reduction of immunity, allergies), and blood circulation system (poor coagulation). Indicators of diseases: blood, urine, and hair.

Magnesium. The daily need is 500–750 mg. It is one of the basic intracellular elements, activates the enzymes necessary for carbohydrate exchange and formation of proteins, adjusts the receipt, storage, and release of energy in adenosine triphosphoric acid, reduces excitation in neurons, and relaxes the heart muscle. Basic causes of deficiency are poor nutrition, alcoholism, diabetes, and long taking of diuretic and antitumor preparations. Organs and consequences: vascular vessels (spasms, hypertension), heart (rhythm infringement), osseous tissue (osteoporosis), kidneys (stone formation), bile system (bile stone illness), thyroid gland (hyperfunctioning), and pancreas (infringement of insulin secretion). Indicators of diseases: blood and hair.

Phosphorus. The daily need is 1500–1600 mg. It plays an especially important role in the activity of head, brain, heart, and skeletal muscles, in the formation of the osseous tissue. In the processes of absorption from the intestine and ossifying, calcium and phosphorus exchanges go simultaneously. Phosphorus participates in transmembrane transport of substances included in the composition of some enzymes. A significant fraction of the energy released at disintegration of carbohydrates and other substances is accumulated in organic compounds of phosphoric acid. Basic causes of deficiency are poor nutrition, alcoholism, drug addiction, liver and kidney illnesses, and bottle feeding of babies. Organs and consequences: the central nervous system (weakness, fatigue), muscular system, including myocardium (weakness, pain), liver (hypofunctioning), and osseous system (osteoporosis). Indicators: blood plasma and hair.

Iron. The daily need is 10–20 mg for men and 20–30 mg for women. A surplus of iron renders a toxic action on functioning of the liver, spleen, head, and brain, strengthens various inflammation processes, and results in copper and zinc deficiency. Basic causes of deficiency are poor nutrition, vitamin C deficiency, bleeding (piles, ulcer, menses), malfunctioning of the thyroid gland, tumors, infections, rheumatism, pregnancy, blood-donoring, and physical loading. Organs and consequences: blood circulation system (anemia), immune system (frequent colds), central nervous system (vertigo, memory reduction, poor attention and concentration), heart (infringement of exchange processes), and muscles (weakness, poor endurance). Indicators: blood and hair.

Zinc. The daily need is 12–50 mg. Basic causes of zinc deficiency are the lack of proteins, alcoholism, drug addiction, kidney diseases, intestinal disbacteriosis, psoriasis, a surplus of some heavy metals (e.g., copper, cadmium, lead, mercury), some oncological diseases, and stress. Organs and consequences: skin (dermatitis, eczema, furunculosis, blackhead rash, trophic ulcers); hair (loss, slow growth); mucous membranes (ulcers, stomatitis); central nervous system (children's development delay, appetite reduction, depression); pancreas (lack of insulin); hypophysis (retardation of puberty in boys); prostate gland (potency reduction, sterility); and cardiovascular system (raised cholesterol level). Indicators: hair, whole blood, and plasma.

Cobalt. The daily need is 15–80 μg. Cobalt is a component of the vitamin B12 molecule and its lack, first of all, is felt in the blood-generating tissues of marrow and nervous tissues. A surplus of cobalt results in irritations, allergies, and diseases of the top windpipe and bronchus. Basic causes of cobalt deficiency are the lack of vitamin B12 and atrophy of the mucous membrane of the alimentary canal. Organs and consequences: blood-generating system (anemia); osseous tissue (osteoporosis); liver (bile stone illness); and central nervous system (memory reduction, inhibited reflexes).

Manganese. The daily need is 2–9 mg. Manganese plays an important role in cellular metabolism and is part of the enzymes protecting the organism from the action of peroxide radicals. A surplus of manganese strengthens the deficiency of magnesium and copper with due negative consequences. Basic causes of manganese deficiency are insufficient consumption of vegetative food and a surplus of copper, iron, calcium, and phosphorus. Organs and consequences: central nervous system (raised fatigue, depression, children's development delay); osseous tissue (frustration of the supporting motive system); pancreas (fatting, diabetes risk); female reproductive system (early climax, malfunctions of ovaries, sterility); and skin (infringement of pigmentation). Indicators: hair, whole blood and urine.

Selenium. The daily need is 20–100 μg. Selenium is one of the major microelements necessary for optimal performance of the organism. The matter is that many external influences and, first of all, abiotic ecological factors of the environment are a cause of the presence of free radicals in the organism. Their high chemical activity allows them to meddle in the course of biochemical reactions and to break their natural course. All this may cause various diseases. Frequently, a rather large surplus of free peroxide radicals is observed, which the organism itself cannot manage. Besides, the formed free radicals "attack" various vital centers, such as cells, enzymes, DNA, and RNA sites. Hence, there appears a need of neutralization of unnecessary (harmful) free radicals, that is, suppression of their chemical activity. For this purpose, there are some physiotherapeutic methods, but natural processes are more optimal and effective. It has appeared that there is a system of antioxidant protection against the action of superfluous quantities of free radicals in the organism. These are vitamins C, E, carotene, and some enzymes. Selenium is a very important part of this protective system. It is a part of glutathione peroxidase—the enzyme capable of suppressing most dangerous free radicals. The idea of the role of selenium in the living organisms (including the man) belongs to V. I. Vernadsky. Selenium has been established to be an antagonist of mercury, arsenic, and cadmium, that is, the less the selenium, the stronger they are accumulated (and vice versa). Basic causes of selenium deficiency are: a low content of proteins and fats, hepatitis, radiation, intestinal bacteriosis, alcoholism, and some tumors. Organs and consequences: immune system (frequent cold and inflammation); cardiovascular system (myocardium dystrophy); liver and thyroid gland (reduction of their functions); skin (dermatitis, eczema); hair (weak growth, loss); nails (dystrophy); and eyes (cataract, glaucoma). Indicators of diseases: blood plasma, hair and nails.

A dependence between selenium deficiency and the frequency of occurrence of lactic gland, intestinal, prostate, and stomach cancer has been established. According to the data of the U.S. National Cancer Institute, among those taking 200 μg of selenium per day within 10 years, the mortality due to these kinds of cancer has decreased by ca. 50%. A very important problem is the establishment of the dose necessary for a particular organism, since a surplus of selenium is as harmful as its deficiency. The need in selenium constantly varies even for the same person, depending on how much free radicals are there in his/her body at this or that time. Thus, it turns out that only the organism itself "knows" how much selenium it needs. At present, essentially new medicinal preparations exist to adjust the concentration of selenium by the organism's demand.

5.8 GLOBALIZATION PROCESSES IN HUMAN SOCIETY AND THEIR POSSIBLE MODELING[56–72]

Etymologically, the term *globalization* comes from the Latin word *globus* meaning a ball. Webster's Dictionary first recorded this term in 1951, explaining it as the act or process of globalizing, as well as the state of being globalized, especially emphasizing the development of an increasingly integrated global economy marked especially by free trade, free flow of capital, and the tapping of cheaper foreign labor markets (http://www.merriam-webster.com/dictionary/globalization). This fixed definition already reflects two factors which will be often repeated in the following definitions: first, not the best for a definition—a tautology, that is, the basic notion is explained in terms of its derivatives, and, second, the economic fullness of the phenomena is rightly noted, which will predominate in many other, more recent scientific and specialized definitions.

Approximately the same can be seen in the Oxford Dictionary definitions of a later period (the late 20th century; http://www.enotes.com/~whencyclopedia/globalization), though it tries to avoid tautology and uses words such as ubiquitous, worldwide, and so forth.

Malcolm Waters in his book on the key ideas of globalization marked the first or one of the first uses of this notion in the press. In April 1959, *The Economist* magazine wrote that the Italian "globalized quota" on machinery imports increased (Waters, 2001). And since the beginning of the 1960s, the world was bombarded with words such as global, globalization, to globalize, globally, and other derivatives. It is believed that the term globalization in its modern sense was first used and scientifically justified by Roland Robertson, professor of sociology, in 1992. And disagreement arises on this point, regarding both the different temporal definition of the notion and its actual interpretation.

How was the term *globalization* evolving and what it means nowadays? First of all, it should be noted that the term itself is rather general in nature, combining a variety of more specific meanings and narrow connotations. The very etymology of the word, understood as something combined, total, complete, universal, and worldwide, often forms the basis of its understanding. In this connection, the relevant literature comprises a very large number of the definitions of globalization.

Wikipedia: "*Globalization (or globalisation) is the process of international integration arising from the interchange of world views, products, ideas and other aspects of culture. Advances in transportation and telecommunications infrastructure, including the rise of the telegraph and its posterity the Internet, are major factors in globalization, generating further interdependence of economic and cultural activities.*" At the present stage, we can say that globalization at least does not prevent the diversity of evolutionary processes in nature and human society. E. Azroyants emphasizes that "the process of globalization should not be treated as a way to achieve homogeneity but as an important source of diversity, gathering diverse items into a whole. Diversity is not a trend but a result determining the viability of the system" (Azroyants, 2002).

It was Roland Robertson who, one of the first people, started to treat the three main areas of human activities, namely, economics, politics, and culture, as a springboard for studying global processes. He is rightly considered a pioneer among the sociologists paying close attention to the processes of globalization and the design of the notion of globalization. His definition of globalization dated back to the end of the 20th century, reads: "Globalization as a concept refers both to the compression of the world and the intensification of consciousness of the world as a whole" (Robertson, 1992).

Antony Giddens was one of the first people who introduced the time–space system into discussions of globalization. He believes that the term globalization is used too universally and, above all, refers to the processes of growth of world unity. Speaking that the globalization of social processes should be understood as transformation of the space and time of modern existence, he further focuses on the impact of globalization processes on everyday life, emphasizes their uneven, sometimes fragmented character. He writes: "The intensification of worldwide social relations which link distant localities in such a way that local happenings are shaped by events occurring many miles away and *vice versa*" (Giddens, 1999, 2003).

A. V. Weber forewarns about excessive generalization: "Globalization is a very broad notion, which allows it to cover various phenomena which have in common only that they reach (or have reached) a global scale. Some people even believe that we should speak of several globalizations rather than one: economic, political, informational, consumer's ("McDonaldization"), environmental threat globalization, drug trafficking, crime, demographic globalization. Moreover, these different processes have different scales and vectors and develop at different rates" (Weber, 2002).

These definitions have already sufficiently expressed, though not quite complete, a more or less general idea of the phenomenon of globalization.

Globalization processes are studied from different scientific positions, and many researchers emphasize the importance of the human factor in these processes. The role of human factor is specific at different stages of globalization development. While single outstanding personalities (rulers, warriors, generals, the great discoverers of new lands, further—great scientists) were advanced to the foreground in the early times, more and more people begin to take direct part in the world's destiny with the development and acceleration of globalization. And what is most important, there appear more and more technical capabilities to change the world and the role of the individual in it. In this regard, of interest is the classification of globalization phases proposed by Thomas Friedman in his book "The World Is Flat: A Brief History of the XXI century." Friedman highlights "three periods of globalization. The first one lasted from 1492—Columbus' voyage, which started the exchange between the Old and New Worlds—till approx. 1800" (Friedman, 2005).

The second period, or "Globalization 2.0," as called by Friedman, lasted roughly from 1800 till 2000, with interruptions to the Great Depression and two world wars. During this period, the world ceased to be medium and became little (Friedman, 2005). During this period, the transnational corporations (TNCs) originated and began to play a key role in the future. New types of more high-speed transport have reduced distances by making it available anywhere in the world. The 19th and 20th centuries were marked by the design of new types of communication. The telegraph, the telephone, then satellite and fiber-optic communication, as well as the emergence of the World Wide Web have laid the foundation for the creation of a common information space of the Earth.

Friedman regards the beginning of the new, 21st century as the beginning of the era of Globalization 3.0. "Globalization 3.0 is shrinking the world from a size small to a size tiny and flattening the playing field at the same time. And while the dynamic force in Globalization 1.0 was countries globalizing and the dynamic force in Globalization 2.0 was companies globalizing, the dynamic force in Globalization 3.0—the thing that gives it its unique character—is the newfound power for individuals to collaborate and compete globally" (Friedman, 2005).

Each period of globalization has its political characteristics, that is, it depends on the political picture of the world on the one hand, and determines it on the other hand. With the development and strengthening of statehood, more powerful states of the world laid the beginning of unification processes from a position of strength. This was expressed in aggressive aspirations, in the struggle for natural and human resources. The political map of the world was determined by the powerful rulers of the more developed and aggressive states.

Then, expansion was enriched by other factors of the world political influence. First of all, the development of science should be noted, providing opportunities to strengthen the state, in that its incarnation as military potential, defensive and offensive capabilities. Scientific advances play an increasingly important role in uniting people of the world, which cannot but be reflected on the political world scenario. The emerging global trends are difficult to rank among one zone of influence, they are woven into human life, transforming it, and the rate of this conversion increases. For example, research and development, which originated in the military–industrial complex, have primarily influenced the degree of openness, accessibility, and rate of information sharing, having given humanity the global information network. The potential for the political activity not only of states and political figures but also of the individual has increased, its capacities have enhanced, their implementation depending on the quality of the individual.

With increasing global trends and their awareness, new actors enter the political arena. The understanding of the impossibility to withstand the challenges of the developing world alone has appeared in humans. Common objectives were the basis for uniting people in various communities, both locally and internationally. First international organizations began to emerge from the middle of the 19th century. The year 1874 can be considered as the date of birth of international organizations, when one of the first interstate organizations, namely, the Universal Postal Union (functioning today), appeared. Then the International Telecommunication Union, the International Union of Railways, and so forth (founded not only on a professional basis) came. By the beginning of the First World War, there were more than 500 various international organizations. They were presented by a broad membership and proclaimed their tasks as improvement of world order in varying degrees, problem-solving based on new principles of interaction.

The establishment of the League of Nations in 1919 was a significant political event. League has been delegated new features previously inherent to individual states only. The 58 participating states jointly solved the issues of global security, the settlement of disputes between States by diplomatic means, and improving the quality of life on our planet. Of course, not all organizations are successful and durable, time and the rapid course of events make their own adjustments, but the process itself continues and is growing rapidly. In the second millennium, there are already more than 23,000 international organizations, 3000 of which are intergovernmental and 20 are nongovernmental. Their organization, membership, charter or the lack thereof, spheres of influence, qualitative and quantitative composition, openness to new members or existing limitations, as well as the goals and objectives of operation are extremely various, reflecting the needs of a changing world and the growing world-order activity of individuals as well as individual states. Among the leading, most capable organizations exerting a marked impact on global development are the World Trade Organization, the International Monetary Fund, the Paris Club, and The Group of Seven (G7, formerly G8).

The United Nations uniting 193 countries in the world (2011) is one of the most influential numerous universal organizations open to all countries in the world. Established during the Second World War, the United Nations continues to be an active and influential actor in the promotion of peace and security, development of cooperation between states, and raising the most pressing global issues.

In April 1968, the Italian manager Aurellio Peccei established the Club of Rome; it was an event that became a key one in the development of global studies (www.clubofrome.org).

Aurellio Peccei together with the Director General of Science OECD (Organization for Economic Cooperation and Development, Paris), professor of physical chemistry, Alexander King, invited about 30 European scientists to Rome—natural scientists, social scientists, and economists.

Thus, an international nongovernmental organization was established to unite scientists and public and political figures from around the world.

The Club of Rome is legally registered in Switzerland. The Club members have put forward two main objectives of their activity. The first one is the awareness of the difficulties faced by humanity, and the second one is finding ways of influencing public consciousness, which should lead to correcting negative situations. It is necessary to emphasize the great importance of the first studies of the Club of Rome for the development of global studies and, more specifically, for the formation and development of global modeling.

The establishment of the Club, the further development of its ideas, as well as scientific advances in the mid-20th century marked the beginning of global modeling of survival prospects of mankind through the use of computer technology. The participants of these studies and other futurists were divided into two main groups: some of them began to develop the so-called neo-Malthusian ideas of "social pessimism" (Jay Forrester, Dennis Meadows, and Robert L. Heilbroner), while the others tried to prove the opportunity to avoid the catastrophe by "optimization" (Alvin Toffler, Mihajlo D. Mesarovic, Ervin Lászl6, and Eduard Pestel). But their combined efforts became that bifurcation point which determined our awareness of the global dangers resulting from human development. They were the first who so convincingly demonstrated and proved the interrelationship and interdependence of global processes and their possible harmful consequences for the entire human community.

Jay Wright Forrester, professor of applied mathematics and cybernetics, Massachusetts Institute of Technology, was the first to introduce a mathematical model under the auspices of the Club of Rome, mimicking the current situation in the world (Forrester, 1971). The purpose of this simulation was to trace the development of crisis global processes in their interaction with society, that is, the man and environment were at the focus of attention. The first model was named World1. It took just 4 weeks for Forrester to design this model, which included about 40 nonlinear equations. The equations described the relationship of five parameters, namely, the demographic parameter, investments, the use of nonrenewable resources, environmental pollution, and food production. The model was considered fairly simple and crude, but its predictions have become staggering: the world was heading for disaster. Soon, Forrester perfected his model, calling it World2.

One saying of the outstanding Soviet (Russian) scientist Academician Nikita N. Moiseyev is of interest: "The work of Professor Jay Forrester

"World Dynamics" was most important known to me. This work was really pioneering. The author made an attempt to describe the basic processes in the economy, demography, pollution growth, and their interdependence on a planetary scale, to describe by means of five variables only. He developed a special language for description, the so-called Dynamo, methods of programming and analysis of the results obtained. This work made a great impression on me. This was not due to its scientism; the majority of the dependencies used was spun out of thin air and could not withstand even benevolent criticism. And not due to its methods, which were very similar to the methods of plus-minus factors used by engineers to calculate electrical circuits back in the twenties. Special impression on me was made by the courage of the author, raised his hand to the problem of a holistic description of the biosphere processes, involving human activity. This work had something in common with those conversations that I had with Timofeyev-Resovsky and was an original response to my doubts. I became firmly convinced of the necessity of such work, though believing that it should be carried out very differently" (http://www.ccas.ru/manbios/kak_daleko_r.html). Indeed, Academician Moiseyev and his team have brilliantly coped with the design of another global model, which, in turn, shocked the world.

Further development of these global mathematical models was due to Dennis L. Meadows and his group, who published the book "The Limits to Growth. Report of the Club of Rome" in 1972 (Meadows et al., 1972).

In 1992, the next book by Meadows, "Beyond Growth," was published (Meadows et al., 1992), and, finally, the book "Limits to Growth. 30 Years Later" appeared in 2000 (Meadows et al., 2000), which used the World3 model. An important thought of the third book's authors is that there is no need to enter the World3 model into a computer and to study it in detail to understand the main conclusions. The most significant conclusions about the likelihood of a global catastrophe stem from a simple understanding of the behavior dynamics of the global system, which is determined by three key factors, namely, the existence of the limits of destruction, the constant striving for growth, and the delay between approaching the limit and the social response to it. The World3 model is based on such cause-and-effect relationships. Since nobody has called off cause-and-effect relationships in the world, it is rather likely that the real world is on the way described in the limits-to-growth scenarios.

Forrester's and Meadows' models predict the inevitable crisis due to resource depletion and increasing pollution, if the current trends are conserved and no measures are taken to prevent this crisis. The crisis shall lead to a sharp drop in industrial production and reduced investments in agriculture. The development of the crisis would lead to a decrease in food production and health-care deterioration, which, in turn, would cause an increase in morbidity and mortality, leading to the population decline. The results of Forrester's and Meadows' calculations have shown that it is absolutely necessary to reduce the consumption of the Earth resources by about eight times to prevent the disaster.

The studies of the Club of Rome have attracted a lot of attention to global issues by convincingly and competently showing their real threat, perhaps for the first time. This is a great merit of the Club of Rome and its members, although it is possible to have different attitudes to the recommendations given by the scientists, in particular, they called to limit the consumption of the Earth resources, to reduce the growth of industrial production, to stop the growth of world population, and thus to achieve "global equilibrium." The authors considered educational activities as one of the factors preventing the catastrophe. We note in passing that in one of the last modifications of Forrester's model, the education level is proposed to be included as a parameter.

Of interest is the dynamics of the Club of Rome policy in developing approaches to address global challenges facing humanity: from the rough World1 model presented in 1971 to developing specific programs, for example, to address energy challenges in 2009. This dynamics and the convincing results of other research groups in the field of global modeling have proven the efficiency of this tool provided that the problem is formulated properly and the parameters of the global process studied are chosen accordingly. With the development of science and, especially, information technology, the process of modeling as a tool for understanding the world is being improved. Although this tool still shows limited capabilities against the background of the increasing variety of global challenges, its targeted and well-tuned use is very fruitful, because mathematical models not only warn of dangers but also help in choosing ways to overcome them.

The Club of Rome continues its activities aimed at finding solutions to global problems in the 21st century. It should be noted that its work has changed both qualitatively and quantitatively. Research activities are supplemented with real projects, which involve experts and capitals from many countries. However, the ideology of the Club remains the same. Here are some examples of the Club of Rome's projects in the new millennium.

In 2003, it founded an international organization called Trans-Mediterranean Renewable Energy Cooperation (TREC), whose purpose was to solve energy problems in the world. TREC includes both scientists and politicians. The power supply concept proposed by TREC is truly global. It provides energy supply in Europe, North Africa, and the Middle East. The Club of Rome, staying true to the principles of conservation of the natural environment, has made this project depending on solar energy. Such an approach would enable getting electric power in large quantities, sufficient even for seawater desalination, that is, another global problem would be solved—the lack of fresh water. Solar energy produces electricity without polluting the atmosphere with greenhouse gases, which is especially important for industrial areas and, finally, it is an alternative to oil and gas. Therefore, the new scientific technologies of this project solve several global issues at once, which will promote nature conservation, improve the ecological situation in the world, and overcome the lack of energy.

At the beginning of the second decade of the 21st century, the Club of Rome brings together more than 30 countries and includes about 100 full members. Noting the positive potential of some global processes and innovative transformations in the economy, the Club of Rome focuses on the development of new strategies to solve global problems such as improving the living conditions and opportunities for the growing world population. In order for mankind to overcome the impending difficulties, a new vision should be created, and a new way of global development should be elaborated. In May 2008, the Club of Rome offered a 3-year program that would run until 2012, which was called "A New Path for World Development," designed to achieve a better understanding of the complex global challenges facing the world today. This awareness should provide the basis for certain actions aimed at improving the prospects for peaceful progress. The difference between this program and others is that not only experts, scientists, or politicians (decision-makers) but also the ordinary citizens of Earth are recruited for its implementation, that is, the program envisages the involvement of the public through various channels. It is called an open-source program.

The new path of world development is an overall conceptual framework for five interrelated clusters of solving world problems. The first one concerns the environment and resources. These include climate change, ecosystems, and water.

The second cluster is called "globalization" and relates interdependence, distribution of wealth and income, demographic change, employment, trade, and finance. Rising inequalities and imbalances associated with the present path of globalization risk the breakdown of the world economic and financial systems.

The third cluster (world development) is relevant to sustainable development, demographic growth, poverty, environmental stress, food production, health, and employment.

The fourth cluster is called "social transformation" and refers to social change, gender equity, values and ethics, religion and spirituality, culture, identity, and behavior. The values and behavior on which the present path of world development is based must change if peace and progress are to be preserved within the tightening human and environmental limits.

The fifth direction (peace and security) is designed to improve justice, democracy, governance, solidarity, security, and peace. Phenomena such as alienation, polarization, violence, and conflict are noted at the present stage of the world development. The preservation of peace is vital itself but is also a precondition for progress and for the resolution of the issues which threaten the future.

The program provides a comprehensive examination of the problems within each domain cluster, that is, one of the striking features of global processes, namely, their interconnectedness, is realized. Each working group sets its task and develops and analyzes very specific, consistently related solvable problems within the framework of the outlined field. In the development of the program, establishing relationships between the trend clusters is expected as well because their larger fields overlap.

The new working program of the Club of Rome, adopted in 2012, is called "Underlying Causes and Remedies: A Key Focus." The program aims to identify key activities which will lead to sustainable and stable development trajectory for the next 40 years. The program consists of four interconnected clusters. The first one is values. "Values lie at the heart of our common future" is said in the program. Can universal values translate into real actions to protect and preserve our planet? Do we care about the future? Is intergenerational equity a basis for long-term action? Here are some of the questions that the researchers have set before themselves. The answers will be sought in a number of studies devoted to searching for common universal values, their development and the impact on people's life, the identification of the role of education in shaping values, ethics in business and consumption, and other similar issues.

The second cluster "Towards a New Economy" examines natural resources, climate change, employment, and real values and their role in the economics, economic growth and "uneconomic growth," new consumption and production paradigms, and so forth. The third cluster considers a range of issues relating to population employment. These include issues of economic growth and employment redistribution, migration, demography, poverty, education and skills of the new generation of workers, their labor rights, agriculture employment, and, finally, the future of the nature of labor in the future global economy. The fourth cluster touches upon the future governance at the highest level of the organization, the foundation of new institutions which would meet the new global challenges and contribute to security and stability. New global governance paradigms, whether they are possible, whether they will work, how local solutions can be effectively woven into the resolution of global problems, how to take into account the public interest, how to create corporate models for research and development, and how to combine values, trust, and new institutions—these are the basic issues of this section of the program.

The Club of Rome report prepared by Jørgen Randers is another significant event of 2012. Kindly do not forget that the spring of 2012 is the 40th anniversary of the book "Limits to Growth," and a new book by Randers appeared at the end of the summer, being forecast for the next 40 years. It caused lively discussions around the world. In his book, Randers reinvents previous experience basing on personal research in various fields of knowledge as well as on the latest research materials from more than 30 research teams. Randers himself says about his new approach to modeling: he writes no scenario, discusses no alternatives, he presents a forecast based on an educated guess. Randers argues that there is no great uncertainty and instability nowadays, that there are forces to inhibit undesirable surges in various areas, citing the population growth in the world and resource consumption as examples.

In order for mankind to be able to overcome global problems, it is not enough to identify, describe, and simulate them, to predict possible consequences, though this is already very much. It is also necessary to educate people for life in our rapidly changing world. The main role should be played by education, which also hits the focus of attention of the Club of Rome researchers as an important component of the proposed strategy for survival. In 1979, the Club of Rome presented the report "No Limits to Learning: Bridging the Human Gap" prepared under the supervision of a team of scientists from many countries (USA, Morocco, and others; Botkin et al., 1979). The researchers noted the widening gap between the knowledge received and the real pattern of the world, considering this phenomenon as one of the major global challenges hindering to overcome the negative trends in the development of mankind. In order to gain an understanding of current world issues in all their complexity, they proposed to reform the education system, to adapt it to the rapidly changing world and to develop new teaching methods to promote the perception and assimilation of new knowledge. In the rapidly developing scientific knowledge promoting the introduction of new technologies into all spheres of human activity, training for acquiring new knowledge and skills goes to the foreground in the education system. The moral aspect of education is not left aside. The capacious moral quality of human dignity is advanced as a basic condition for the survival of mankind.

One of the global challenges of the modern world, the uneven levels of development of diriment countries, has attracted the attention of politicians and academics. The world divided by the so-called developed and developing regions is fraught with the danger of causing local conflicts with far-reaching consequences. The 21st century was already sadly marked by the September 11, 2001 disaster, one of whose root causes lies in the economic and political inequality of nations and people. Unfortunately, this global problem remains one of the most critical to the present day, and its solution is an urgent prospect of the future. Scientists study it under the auspices of the Club of Rome, as mentioned below, and in other research teams in many countries. That is, the study of global processes and their impact on global development does not remain the Club of Rome's prerogative.

In one of his works, J. M. Gvishiani shows other models of global development in the 1970s, the time when modeling of the global process was gaining momentum, offering more and more new versions of the models themselves and their thematic content (Gvishiani, 1978). For example, a team of Argentine scientists led by Professor Amulcar O. Herrera offered its own Latin American model of global development in 1974, whose aim was to find ways to achieve "adequate living standards" for developing countries. The model provides the division of the whole world into four regions, namely, Asia, Africa, Latin America, and other developed countries. The goal is achieved by controlling the development of the regions through centralized redistribution of capital, that is, at the expense of economic assistance provided by developed countries. According to the authors, most developing countries, where the population is starving, with a high infant mortality rate, are already affected by the crisis. The authors believe that the causes of the crisis phenomena lie in the flawed system of values and are due to the uneven distribution of wealth and power between countries rather than by depletion of nonrenewable resources, environmental pollution, or demographic causes.

The results of the model are reduced to two scenarios. In the first scenario, no economic assistance to developing countries is provided. As a result, the population of Latin America will reach "adequate living standards" in about 40 years, but Asia and Africa will not achieve them under any circumstances, and the situation by all vital parameters will only get worse there. The second scenario envisages the provision of assistance to developing countries since 1980. The aid is expected for 10 years, starting from 0.2% of the annual total final product of the developed country region and increasing it up to 2%. In this case, the period of achieving the level of "adequate living standards" for the people of Asia and Africa will be 57 and 65 years, respectively.

This type of model is interesting, first of all, as an example of searching for solutions to overcome the threats of global processes, as well as in terms of the choice of parameters, obtaining and describing the results. One may treat the recommendations ambiguously, but it should be noted that the problem of inequality of regions is urgent for the world community and is still waiting for solution.

Even this brief overview of global modeling and its initial development gives an idea of the change in the attitude of mankind to global processes. The awareness of the danger posed by a few of them continues in the world. First of all, they include demographic changes toward growth of the world population, the lack of food, the depletion of nonrenewable natural resources, the growing demand for energy resources, the uneven economic and political development of the regions on our planet, environmental pollution and its consequences impacting the global climatic and evolutionary processes, culture dehumanization, and the design and build-up of weapons of mass destruction. Global modeling has demonstrated the trend to identify and study such global processes which had a negative effect of their impact. Accordingly, the concept of the inevitability of the global catastrophe at the current trends in the development of society has become one of the leading concepts.

The process of developing global simulation continues, attracting more and more participants from around the world, contributing to the formation of new creative teams of scientists from different areas of knowledge. We can say that this process is globalized together with the tasks posed. Structuring of global processes is an important factor determining the success of global modeling. J. M. Gvishiani suggests an approach to this problem (Gvishiani, 1978). He offers a multilevel view of the world and the corresponding description of global processes going therein. The first level is the level of interactions between regions and countries with their complex of global economic (e.g., world market), political (international agreements, etc.), scientific, environmental, and cultural processes. The next level of description and study of global processes is the level of a separate region or country with their quality priorities. As priority processes at this level, Gvishiani offers the following ones: the demographic ones with the whole complex of social problems; the production ones, taking into account natural resources and scientific and technological progress; and food production. The processes related to food production, natural resource stocks, taking into account their impact on technological processes and the degree of environmental pollution, as well as problems of scientific and technical progress are highlighted by him separately. He also considers it important to take into account the human impact on the environment, social processes, and various government control mechanisms. In fact, it is proposed to study global processes through their local consideration.

By analyzing the principle of choice of global processes to build global models, as well as their different definitions, one can propose another approach to the design of a classification based on the objectification of global processes in the overall process of globalization.

1. The processes resulting from technical, scientific, and technological progress. The basis for such processes is achievements in the field of science and technology, bringing together scientists in the framework of international research programs (e.g., the Large Hadron Collider), information technology, just uniting the world community in all major areas of life, new vehicles, health-care development, and modernization of living standards determining the level of quality of life. This may also include global threats arising from the increasing military means of destruction.

2. Production processes marked by an increased scale of production, its increased productivity, the establishment of TNCs, modernization processes, the world distribution of manufactured brands, products, services, and resources, and globalization of the labor market.
3. Economic global processes expressed in the expansion and strengthening of world economic relations, the free movement of capital, strengthening the global impact of TNCs, the global movement in the labor market.
4. Environmental global processes associated with the use of natural resources, plant emissions of pollutants, and other human impacts on the environment. Just the realization of the harm caused to the environment was one of the first unifying factors for the international community to combat environmental problems.
5. Political processes that manifest themselves in the new reality of the interaction among countries (United Europe), the weakening of state frontiers, increasing the movement freedom of people, capital, goods, and services; the emergence of interstate world nongovernmental organizations, strengthening their role in world politics, the influence of TNCs, the development of new forms of global policy-making, greater openness of world politics due to the global Internet network.
6. Social processes often occurring at the local level and growing into global social processes, such as social movements for environment protection, bringing people together into a unified global network of communication, expansion of the cultural, spiritual, and cognitive horizons, increased tolerance, gaining greater economic, political, and spiritual freedom, the adoption of common standards in the lifestyle (e.g., we can talk about the world fashion, trends in nutrition and healthy lifestyles, etc.). One cannot ignore protest mood and terrorist threats.
7. The group unites all the natural phenomena having a global impact on animate and inanimate nature and, of course, on the man.

Certainly, the presented classification does not include all the quantitative and qualitative diversity of global processes. It only roughly outlines possible ways of classifying a certain global process into one group or another, thereby contributing to its understanding at a new level. It once again shows the relationship and impacting variability of global processes aimed at the establishment of the common, united, interdependent world. This quality is especially brightly seen in global information processes implemented in almost all spheres of human activity.

New challenges of globalization put the man in completely new-for-him living conditions, changing his views, mentality, and behavior. One of these most striking manifestations is the creation of a single world space using new communication means, which, in turn, are developing rapidly, giving humanity more and more new capabilities. Time and space begin to perform entirely different social functions. People have acquainted the ability to simultaneous present in different places at once; at least, the simultaneous awareness of what is happening in different places. Involvement in its timing certainly affects the man and his social behavior, which still needs a special study.

Globalization presents another social requirement to the man associated with trust and risk. The life of society in our rapidly globalizing world constantly proceeds under the risk–trust dyad, which is realized in a variety of spheres of life. Any person everyday has to trust strangers, some impersonal forces (market relations) and standards developed for him (human rights), virtual exchange commissions, which are outside the control of an individual or a group of people. Therefore, a person is consciously or puts himself at the mercy of other people with a certain degree of risk. Under these conditions, high morals and conscious contribution to the trust of all parties is the key to the well-being of each person.

KEYWORDS

- **biospheric studies**
- **global climate change**
- **biodiversity**
- **evolution of species**
- **noosphere**
- **immunochemical techniques**
- **biosensor technologies**
- **detoxification**
- **minerals**
- **globalization**

REFERENCES

1. Alexeyev, A. S. Mass Extinctions in the Phanerozoic. Dissertation, Moscow State University, 1998. [Russ].
2. Asimov, I. I. *Robot*; Gnome Press: New York, 1950. [Russ].
3. Alexeyev, A. A. *Current State of the Extinction Problem. Current Status and Main Directions in the Study of Brachiopods*; Dokl. IV Int. School, Zvenigorod, October 28–November 3, 1991. Moscow: PIN RAN; pp 21–50. [Russ].
4. Borinskaya, S. A. *Evolutional Principles in Nature and Society*; Max-Press: Moscow, 2002. [Russ].
5. Budyko, M. I. *Evolution of the Biosphere*; Hydrometeoizdat: Leningrad, 1984. [Russ].
6. Gabdullin, R. R.; Il'in, I. V.; Ivanov, A. V. *Evolution of the Earth and Life*; MGU Press: Moscow, 2005. [Russ].
7. Ebeling, W.; Engel, A.; Feistel, R. *Physik Der Evolutionsprozesse*; Wiley-VCH Verlag GmbH: Berlin, 1990.
8. Goldfein, M. D.; Ivanov, F. V.; Kozhevnikov, N. V. *Fundamentals of General Ecology, Life Safety and Environment Protection*; Nova Science: New York, 2010. [Russ].
9. Goldfein, M. D.; Ivanov, A. V.; Malikov A. N. *Concepts of Modern Natural Sciences. A Lecture Course*; Goldfein, M. D., Ed., 2009. [Russ].
10. Goldfein M. D.; Ursul, A. D.; Ivanov, A. V.; Malikov, A. N. *Fundamentals of the Natural-Scientific Picture of the World*; Goldfein, M. D., Ed.; SI RGTEU Press: Saratov, 2011. [Russ].
11. Goldfein M. D.; Ivanov, A.V. *Modern Concepts of Natural Science*; Nova Publishers: New York, 2013.
12. Leonovich, A. A. *Physical Kaleidoscope*. Bureau Quantum. (Appendix to the Quant Journal, No. 2). 1994.
13. Bak, P. *How Nature Works: The Science of SelfOrganized Criticality*; Copernicus: New York, 1996.
14. Bak, P.; Sneppen, K. Punctuated Equilibrium and Criticality in a Simple Model of Evolution. *Phys. Rev. Lett.* **1993,** *71,* 4083.
15. Hoffman, A. A.; Parsons, P. A. *Evolutionary Genetics and Environmental Stress*; Oxford University Press: Oxford, 1991.
16. Newman, M. E. J. A Model of Mass Extinction. *J. Theor. Biol.* **1997,** *189,* 235.
17. Newman, M. E. J. Simple Models of Evolution and Extinction. *Comput. Sci. Eng.* **2000,** *2,* 80.
18. Newman, M. E. J.; Eble, G. J. Power Spectra of Extinction in the Fossil Record. *Proc. R. Soc. Lond.*, **1999,** *266,* 1267.
19. Newman, M. E. J.; Palmer, R. G. *Modeling Extinction*; Oxford University Press: Oxford, 2003. См. Также Electronic Preprint in Internet http://arxiv.org/abs/adap-org/9908002v1.
20. Newman, M. E. J.; Roberts, B. W. Mass Extinction: Evolution and the Effects of External Influences on Unfit Species, *Proc. R. Soc. Lond.*, **1995,** *260,* 31–37.
21. Newman, M. E. J.; Sibani, P. Extinction, Diversity and Survivorship of Taxa in the Fossil Record. *Proc. R. Soc. Lond.*, **1999,** *266,* 1593.
22. Newman, M. E. J.; Sneppen, K. Avalanches, Scaling and Coherent Noise. *Phys. Rev. E.* **1996,** *54,* 6226.

23. Ray, T. S. *Evolution, Complexity, Entropy, and Artificial Reality. Phys. D.* **1994,** *75,* 239–263; see also Ray, T. S. *Artificial Life.* Electronic preprint in the Internet http://www.hip.atr.co.jp/~ray/pubs/fatm/fatm.html.
24. Roberts, B. W.; Newman, M. E. J. A Model for Evolution and Extinction. *J. Theor. Biol.* **1996,** *180,* 39.
25. Wilke, C.; Martinetz P. Simple Model of Evolution with Variable System Size. *Phys. Rev.* **1997,** *E 56,* 7128.
26. Rogers, K. R.; Gerlach, C. L. Update Environmental Biosensors. *Environ. Sci. Technol. News.* **1999,** *1,* 500–506.
27. Zyryanov, V. V.; Goldfein, M. D. *Trends in the Application of Immunoassay and Biosensors Technology to the Analysis of Environmental Objects.* International Chemical Congress of Pacific Basin Societies. Honolulu, Hawaii. Books of Abstracts. 2000. Part 1, Agrochemistry, section 265.
28. Van Emon, J. M. Immunochemical Application in Environmental Science. *J. Assoc. Off. Anal. Chem. Int.* **2001,** *84,* 125–133.
29. Chuang, J. C.; Chou, Y. L. Comparison of Immunoassay and Gas Chromatography/Mass Spectrometry for Measurement of Polycyclic Aromatic Hydrocarbon in Contaminated Soil. *Anal. Chim. Acta.* **2003,** *486,* 31–39.
30. Lobanova, A. Y.; Shutaleva, E. A. Definition of Polyaromatic Hydrocarbons in Food and Objects of Environment with a Method of the Polarising of Fluorescent Immunoanalysis. *Biomed. J.* **2007,** *8,* 169–83. [Russ].
31. Lu, C.; Anderson, L.; Morgan, M.; Fensce, R. *Saliva Biomonitoring for Pesticides Exposure Among Application using ELISA*; International Chemical Congress of Pacific Basin Societies: Honolulu, Hawaii. Books of abstracts. 2000. Part 1. Agrochemistry, section 267.
32. Lee, N., Allan, R.; Kill, A., *Thin B. Monitoring Mycotoxins and Pesticides in Grain and Food Production Systems for Risk Management in Vietnam and Australia*; Pacific Basin Societies: Honolulu, Hawaii. Books of abstracts. 2000. Agrochemistry, section 281.
33. Dzantiev, B. B., Zherdev, A. V. *Rapid Enzyme Immunoassay for Pesticides Detection with the Use of Bacillary-Amylase as Label*; Pacific Basin Societies: Honolulu, Hawaii. Books of abstracts. 2000. Agrochemistry, section 268.
34. Sanchez, F. G.; Diaz, A. N.; Development of Homogenious Phase-Modulation Fluoroimmunoassay for 2,4-Dichlorophenoxyacetic Acid. *Analyt. Chim. Acta.* **1999,** *395,* 132–142.
35. Biagini, R.; Striley, C.; Murphy, D. *Use of Fluorescence Microbead Immunosorbent Assay for Pesticide Biomonitoring*; Pacific Basin Societies: Honolulu, Hawaii. Boors of abstracts. 2000. Agrochemistry, section 263.
36. Van Emon, J. M.; Lopez-Aliva, V. *Immunoaffinity Extraction with On-Line Liquid Chromatography Mass-Spectrometry.* Environmental Immunochemical Methods, ACS Symposium. Series 646. pp 74–88.
37. Krosaviani, M.; Blake, D. A. Binding Properties of a Monoclonal Antibody Directed Toward Lead-Chelate Complexes. *Bioconjugate Chem.* **2000,** *11,* 267–277.
38. Zyryanov, V. V.; Goldfein, M. D. *Biosensor Definition of Organic Compounds in Environment and Organism Liquids. Works of International Symposium on problems of science technique and education*; Moscow, 2000; Vol. 2, p 82. [Russ].

39. Spier, C. R.; Vadas, G. G.; Kaattari, S. L; Unger, M. A. Near Real-Time, On-Site, Quantitative Analysis of PANs in the Aqueous Environment using an Antibody-Based Biosensor. *Environ. Toxicol. Chem.* **2011,** *30*(7), 1557–1563.
40. Shimomura, M.; Nomura, Y.; Zhang, W.; Sakino, M. Simple and Rapid Detection Method Using Surface Plasmon Resonance for Dioxins, Polychlorinated biphenyls and Atrazine. *Anal. Chim. Acta.* **2001,** *134,* 223–230.
41. Zaalishvili, G. V.; Khatisashvili, G. A.; Ugrekhelidze, D. Sh.; Gordeziani, M. Sh.; Kvesitadze, G. I. Detoxication potential of plants. *Appl. Biochem. Microbiol.* **2000,** *36*(5), 515–524.
42. Kvesitadze, G. I.; Khatisashvili, G. A.; Sadunishvili, T. A.; Yevstigneyev, V. G. *Metabolism of Anthropogenous Toxicants in Higher Plants*; Nauka: Moscow, 2005; 199 p. [Russ].
43. Pomerantz, S. H.; Murthy, V. V. Purification and Properties of Tyrosinases from *Vibrio tyrosinaticus*. *Arch. Biochem. Biophys.* **1974,** *160*(1), 73–82.
44. Leonowicz, A.; Grzywnowicz, K. Quantitative Estimation of Laccase Forms in Some White-Rot Fungi Using Syringaldazine as a Substrate. *Enzyme Microb. Technol.* **1981,** *3,* 55–58.
45. Niku-Paavola, M.-L.; Karhunen, E.; Salola, P.; Paunio, V. Ligninolytic Enzymes of the White-Rot Fugus *Phlebia radiate*. *Biochem. J.* **1988,** *254,* 877–883.
46. Korolyova, O.V.; Yavmetdinov, I.S.; Shleyev, S.V.; Stepanova, S.V.; Gavrilova, V.P. Isolation and Studying of Some Properties of Laccase from the Basidial Mushroom *Cerrena maxima*. *Biochemistry*. **2001,** *66*(6), 762–767.
47. Bradford, M. M. A Rapid and Sensitive Method for the Quantitation of Microgram Quantities of Protein Utilizing the Principle of Protein-Dye Binding. *Anal. Biochem.* **1976,** *72*(1), 248–254.
48. Lukner, M. *Secondary Metabolism in Microorganisms, Plants, and Animals*; Mir: Moscow, 1979; 548 p.
49. Pozdnyakova, N. N.; Rodakiewicz-Novak, J.; Turkovskaya, O. V.; Haber, J. Oxidative Degradation of Polyaromatic Hydrocarbons Catalyzed by Blue Laccase from *Pleurotus ostreatus* D1 in the Presence of Synthetic Mediators. *Enzyme Microb. Technol.* **2006,** *39,* 1242–1249.
50. McEldoon, J. P.; Pokora, A. R.; Dordick, J. S. Lignin Peroxidase-Type Activity of Soybean Peroxidase. *Enzyme Microb. Technol.* **1995,** *17,* 359–365.
51. Sanches-Ferrer, A.; Rodrigues-Lopez, J. N.; Garcia-Canovasm, F.; Garcia-Carmona, F. Tyrosinase: A Comprehensive Review of its Mechanism. *Biochim. Biophis. Acta.* **1995,** *1247,* 1–11.
52. Tumaykina, Yu. A.; Turkovskaya, O. V.; Ignatov, V. V. Destruction of Hydrocarbons and Their Derivatives by a Vegetative and Microbic Association on the Basis of *Elodea Canadensis*. *Appl. Biochem. Microbiol.* **2008,** *44*(4), 422–429. [Russ].
53. Günther, T.; Sack, U.; Hofrichter, M.; Lätz, M. Oxidation of PAH and PAH-Derivatives by Fungal and Plant Oxidoreductases. *J. Basic Microbiol.* **1998,** *38*(2), 113–122.
54. Goldfein, M. D.; Kozhevnikov, N. V.; Ivanov, A. V., Kozhevnikova, N. I.; Malikov, A. N.; Timusy, L. G. *Fundamentals of General Ecology, Life Safety and Economic and Legal Regulation of Natural Usage*; Economical State University: Moscow, 2006. [Russ].

55. Goldfein, M. D.; Kozhevnikov, N. V.; Ivanov, A. V.; Malikov, A. N. *Life Safety and Ecological and Economic Problems of Natural Usage*; Economical State University: Moscow, 2008.
56. Arzoyants, E. *Globalization: A Disaster or a Path to Development*?; Novyi Vek: Moscow, 2002.
57. Weber, A. B. Neoliberal Globalization and Its Opponents. *Polity*. **2002,** *2*(25), 22–37. Moscow: Leto.
58. Gvishiani, D. M. (1978). Methodological Problems in the Modeling of Global Development. *Russ. Probl. Philos*. *17*(2), 14–28.
59. Giddens, A. *Sociology* (7th ed.); Polity: Cambridge, 2013.
60. Meadows, D. H.; Meadows, D. L., et al. *The Limits to Growth: A Report to the Club of Rome*; Universe Books: New York, 1972.
61. Meadows, D. H.; Meadows, D. L.; Randers, J. *Beyond the Limits*; Chelsea Green Publishing Company: White River Junction, Vermont, 1992.
62. Meadows, D.; Randers, J.; Meadows, D. L. *Limits to Growth: The 30-Year Update*; Chelsea Green Publishing Company and Earthscan: White River Junction, Vermont, 2004.
63. Moiseyev, N. N. *To Be or Not to Be the Humankind*?; Ulyanovsk Printing House: Moscow, 1999.
64. Peccei, A. *The Human Quality*; Pergamon Press: Oxford, 1977.
65. Forrester, J. *World Dynamics*; Wright-Allen Press: Cambridge, 1971.
66. Friedman, T. *The World Is Flat: A Brief History of the Twenty-first Century*; Farrar, Straus and Giroux: New York, 2005.
67. Botkin, J.; Elmanjra, M.; Malitza, M. *No Limits to Learning. Bridging the Human Gap. A Report to the Club of Rome*; Pergamon Press: Oxford, 1979.
68. Giddens, A. *Runaway World: How Globalization Is Reshaping Our Lives*; Profile: London, 1999.
69. Robertson, R. *Globalization: Social Theory and Global Culture*; Sage: London, 1992.
70. Simon, J. *The Ultimate Resource*; University Press: Princeton, 1981.
71. Waters, M. *Globalization. Key Ideas*; L. N. Y. Routledge: New York, 2001.
72. Ilyin, I. V.; Trubetskov, D. I.; Ivanov, A. V. (Eds.). *Nonlinear Dynamics of Global Processes in Nature and Society*; Moscow University Press: Lomonosov Moscow State University, Faculty of Global Studies, 2014.

CHAPTER 6

PHENOMENON OF THE LOGISTIC EQUATION AS A BASIC MODEL FOR DESCRIBING GLOBAL PROCESSES IN SOCIETY

CONTENTS

6.1 The Disease Spread Problem 194
6.2 The Simplest Models of Population Growth 199
6.3 A Phenomenological Model Describing the Earth Population Growth 204
6.4 Growth Dynamics of World Urbanization 213
6.5 Migration Patterns 216
Keywords 224
References 224

ABSTRACT

The logistic equation to describe the processes of globalization is considered. The problem of the spread of diseases is considered. The models of population growth are listed. A phenomenological model of the Earth's population growth is analyzed. The problems of urbanization and migration are considered.

In the analysis of the models of global processes, special attention should be paid to the models based on the equations of logistic type. The logistic behavior is typical for many processes in animate nature and human society; it provides both qualitative and quantitative agreement with the known statistics of global human development.

This chapter provides a number of specific examples of the application of the logistic equation.

6.1 THE DISEASE SPREAD PROBLEM

We begin with a short story about a severe outbreak of cholera in London in 1854, where the use of the natural-scientific approach proved useful from both practical and theoretical viewpoints. No causes and propagation ways of cholera were known at that time, so this sudden outbreak of this terrible disease was a disaster for people, taking away many lives. For example, only for three days in 1854 (from August 31 to September 2), 83 cases of death from cholera were recorded. The English physician John Snow invented to map all the registered deaths from cholera onto the map of London (Fig. 6.1).

It was found that the greatest number of deaths had occurred in the vicinity of Broad Street, and J. Snow came to the conclusion that the public water pump located on this street was the source of infection. Then he decided that it was necessary to remove the handle of this pump to make impossible any use of this bad water; the local residents were thus forced to go to other water sources, which immediately led to the termination of the outbreak. Therefore, in addition to this quick termination, contaminated water was found to be the source of cholera.

FIGURE 6.1 Map of London with marked cases of reported deaths from cholera (dark thick lines).

To fight against epidemics in the human society, one needs to know not only the mechanism of infection on the human body, but also how diseases are spread. While medicine should answer the first question, the second one is for nonlinear dynamics. Consider the simplest case, when a disease spreads among a group of susceptible individuals, with no removal of them from the population due to death, recovery, or isolation. Such a formulation of the problem, in particular, may be acceptable for the initial stages of some diseases of the upper respiratory tract, since a long time may have elapsed for such diseases before the source of infection will be removed from the population. Let us also assume that the group of individuals is rather numerous (so we can use the mathematical formalism of differential calculus), but, at the same time, this group is compactly concentrated within a certain area, so that any spatial aspects of the epidemic can be neglected.

Suppose there are n individuals susceptible to the disease and a source of infection is introduced into the group at time $t = 0$. Consider a uniformly mixed group consisting of $(n + 1)$ individuals. Suppose that this group contains x susceptible individuals and y infection sources at time t, that is,

$$x + y = n + 1. \tag{6.1}$$

It is reasonable to assume that the average number of new disease cases occurring within the interval dt is proportional to the numbers of both the infection sources and susceptible individuals. If the frequency of contacts between the members of this group is b, the average number of new disease cases occurring within dt is equal to $bxydt$, that is,

$$dx = -bxydt. \tag{6.2}$$

The sign minus in the right-hand side of eq 6.2 indicates that the number x of susceptible individuals decreases over the time interval dt by $(bxydt)$. If we introduce a dimensionless time $\tau = bt$, then the evolution equation becomes

$$\frac{dx}{d\tau} = -x(n - x + 1) \tag{6.3}$$

with the initial condition

$$\tau = 0, x = n \tag{6.4}$$

In the design of this model, it is believed that as soon as an individual gets infected himself, he becomes infectious to other susceptible individuals, that is, the latent period is zero.

The solution of the resulting equation is

$$x(\tau) = \frac{(n+1)n}{n + e^{(n+1)\tau}} \tag{6.5}$$

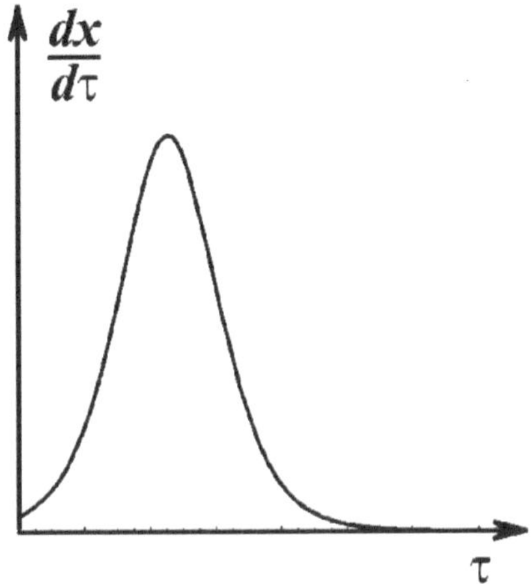

FIGURE 6.2 Epidemic curve.

In practice, the number of new cases occurring per day or week is usually registered during an epidemic, rather than the total number of cases. It is thus more convenient to consider the dynamics of growth of the number of new cases, described by the so-called epidemic curve. The corresponding equation is

$$\frac{dy}{d\tau} = -\frac{dx}{d\tau} = x(n - x + 1) = \frac{n(n+1)^2 \, e^{(n+1)\tau}}{\left(n + e^{(n+1)\tau}\right)^2} \tag{6.6}$$

This is a symmetric, single-humped curve with one maximum. We thus obtain a characteristic of epidemics: at first, the number of new cases increases rapidly, reaches a maximum at some time, and then decreases to zero. This shape of the epidemic curve is a purely mathematical consequence of the accepted assumption that the average number of new cases is proportional to both the numbers of susceptible individuals and infection sources (Fig. 6.2).

It should be noted that even such a seemingly simple model provides a very good correlation with the actual statistical data (Fig. 6.3), describing acquired immunodeficiency syndrome (AIDS) spread over the territory of the United States.

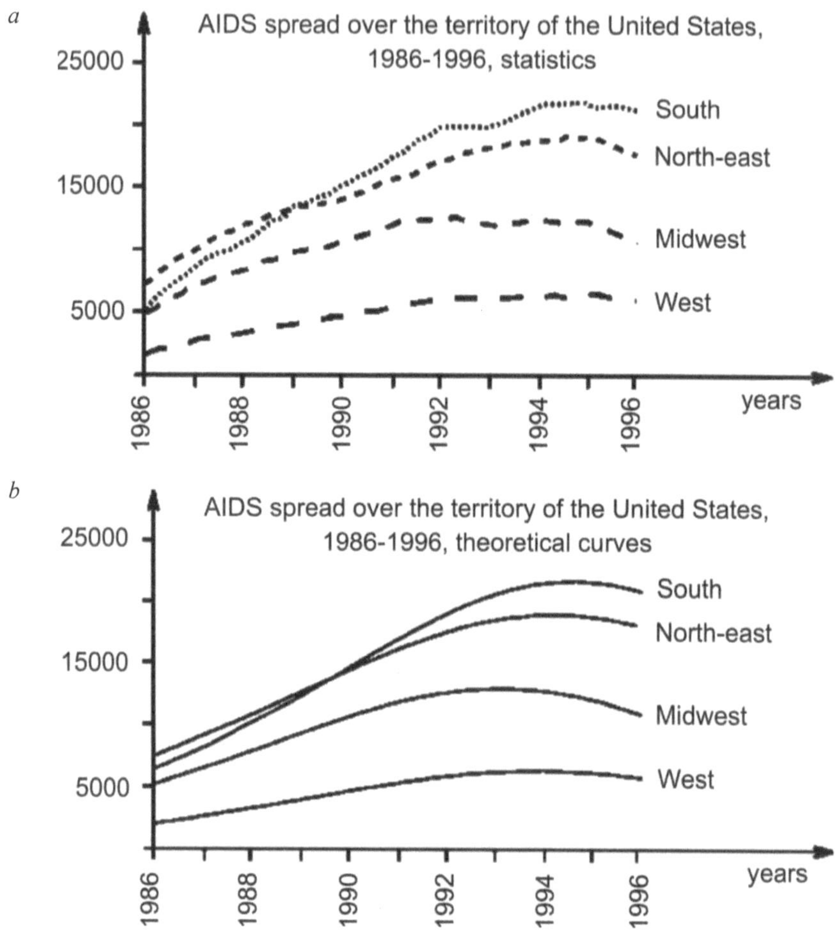

FIGURE 6.3 Epidemic curves of AIDS spread over U.S. regions: (a) the actual statistics (according to the Centers for Disease Control and Prevention, http://thebody.com/cdc/cdcpage.html); (b) the analytical epidemiological curves $dy/d\tau$ plotted in the relevant dimensional units.

In this case, it is very important that the basis of this simple epidemiological model is a reference object of nonlinear dynamics, the logistic equation is generally written as (Koronovsky et al., 2003)

$$\frac{dN}{dt} = \frac{\mu N(\mathrm{K} - N)}{\mathrm{K}} \tag{6.7}$$

where N is a quantity characterizing the process, $K = N$ at $dN/dt = 0$, t is time, and μ is a constant coefficient.

6.2 THE SIMPLEST MODELS OF POPULATION GROWTH

Let us derive the logistic equation (eq 6.7) from the population growth dynamics problem.

Thomas Malthus authored one of the first models of the population growth dynamics. This model described in "An Essay on the Principle of Population" (Malthus, 1798) is one of the earliest and simplest models that exist in biology (though it has "predecessors," for example, the 1202-dated model by Fibonacci, describing the dynamics of a rabbit population (Vorobiev, 1978)).

Malthus' model was obtained under the following assumptions. First, despite the population size being discrete (i.e., can change by a whole number of units only), it is assumed to be continuously changing in time. This is only justified if the population size is large enough—then its change for few whole units can be considered infinitely small as compared with the total population, and, accordingly, it is possible to describe the dynamics in the language of derivatives. Second, we assume that the change of the population per time unit is directly proportional to the number of individuals in this population, that is, the greater the population size, the more individuals die and, accordingly, are born per unit time. The proportionality coefficient is assumed constant (in Malthus' paper, this assumption is supported by his analysis of demographic information). Furthermore, we assume that the resources (space habitat, food, etc.) necessary for the life of individuals in the population are unlimited. Then the mathematical description of the model looks as

$$\frac{dN}{dt} = \mu N, \tag{6.8}$$

where N is the population size, μ the difference between the birth and death rate constants. The solution of eq 6.8 predicts exponential growth of the population $N(t) = N(0)e^{\mu t}$ known as the Malthusian one, provided that $\mu > 0$ (the birth rate exceeds the death rate); if $\mu < 0$ (more deaths than births), there will be exponential decay of the population to zero (Fig. 6.4). Indeed, if μ is positive, then the population increases with time, the population growth rate rises with this size, and the population thus grows faster and faster, without any restrictions.

Malthus, the first to get similar results with regard to the human community, came to the conclusion that the growth of human society would be ahead of the growth rate of food stocks and, therefore, fierce competition among the people is inevitable ("for a place under the sun"). It is clear that no biological population can grow indefinitely in real life since any ecological system is stabilized by natural resource constraints. That is why Pierre François Verhulst suggested a logistic model in 1838, which was more reliable than the Malthusian model (eq 6.8) to describe the dynamics of populations (Verhulst, 1838). We note in passing that Verhulst's equation also has its "predecessor," the 1825 Gompertz model of population dynamics

$$\frac{dN}{dt} = \mu\left(\ln \mathrm{K} - \ln N\right), 90^{\circ} \tag{6.9}$$

which exhibits saturable increase (Gompertz, 1825), that is, the number of population stabilizes at a certain level over time and then no longer changes. The parameter K in this equation has the sense of the maximum possible population. However, this model was not widely used because the logarithmic terms in the right-hand side of eq 6.9 had no justification. In addition, the curves corresponding to real time dependences of the population size go to a steady-state level, faster than expected from Gompertz' model.

In this regard, the model proposed by Verhulst was more successful. Verhulst suggested that the resources needed for the life of individuals in a population are limited, and no more than K individuals can simultaneously live in the area under consideration. The value K is time independent and a parameter of the model called the capacity of the medium. Then the factor responsible for the rate of changes in the population size should depend on the number N of individuals in the population, and for small N values, when the population size is much smaller than the capacity of the medium ($N \ll \mathrm{K}$), the limited natural resources will not be noticeable, and hence both equation and its solution should differ very little from eq 6.8 and its solution, respectively. Consequently, the change rate of the population should be close to μN for small N. On the other hand, the limited nature of natural resources begins to manifest itself with an increase in the population, and the factor responsible for the growth rate should decrease, reaching zero when $N = K$. The equation taking into account all of the above is called Verhulst's equation and coincides with the logistic equation (eq 6.7) by form.

The Verhulst equation solution is shown in Figure 6.4: when the population size is small ($N \ll K$), eq 6.7 is identical to eq 6.8 and the population size increases exponentially (provided, of course, that $\mu > 0$). But, as N increases, the limited nature of natural resources begins to be displayed, the population growth rate reduces, and ($t \to +\infty$) the population size N tends to K, and the population growth rate N vanishes.

It is easy to show that the solution of eq 6.7 (assuming that at time $t = t_0$, the population size is $N(t_0) = N_0$) can be written as

$$N(t) = \frac{KN_0}{N_0 + (K - N_0)e^{\mu(t_0 - t)}} \tag{6.10}$$

The curve described by eq 6.10 is often called the logistic curve.

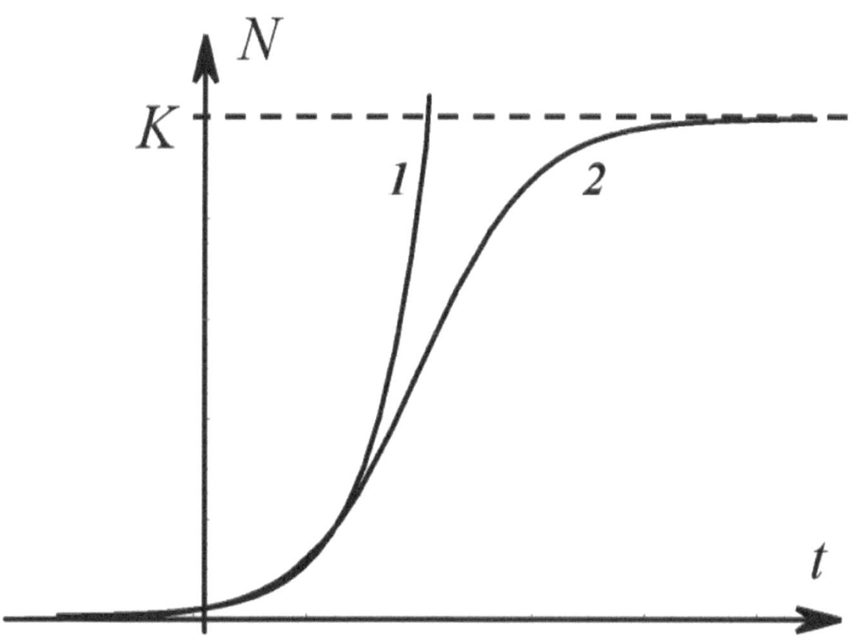

FIGURE 6.4 Solutions of eq 6.8 and of the logistic equation (eq 6.7): *1* unrestricted (exponential) growth of N if $\mu > 0$ for eq 6.8 (Malthus' equation); *2* solution (eq 6.7) of the logistic equation (Verhulst's equation).

It is important to note that Verhulst's equation fairly well describes the dynamics of simple biological systems, such as bacterial colonies. Figure 6.5 (taken from Smith, 1976) shows the time dependence of the numbers of a bacterial colony and the corresponding solution of the logistic equation (eq 6.7).

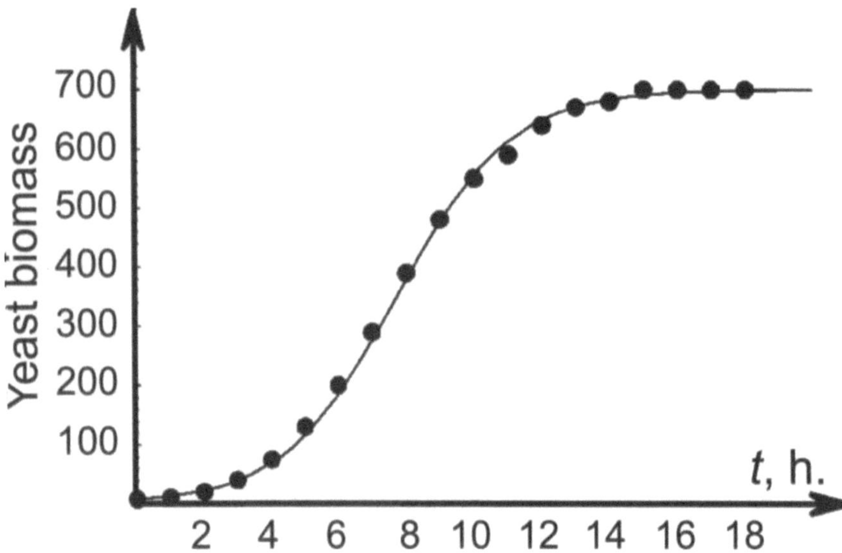

FIGURE 6.5 Time dependence of the numbers of a bacterial colony, with the logistic curve superimposed (Smith, 1976).

Clearly, in the process of a long-term evolution, the system parameters may change (e.g., under the influence of climatic fluctuations).

There were attempts to describe the population dynamics using the logistic equation. One of the first attempts to compare the results predicted by the solution of the logistic equation (eq 6.7) and statistical demographic data was made by Raymond Pearl for the United States of America in the 1920s (Gilyarov, 1998). At the time of the survey (the early 20th century), it seemed that Verhulst's model provided a good description of the population dynamics. However, as it turned out later, the real statistics is well fitted by the logistic curve before the 1940s only. Further, in accordance with Verhulst's equation, stabilization of the numbers should be observed, which not the case is—the population continues to grow. The results by

Pearl are shown in Figure 6.6 with a dotted line, the points marked are the real statistics of the U.S. population.

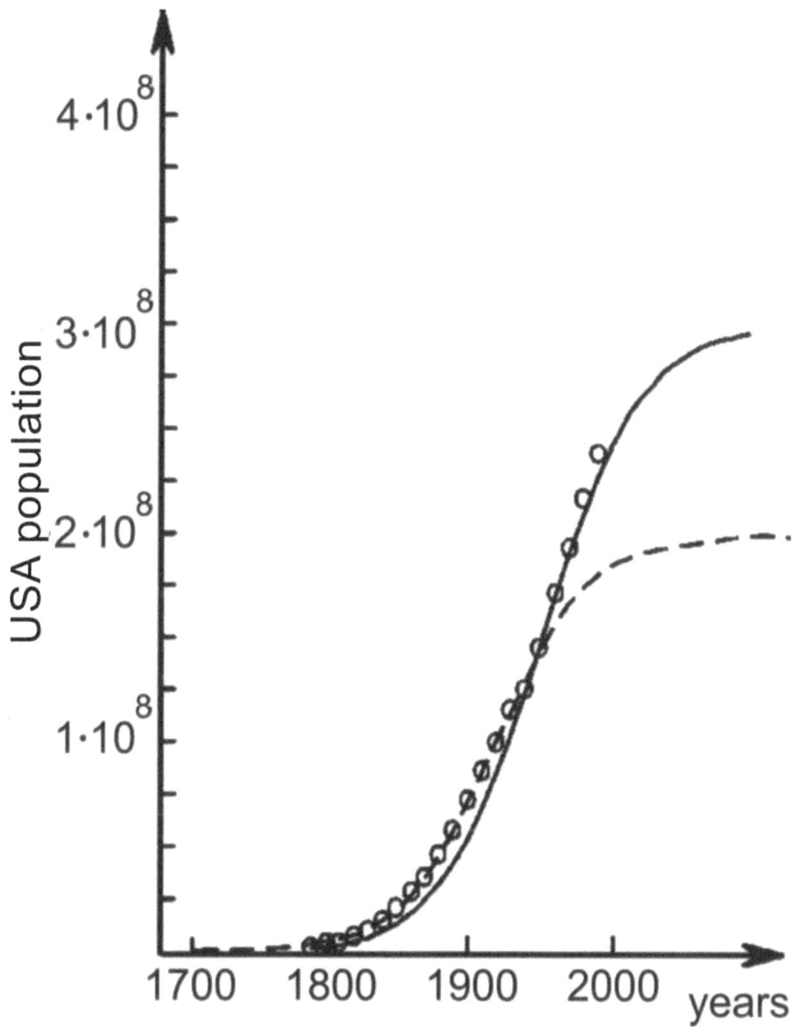

FIGURE 6.6 The U.S. population (circles are real statistics), the results of the logistic curve fitting the data by Pearl (dashed line) at the beginning of the 20th century (Gilyarov, 1998), and the logistic curve plotted on the basis of the U.S. population data before 1990 (continuous line).

Trying to fit the parameters in the logistic equation so that its solution is in accordance with today's statistics seems unsuccessful at the first glance (Fig. 6.6, solid line). However, some researchers of this problem have shown that, by taking into account a number of important factors and some modifications of the Verhulst model, satisfactory results in good agreement with the known demographics can be obtained.

6.3 A PHENOMENOLOGICAL MODEL DESCRIBING THE EARTH POPULATION GROWTH

The population growth problem seems to be one of the most important global issues facing mankind. It closely relates to other challenges faced in all countries, for example, migration and agriculture problems. The increasing globalization of the modern economy, open frontiers, the high mobility of individuals and, at the same time, the increasing social tension while interaction of different cultures (especially vividly exemplified in Western Europe, where migrants from Africa, the Middle East, India, and so forth begin to exert a strong impact on the traditional way of life and the system of European values) requires a special analysis of economic, demographic, social, psychological, and other processes. It should also be noted that agriculture which has been evolving quite conservative and stable and is determined by the local climatic and geographical conditions over the centuries, has experienced the impact of population growth in recent decades. This is due to the emergence of new technologies providing such capabilities.

Population growth and migration are usually treated by demographic statistical methods developed to describe the population of individual countries or regions (Borisov, 2004; Korotayev et al., 2007). These statistical data serve the basis of quantitative laws, often suitable for describing the global dynamics of population growth in the world as well as in individual regions and countries.

This section presents issues of the application of interdisciplinary methods and approaches of nonlinear dynamics for qualitative and quantitative analysis of the global processes associated with the Earth population growth. The theory (model) proposed by Sergei Kapitsa is the most advanced in this respect and considers the population of the world as a global system, representing a single and closed object, whose all parts interact with each other. Just this interconnectedness and interdependence of the modern world is due to transport and trade links, migration, and information flows. In the distant past, when there were less people and the world was largely divided, its populations however were slowly but surely interacting. It would suffice to recall the well-known history of the Crusades, the Silk Road trade, and the colonization of Africa and America. Currently, due to the advances in modern telecommunications and transport network, the world should be certainly treated as a single system.

It should also be stressed that for the population of the world as a whole, migration cannot be taken into account, as playing a certain role for separate countries or regions only: it is nowhere to migrate on a global scale. It is also important that all people biologically belong to one species (*Homo sapiens*), they have the same number of chromosomes (46), different from all other primates, and all races are able to mixing and social exchange. Moreover, mankind is a relatively uniform species and exceeds the number of animals by about five orders of magnitude. There is evidence to state that over the last million years, human has biologically changed little, and the mainstream development and self-organization of mankind took place in the social sphere (Cambridge Encyclopedia, 2004). Just these processes of social, technological, and economic development are quantitatively described by Kapitsa's phenomenological mathematical model, interpreting demographic data through a systematic approach and synergy.

Much attention is currently drawn to the rapid growth of the world population. The growth rate is ever-increasing and so high (approaching 100 million per year), that it is characterized as a population explosion. This increase in the world population requires the ever-increasing food and energy production, the consumption of mineral resources, and leads to the increased pressure upon the biosphere (Fig. 6.7), in particular, to the much-discussed problem of global warming.

The picture of the ever-increasing and uncontrolled population growth, naively extrapolated to the future, leads to alarming forecasts and even apocalyptic scenarios for the future of mankind. Despite the dramatic population explosion and often very emotional reactions it causes, it is essential that mankind is experiencing some demographic transition (Fig. 6.7). This phenomenon is in a sharp increase in the rate of population growth, subsequently replaced by an equally rapid decrease of it; then the population will stabilize in numbers. This transition has already been passed by developed countries, and the same process now takes place in developing countries. The demographic transition is accompanied by the growth of productive forces and the movement of large masses of people from rural to urban areas.

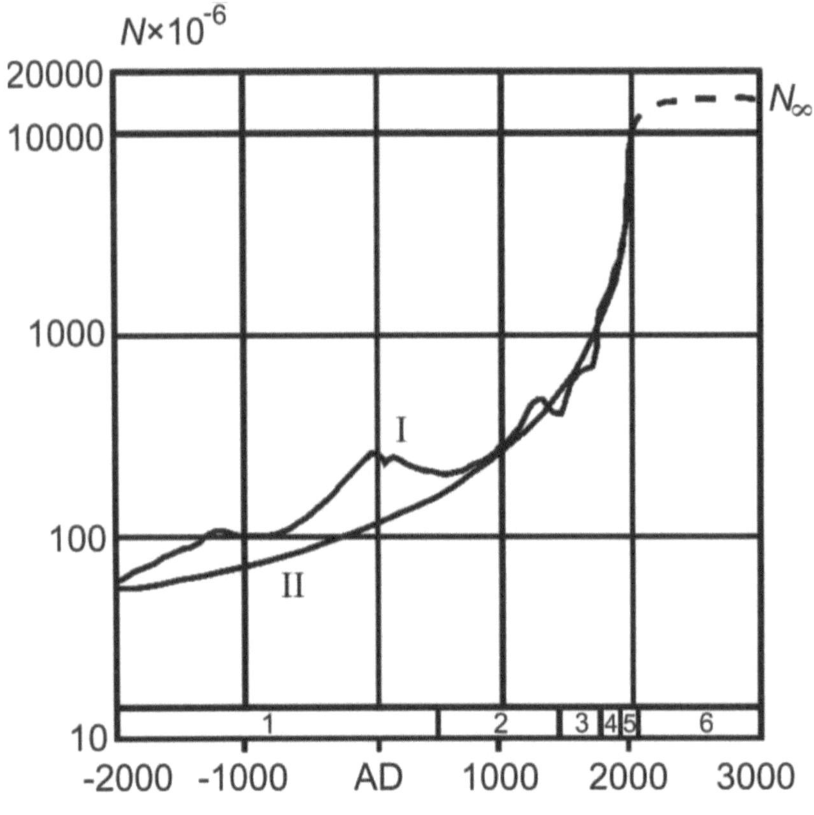

FIGURE 6.7 World population: *I* demographics, *II* the blow-up regime. Time intervals: *1* Ancient World, *2* Middle Ages, *3* early modern period, *4* recent history, *5* a demographic transition, *6* stabilization of the population (Kapitsa, 1999)

Upon completion of this transition, a sharp change in the age composition of the population occurs. The population distribution by age and sex is usually represented as diagrams, to be clearly seen how the population composition changes with age and how the evolution of the system proceeds in the case of a nonstationary state. These distributions have changed little, and only on passing the demographic transition, a qualitative change in the age distribution of population is observed. From the pyramid, characteristic of the period of growth, a rapid transition occurs to a columnar distribution, where the population growth is almost stopped (Fig. 6.8).

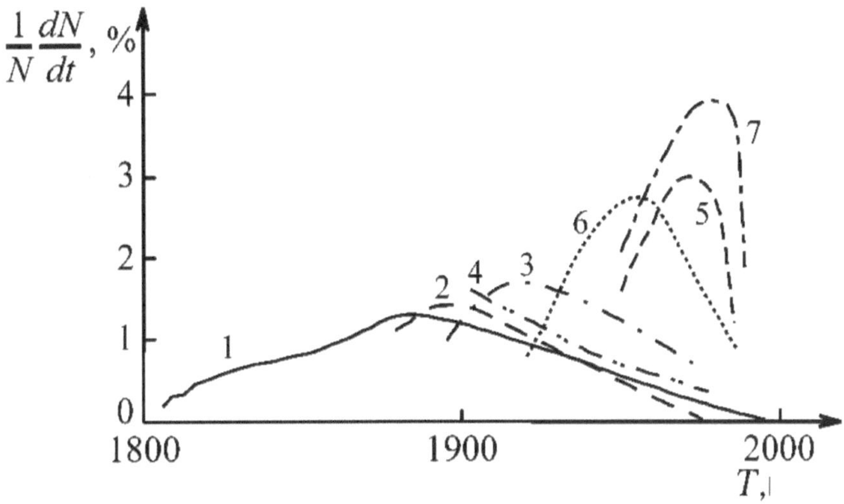

FIGURE 6.8 Passage of the demographic transition by various countries: *1* Sweden, *2* Germany, *3* Soviet Union (Russia), *4* United States, *5* Mauritius, *6* Sri Lanka, and *7* Costa Rica. The curves are smoothed (Kapitsa, 1999).

Let us now consider the phenomenological mathematical model of the hyperbolic growth of the world population in more detail (Kapitsa, 1996; Kapitsa, 1999; Kapitsa, 2010).

Let the world population at some time T be characterized by the number of people $N(T)$. The growth process will be considered over a considerable time interval, that is, over a very large number of generations. The very length of a generation and the distribution of people by age and sex are not included in the calculations.

In system analysis, when a multifactorial statistically stationary process goes with many degrees of freedom, one could expect that the growth will be self-similar. Under these conditions, the simulation is based on the assumption of the time-homogeneous self-similar nature, resulting in a scaling process.

The meaning of this basic hypothesis is in the constancy of the relative growth rate of the system

$$\lim_{\Delta N, \Delta T \to \infty} \frac{\Delta N}{N - N_1} \frac{T - T_1}{\Delta T} = \frac{d\left(\ln\left|N - N_1\right|\right)}{d\left(\ln\left|T - T_1\right|\right)} = \alpha, \qquad (6.11)$$

where N_1 and T_1 are some reference values of the numbers and time. In most cases, we can assume that $N_1 = 0$.

Equation 6.11 leads to the fact that the self-similar growth should be (with need) described by power laws like $N = C(T_1 - T)^{\alpha}$, where C and α are constants. The constancy of the population–time ratio manifests itself in such self-similar processes. The simplest but not unique example of such a kind is linear increase when $\alpha = 1$. The growth in geometric progression (exponential growth) or development by the logistic law is not satisfied by this condition, because an internal time scale appears, namely, the time of doubling and scale invariance.

Searching for a self-similar solution is the first step in designing a model, and Foerster (1960) proposed an empirical formula

$$N = \frac{179 \times 10^9}{\left(2027 - T\right)^{0.99}} \qquad (6.12)$$

to describe the world population growth, where the respective values of the constants

$$C = (179 \pm 0.14) \times 10^9,\ T_1 = 2027 \pm 5,\ \alpha = 0.99 \pm 0.009.$$

were obtained by statistical processing (least squares) of a large amount of data on the world population from Christ till 1960. The accuracy of the proposed values for α seems somewhat overestimated. In what follows, the value $\alpha = 1$ will be accepted. We note that Horner (1975) later proposed a similar expression

$$N = \frac{C}{T_1 - T}, \qquad (6.13)$$

where $C = 200 \pm 10^9$, $T_1 = 2025$, and $\alpha = 1$. The above formulae are capable of describing the world population growth for hundreds and even thousands of years.

Equation 6.13 describes growth as a process of self-similar development, obeying the hyperbolic law of evolution, the so-called *blow-up regime* (Fig. 6.7), studied for many nonlinear phenomena in detail. Therefore, the population explosion appears as a global systemic instability characteristic of the growth of clearly nonlinear systems.

However, eqs 6.12 and 6.13 are fundamentally limited to a certain area, where applicable. First, as $T_1 = 2025$ is approached, the world population will tend to infinity. This conclusion makes some people consider 2025 as the Day of Judgment. Second, an absurd result is derived for the distant past: 10 people existed 20 billion years ago, at the creation of the Universe.

Therefore, the self-similar solution eq 6.12 is limited from both the future and the past, and we can inquire of its applicability limits. Indeed, the theory of self-similar processes indicates that these power solutions are valid only as intermediate asymptotic ones. In other words, the proposed expressions apply only within a certain range to be found.

To do this, we turn to these 10 "elders" who presumably lived 20 billion years ago. If such phenomena would have existed, they would have lived a billion years, which is just as absurd as the fact of their existence. On the other hand, the rapid growth of mankind as T_1 is approached means that the whole population size will be doubled every year since 2024, which is also impossible, because the maximum rate of population growth is limited.

In other words, there is a factor to be taken into account, namely, the time characterizing human life, including the reproductive capacity and life expectancy. In the initial period of growth, when N is small, at T_0 (the beginning of population growth), the minimum rate of the world population growth cannot be less than the appearance of an individual for a characteristic time τ. Analytically, it can be written as

$$\left(\frac{dN}{dT}\right)_{\min}\Bigg|_{\substack{N\to 1\\ T\to T_0}} \geq \frac{1}{\tau} \tag{6.14}$$

This condition should be regarded as an approximate representation of continuity, suitable for describing the initial development. On the other hand, as the demographic transition is passed through and T_1 is approached, the maximum growth rate should be limited, when the increment of the world population over the characteristic time τ cannot be larger than the population itself:

$$\frac{N}{\tau} \geq \left(\frac{dN}{dT}\right)_{\max}\Bigg|_{T \to T_1} \tag{6.15}$$

Equations 6.14 and 6.15 indicate that the passage to the limit (eq 6.11) is impossible, since the continuity of the number of people is limited by integers (by one person in the limit), and time is limited by the characteristic time interval, the duration τ of human life, having the dimension of [time/ind].

In order to overcome these limitations and mathematical features of the self-growth development, Sergei Kapitsa proposes to divide the entire period of the development of mankind into three epochs (which he called epochs A, B, and C, respectively). During epoch B, the change in the world population is described by eq 6.13 (the numerical values of its parameters are somewhat different from what suggested by Horner). Differentiating eq 6.13 with respect to time T, we find that the rate of global population growth in this period will be described by the relation

$$\frac{dN}{dT} = \frac{\mathrm{C}}{(T_1 - T)^2} = \frac{N^2}{\mathrm{C}} \tag{6.16}$$

In order to satisfy the required condition (eq 6.14) during epoch A, the formula of the growth rate of the world population (eq 6.16) should be corrected:

$$\frac{dN}{dT} = \frac{N^2}{\mathrm{C}} + \frac{1}{\tau} \tag{6.17}$$

Similarly, with regard to eq 6.12, the growth rate for epoch C is corrected:

$$\frac{dN}{dT} = \frac{\mathrm{C}}{(T_1 - T)^2 + \tau^2} \tag{6.18}$$

Therefore, after the introduction of the characteristic time of human life as a microscopic phenomenological parameter into the model, it is possible not only to determine the applicability limits of eq 6.11 but also to extrapolate the solution to both the past, covering the history of mankind from T_0 till T_1, and the foreseeable future. Epochs A, B, and C sequentially follow each other, and at the boundaries of these epochs, the corresponding solutions of eqs 6.16–6.19 (first, the solutions of eqs 6.16 and 6.17 at the border between epochs A and B, then, the solutions of eqs 6.17 and 6.18 at the boundary between B and C) must be joined. The values of the constants determining the solution are obtained on the basis of a comparison of modern demographic data with the calculations by the formula obtained by integrating eq 6.18:

$$N = \frac{\mathrm{C}}{\tau}\operatorname{arcctg}\left(\frac{T_1 - T}{\tau}\right) \tag{6.19}$$

Equation 6.19 describes the demographic transition and the future of mankind (epoch C). For the constants in eqs 6.16–6.18, Kapitsa (1999) calculated the following values by least squares:

$$\mathrm{C} = (185 \pm 1) \times 10^9,\ T_1 = 2005 \pm 1,\ \tau = 45 \pm 1,$$
$$\mathrm{K} = \sqrt{\mathrm{C}/\tau}\ 64100 \pm 1000, \tag{6.20}$$

where K is a dimensionless growth constant, which plays an important role in this model. In fact, all quantitative estimates resulting from this model can be expressed in terms of two quantities, namely, the constant K of growth and the characteristic time τ of human life.

In principle, one could introduce two values for τ, one for the initial period and another for the demographic transition epoch. But, as the author of the model believes, the estimate of the origin of development $T_0 = 5.4$ million years ago obtained with the value of characteristic time scale $\tau =$ 45 years, satisfactorily meets paleoanthropological data. Therefore, there is no point in introducing any second value for the phenomenological constant and we can limit ourselves to one value for τ.

Thus, the whole history of human development and the world population growth, according to the model of Kapitsa, can be represented as follows (Fig. 6.9): the earliest and longest epoch of linear growth (epoch A) started at time $T_0 = T_1 - \pi \mathrm{K}\tau/2 = 4.5$ million years ago, its duration was $\Delta T_A = \mathrm{K}\tau = 2.9$ million years and the world population was $N_{A,B} = \mathrm{K}\,\mathrm{tg}1 = 10^5$ individuals by its completion. The next epoch B of hyperbolic growth lasted $\Delta T_B = (\pi/2 - 1)\mathrm{K}\tau = 1.6$ million years and ended $\tau = 45$ years before the critical date $T_1 = 2005$ in 1960, when the world population was $N_{BC} = \pi K^2/4 = 3.22$ billion. The demographic transition takes $2\tau = 90$ years and will be over, respectively, in $T_1 + \tau = 2050$. From the demographic transition, epoch C (transition to a stable state) begins. As can be seen from the model, the population number in such a stable state is $N_\infty = \pi K^2 = 13$ billion.

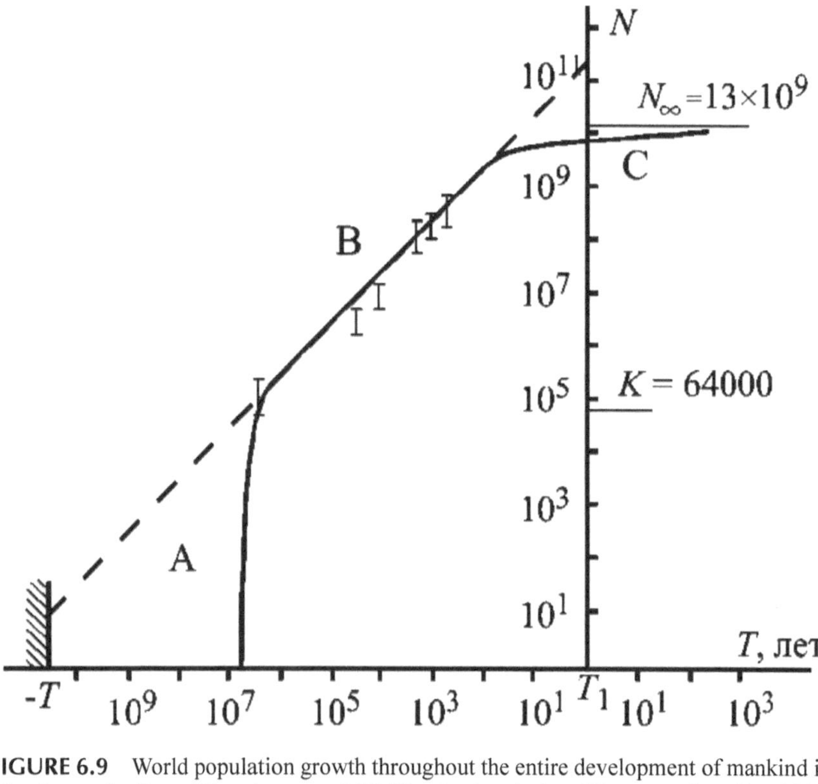

FIGURE 6.9 World population growth throughout the entire development of mankind in a log-log scale. Dotted line shows eq 6.15. The abscissa is time T in years since $T_1 = 2005$, the vertical axis is the number N of people from one person (Kapita, 1999).

It should be noted that the evaluation of the stable population as $N_\infty = \pi K^2 = 13$ billion obtained in the framework of the described model is in good agreement with the UN prediction, as well as with another model of the world population dynamics, based on the nonlinear diffusion equation and logistic equation (Koronovsky et al., *2000*).

6.4 GROWTH DYNAMICS OF WORLD URBANIZATION

Interestingly, the global growth dynamics of the population living in urban areas (Korotayev et al., 2010) correlates with the *blow-up regimes*. Korotayev et al. (2007) proposed a simple model describing the global dynamics of urbanization, based on a modification of the logistic equation:

$$\frac{du}{dt} = \mathrm{k}S(u)u(u_{max} - u), \tag{6.21}$$

where u is the proportion of the urban population, $S(u)$ the "excess" product produced at the given level of the technological development of the society per person, k a constant, u_{max} the maximum possible proportion of the urban population. As shown by Korotayev et al. (2010), this model predicts that the blow-up regimes are observed in the epoch of rapid technological development, when the hyperbolic growth of world urbanization is accompanied by a quadratic hyperbolic growth of the urban population in the world. This is supported by well-known statistical data and empirical estimates.

Figure 6.10 shows the dynamics of the world urban population (millions) for cities and towns (with populations of at least 10,000 people) for the period from 5000 BC till 1990 AD. The dots in the figure correspond to the empirical estimates by Modelski (2003), Grubler (2006), and UN population division (2010). The solid line corresponds to the empirical relationship

$$u(t) = \frac{\mathrm{C}}{(T - t)^2}, \tag{6.22}$$

where the constants C = 7,705,000 and $T = 2047$ have been determined by least squares.

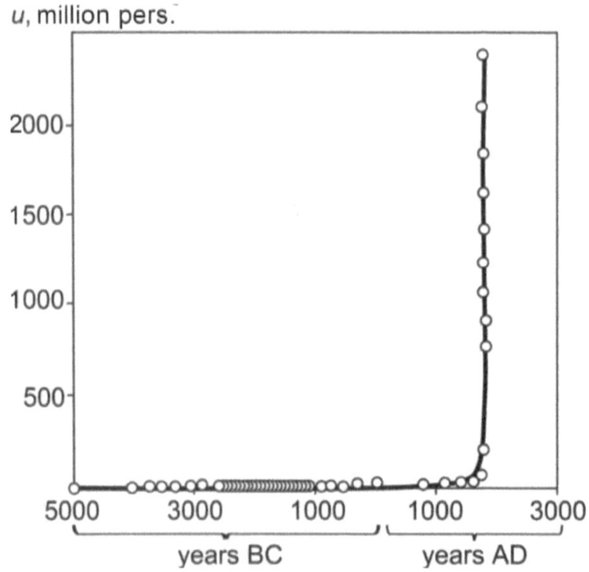

FIGURE 6.10 Urban population dynamics for cities and towns (with populations of at least 10,000 people) in the period from 5000 BC till 1990 AD (Korotayev et al., 2010).

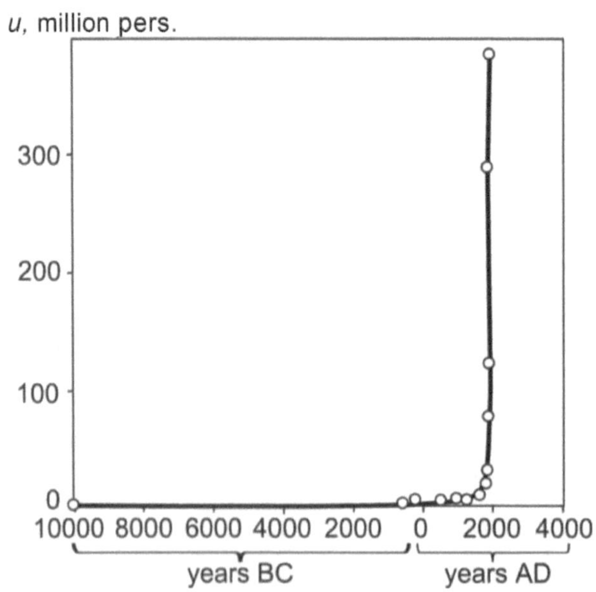

FIGURE 6.11 Urban population dynamics of large cities (with populations of at least 250,000 people) for the period from 10,000 BC till 1960 AD (Korotayev et al., 2010).

A similar behavior is demonstrated by big cities as well (with populations of over 250,000 people). Figure 6.11 shows the corresponding known statistical demographic data (White et al, 2007; Chandler, 1987; UN population division, 2006) and the approximation of these data by eq 6.22. The parameters C = 403 and T = 1,990 were also determined by least squares.

If we consider the population growth dynamics of the largest city in every epoch in the world from 10,000 BC till 1950 AD, we will again arrive at the dependence showing a quadratic hyperbolic growth—eq 6.22 (Fig. 6.12), where C = 104,020,000 and T = 2040. The statistics are also taken from Modelski (2003), known statistics (UN population division, 2006) were also used.

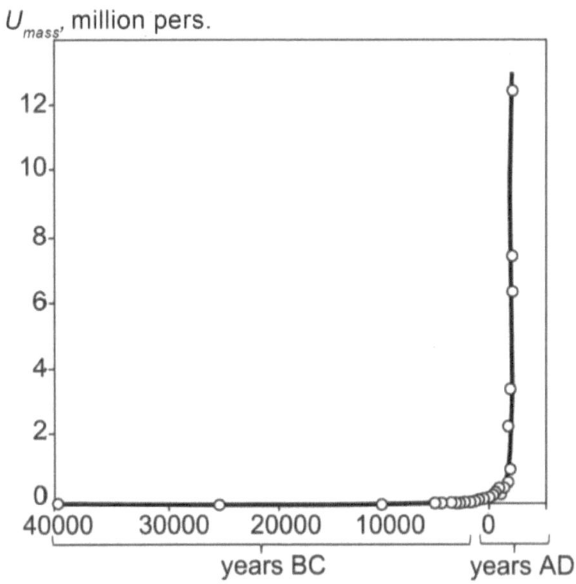

FIGURE 6.12 Growth dynamics of the largest (at any one time) city in the globe during the period from 10,000 BC till 1950 AS (Korotaev et al., 2010).

Importantly, studies of urbanization are directly relevant to understanding the changes and growth of the social development of mankind. For example, as has been shown by anthropologists (Naroll et al., 1976; Levinson et al., 1980), the size of the largest settlement in the pre-agrarian, agrarian, and early industrial societies is an important indicator of the general level of social and cultural complexity of the corresponding system. The latter allows assuming that this important indicator of human development increases by a quadratic hyperbolic law in the epoch of development in the blow-up regime.

In conclusion, we emphasize that there are different methods and approaches to modeling the global process of population growth in today's demographics, among which some interdisciplinary methods developed in nonlinear dynamics come into play. It is well known that demography is a science with its effective well-established traditional methods; at the same time, analysis of global demographic systems with new nonlinear methods developed in the science of oscillations and waves, synergetics, the theory of dynamic systems, is very efficient, allowing to detect a number of new aspects and regularities in the global population dynamics of separate regions and countries, as well as throughout the globe as a whole. As noted by Sergei Kapitsa in his monograph (Kapitsa, 1999), the methods of nonlinear dynamics and synergetics—the science of the complex—are continually expanding the scope of their application, providing new capabilities for the study of phenomena, which, because of their complexity, require a variety of new techniques. This interdisciplinary way will ultimately involve new concepts and ideas into the orbit of the humanities, especially history, sociology, psychology, demography, and so forth promote integration of the sciences called the exact and natural ones, with the sciences of human and society.

6.5 MIGRATION PATTERNS

Considering the design of models to describe the dynamics of demographic systems, we cannot but dwell on the related issues of accounting and modeling migration. The history of analysis of migration processes and modeling migration also begins with the 20th century, when the English scientist Ernst Georg Ravenstein formulated seven migration laws on the basis of migration statistics in European countries and the United States (Ravenstein's Laws of Migration) (Ravenstein, 1885; Lee, 1966):

1. Most migrants only proceed a short distance, and toward centers of absorption.
2. As migrants move toward absorption centers, they leave "gaps" that are filled up by migrants from more remote districts, creating

migration flows that reach to "the most remote corner of the kingdom."

3. The process of dispersion is inverse to that of absorption.
4. Each main current of migration produces a compensating counter-current.
5. Migrants proceeding long distances generally go by preference to one of the great centers of commerce or industry.
6. The natives of towns are less migratory than those of the rural parts of the country.
7. Females are more migratory than males.

The most important generalization by Ravenstein is associated with his conclusion that economic factors are decisive in the decision-making process of migration (Ravenstein, 1885; Lee, 1966): "*Bad or oppressive laws, heavy taxation, an unattractive climate, uncongenial social surroundings, and even compulsion (slave trade, transportation), all have produced and are still producing currents of migration, but none of these currents can compare in volume with that which arises from the desire inherent in most men to 'better' themselves in material respects.*" Most of Ravenstein's laws have not only been preserved but even increased their value to the present time. These are the regularities formulated by him which underlie many modern concepts and models describing migration flows.

Consider some most important migration models. Traditionally, two types of models are resolved in the theory of migration, namely, macrolevel and microlevel models. Macrolevel models reflect the patterns of migration movements of the entire population or large aggregate sociodemographic groups (e.g., the elderly, young couples, etc.) as part of some territorial system and are based on census or current accounts. Indicators of migration processes are forecasted with the help of these models (e.g., a forecast of the migration of people from one area to another), and possible scenarios for the development of migration processes are analyzed. Microlevel models consider the migratory behavior of an individual (a family, a household). It is obvious that no specific particular individual is meant, selected in any way from a given set, but the representative individual, subordinate to the general laws of migration behavior, characteristic of the given social group or territory. These models are important for understanding the social and psychological aspects of migration, and analysis of an individual's decision-making processes on the fact that and direction of migration is carried out on their basis.

Let us first dwell on some dynamic macrolevel migration models, used to predict migration processes.

Interactional model or models of spatial interaction historically appeared first in the theory of migration. They are based on Ravenstein's empirically established "migration law of gravity," according to which the majority of migrants move only a short distance, and such migration flows tend to big cities.

In the most general form, this type of model can be represented by the equation

$$M_{ij} = \mathrm{k}\frac{f\left(R_i, A_j\right)}{\varphi\left(D_{ij}\right)}, \tag{6.23}$$

where M_{ij} is the interaction parameter (the number of migrants per unit time from the area i to the area j), $f(R_i, A_j)$ is a function of the factors R_i (the parameter characterizing the factors of "pushing" from the area i) and R_i (the parameter characterizing the "pull" factors to the area j), $\varphi(D_{ij})$ is a function of the distance D_{ij} between the areas i and j, and k is a coefficient of proportionality. It is generally believed that $f(R_i, A_j) = R_iA_j$ and $R_i = P_i$, $A_j = P_j$, where P_i and P_j are the populations of areas i and j. Equation 6.23 can thus be rewritten as

$$M_{ij} = \mathrm{k}\frac{P_iP_j}{f\left(D_{ij}\right)} \tag{6.24}$$

If we consider a territorial system consisting of n elements, the total flow of migration from i will be $M_i = \sum_{j=1}^{n} M_{ij}$, and the total migration flows to j, respectively, will be $M_j = \sum_{i=1}^{n} M_{ij}$.

One of the main problems associated with the use of interactional patterns of migration is the need for a sufficiently large level of population aggregation. For example, the model designed can well reflect the migration patterns for some region as a whole, whereas no adequate interactional model can be built for its separate areas. An important condition for the design of an interactional model is the comparability of the interacting territorial units. Depending on the features of the modeling of interaction of a territorial system, gravitational models, models of "intervening opportunities," and models based on Alonso's theory of spatial mobility are counted among the interactional models.

The so-called gravity models of migration are most numerous among the interactional models. These models consider the territorial mobility of the population by analogy with Newton's law of gravitation as the interaction of two spatially separated elements (territories). One of the most widely used types of gravity models was proposed in the early 1940s by the American sociologist George Kingsley Zipf (Zipf, 1946):

$$M_{ij} = \mathrm{k}\frac{P_i P_j}{D_{ij}} \quad (6.25)$$

Analyzing the intercity migration in the United States in the 1930s with his model, Zipf showed that it could be used to analyze various types of movement of individuals between settlements (Zipf, 1946).

This model was later complicated by introduction of some correction factors ("specific codes") for the population figures of the interacting regions (Dodd, 1950). Such indicators as the male-to-female population ratio, the income level per capita, the population share of working age, the educational level of population, the confessional structure of population, and so forth can act as weights (possibly several ones). In this case, the generalized gravity model eq 6.25 takes the form

$$M_{ij} = \mathrm{k}\frac{\left(\sum I_n\right) P_i^{\beta} \left(\sum J_m\right) P_j^{\gamma}}{D_{ij}^{\alpha}}, \quad (6.26)$$

where α, β, γ are the exponents of the populations of the areas i and j, $\sum I_n$ and $\sum J_m$ are the summarizing weights for the population of areas i and j.

The advantages of this type of models are, primarily, the relative simplicity of their construction and the availability of statistical information for almost any level of analysis (cross-country, interregional, interregional migration, etc.). The main disadvantage is the assumption of the symmetry of migration flows ($M_{ij} = M_{ji}$), which is almost never seen in reality. The definition of the "distance" indicator also seems ambiguous. The physical distance used in the gravity models, measured in units of length or time (taking the time to move as a measure of distance) can be perceived in different ways by different migrants. This is a bright example of the human mind's nonlinearity. For example, G. Olsson, analyzing the internal migration in Sweden, noted that migrants from various regions of the country differently perceive physical distance. According to him, a 300 km distance for a resident of the vast but sparsely populated northern part of the country corresponds to approximately 100 km in the perception of a resident of southern Sweden (Jagielski, 1980). Therefore, it is advisable to use a modified measure of the distance between regions.

With this in mind, somewhat more complex models were proposed, which take into account no direct dependence between spatial mobility and physical distance. The model of Samuel Andrew Stouffer, called the theory of intervening opportunities, is most famous among these models. The basis of this model is the concept of intervening opportunities, so that "the number of persons going a given distance is directly proportional to the number of opportunities at that distance and inversely proportional to the number of intervening opportunities" (Stouffer, 1940). The term "intervening opportunities" is to be defined for each specific case: it can be free housing, free workplaces, shopping places, and so forth in different areas of a city far from the city center, and so forth. For instance, if free housing is considered as opportunities, all the available apartments which every migrant from one district to another district could take in other places on his way to the destination and directly at this destination, are called intermediate opportunities. As the moving distance increases, the "alternative cost" of moving at this distance also increases, due to an increasing number of "intermediate opportunities" available to our migrant. Trying to eliminate physical distance has led to the introduction of the notion of "social distance" measured by social and economic indicators.

The modified model of intermediate opportunities can be represented as

$$M_{ij} = \mathrm{k}\frac{X_M^{\,A}}{X_B^{\,B} X_C^{\,C}}, \tag{6.27}$$

where X_M is a scale indicator, defined as the sum of the total number of entrants for the period into the area *j* and having left the area *i*, X_B the number of intervening opportunities, X_C the number of "competing migrants," *A, B,* and *C* are model parameters determined empirically. The number of "competing migrants" is defined as the ratio of the number of migrants in the areas around the immigration centers under comparison with the same radius of influence (equal to the distance between the competing cities).

Stouffer's model was empirically tested by Wadycki (1975) on an example of migrations between U.S. states. As a result, he concluded that the "social distance" measured in intervening opportunities and competing migrants to identify spatial mobility is better than the physical distance, and the factor "competing migrants" also has a significant impact on limiting the migration flows. The study has found that in most cases, the model of intervening opportunities explains migration flows better than the simple gravity model.

Finally, let us dwell on a microlevel model based on the theory of spatial diffusion. Recall that in the modeling of migration at the microlevel, the object of study is the migratory behavior of individuals, families, or households, who make decisions on two major issues, namely, "whether to migrate or not?" and "where to migrate?"

The diffusion migration model was developed by the Swedish researcher Torsten Hägerstrand (Hägerstrand, 1967). It treats migration as a diffusion process, that is, the process of propagation of information and material objects around a certain primary source (Fig. 6.13).

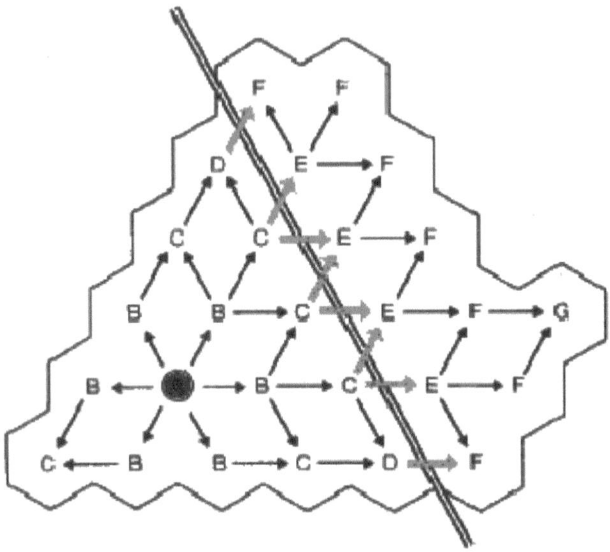

FIGURE 6.13 Hägerstrand's process of spatial diffusion: point A is the source of information, → denotes "normal" diffusion (propagation occurs in one period), ‖ a "contact" barrier, ⇨ is "slow" diffusion (propagation occurs in the two periods) (Hägerstrand, 1967)

Consider an example of the operation of this diffusion model, which has some similarity with the above model of lattice gas. Any movement can be seen as a result of two successive random events: (i) the identification from all the inhabitants of this area, one who will become a migrant (the probability of getting "information"); and (ii) the selection of the direction and distance of movement (each direction and each distance is characterized by a certain probability). Both probabilities are functions of individual characteristics, the characteristics of places, and so forth and should be estimated a priori on the basis of other studies or working hypotheses (e.g., the migration movement length is determined by an exponential function of the distance).

At the first stage of modeling, the population distribution map is covered by a square or hexagonal grid on a layout, which corresponds to the studied area by shape, and the dimension of cell is selected in such a way that the central cell corresponds to the region of the source of information. The resulting grid is a spatial ordering of the spread of information or contacts, that is, each cell is spatially linked to the central one, and the probability of this link corresponds to the distribution described by some function (e.g., $f(x) = \mathrm{k}D_x^{-b}$, where D is the distance from the information source to the region x, k, and b are parameters of the model). This function shows that the probability of contact symmetrically decreases with the distance from the source (the central cell); the extreme cells will have the lowest probability of contact. The sum of the probabilities inscribed in all cells is equal to unity. Therefore, it is possible to obtain a matrix of the probabilities of direct contacts of the source with a potential migrant, whose place of residence coincides with one of the cells (Table 6.1). This matrix is called mean information field (MIF).

Then, groups of random numbers (e.g., 1–10,000) are assigned to each probability value in individual cells, and these groups are inscribed into the respective cells of the layout (e.g., cell C1 corresponds to the random numbers 37–40). The filled grid is superimposed on the map of the population distribution, and taking some number from a table of random numbers, one can get a place (a cell on the layout), at which a potential migrant is now present.

At the second stage of modeling, the direction and path length of the migrant are selected. If the probability distribution of contacts used at the first stage is accepted as determining the direction and distance of territorial movement, then we can use the same grid to determine the distance of migration. Otherwise, a new grid with new distributions of random numbers should be built. This new layout is centered by a randomly selected cell (i.e., by the identified migrant). Then, taking any random number (a group of numbers) from a table of random numbers, we find the destination cell, which the migrant should go to (Table 6.2). It may happen that a random number will be selected corresponding to the cell in which the potential migrant lives. This can be treated as a change of residence within the settlement or a short-distance migration.

TABLE 6.1 A Layout to Simulate Migration. Potential Migrant Placement Matrix

Matrix of probabilities					Matrix of random numbers				
0.01	0.02	0.04	0.02	0.01	1	2–3	4–7	8–9	10
0.02	0.06	0.10	0.06	0.02	11–12	13–18	19–28	29–34	35–36
0.04	0.10		0.10	0.04	37–40	41–50		51–60	61–64
0.02	0.06	0.10	0.06	0.02	65–66	67–72	73–82	83–88	89–90
0.01	0.02	0.04	0.02	0.01	91	92–93	94–97	98–99	100

TABLE 6.2 A Layout to Simulate Migration. Selection of the Migration Path

0.022	0.028	0.032	0.028	0.022	000–021	022–049	050–081	082–109	110–131
0.028	0.045	0.063	0.045	0.028	132–159	160–204	205–267	268–312	313–340
0.032	0.063	0.127	0.063	0.032	341–372	373–435	436–563	564–626	627–658
0.028	0.045	0.063	0.045	0.028	659–686	687–731	732–794	795–839	840–867
0.022	0.028	0.032	0.28	0.022	868–889	890–917	918–949	950–977	978–999

A major shortcoming of such diffusion models is that the exploration of real migration processes with their usage entails huge technical difficulties, so well-known examples are only for small populations and relatively small areas. Models of this type are also unsuitable for predicting migration flows. However, they enable to analyze quality consequences and the most preferred migration destinations for individuals (Jagielski, 1980).

The design and use of complex nonlinear dynamic models to describe and predict global migration processes are just beginning. Most currently developed migration models are aimed at the description of only some forms and aspects, are local in nature and limited in scope. Many of the designed models are applicable only to a specific area (e.g., intra-regional migration in the United States), and they cannot be automatically extended to analysis of migration processes in another territorial system, or to describing the global migration on Earth. It should also be noted that the modeling of migration processes is based on a very poor information base, which does not allow us to fully test many of the theoretically developed models or forces their simplification, consciously and to the detriment of the quality of these models, to put them into practice. Thus, the design of effective models of the global migration is now an important task and, no doubt, the active involvement of interdisciplinary methods and approaches in solving this problem will allow us to go to a new level.

KEYWORDS

- **logistic equation**
- **globalization**
- **the problem of the spread of diseases**
- **population growth**
- **phenomenological model**
- **urbanization**
- **migration**

REFERENCES

1. Plotinsky, Yu. M. *Mathematical Modeling of the Dynamics of Social Processes*; MSU Press: Moscow, 1992. [Russ].
2. Plotinsky, Yu. M. *Theoretical and Empirical Models of Social Processes*; Logos: Moscow, 1998. [Russ].
3. Koronovsky, A. A.; Trubetskov, D.I. Usage of the modified Weidlich equations for modeling of social processes. *Appl. Nonlinear Dyn.* **1996,** *4*(3), 31. [Russ].
4. Meadows, D.; Randers, J.; Meadows, D. *Limits to Growth: The 30 Year Update*; Chelsea Green Publishing Company: White River Junction, Vermont, 2002.
5. Borisov, B. A. *Demography*, 4th ed.; Nota Bene: Moscow, 2004. [Russ].
6. Kapitsa, S. P. *Paradoxes of Growth: The Laws of Human Development*; Alpina Non-fiction: Moscow, 2010. [Russ].
7. Korotayev, A. V.; Khalturina, D. A.; Bozhevol'nov, Yu. V. *The Laws of History Secular Cycle and Millennial Trends. Demography, Economics, Wars*, 3rd ed.; LKI Press: Moscow, 2010. [Russ].

8. Haken, H. *Synergetics, an Introduction: Non-equilibrium Phase Transitions and Self-Organization in Physics, Chemistry, and Biology*, 3rd ed.; Springer-Verlag: New York, 1983. [Russ].
9. Kapitsa, S. P. A phenomenological theory of world population growth. *Adv. Phys. Sci.* **1996**, *166*(1), 63. [Russ].
10. Abylgaziev, I. I.; Ivanov, A. V.; Il'in, I. V.; Sayamov, Yu. N.; Ursul, A. D.; Sheshnyov, A. S.; Yashkov, I. A. *Global Systems of Cities*; Max-Press: Moscow, 2012. [Russ].
11. Likhachova, E. A.; Timofeyev, D. A., et al. *Essays on the Geomorphology of the Urbosphere*; Media-PRESS: Moscow, 2009. [Russ].
12. Semenov, V. T.; Schtompel, N. E. *Formation of Sustainable Development of Megacities. Urbanistic Aspects*; Kharkov National Academy of Urban Economy: Kharkov, 2009.
13. Shuper, V. A. *A Synergetic Approach to the Russian Urbanization/Synergetics: The Future of the World and Russia*; URSS: Moscow, 2008. [Russ].
14. Butty, M. *Cities and Complexity: Understanding Cities with Cellular Automata, Agent-Based Models, and Fractals;* The MIT Press: Cambridge, 2007.
15. Butty, M.; Longley, P. *Fractal Cities: A Geometry of Form and Function*; Academic Press: New York, 1994.

PART 2

CHAPTER 7

THEORY AND PRACTICE OF ORGANIC PARAMAGNETICS

CONTENTS

7.1 Classification and Identification 230
7.2 Magnetism and Organic Paramagnets 256
7.3 A Brief Description of the Chemistry of Paramagnetic Substances as Antioxidants 284
7.4 Stable Radicals (Paramagnets) in the Polymerization Kinetics of Vinyl Monomers 295
7.5 Usage of Stable Radicals for Studying Biological Systems 312
Keywords 325
References 325

ABSTRACT

The theory of organic paramagnetic materials and their applications are described. Their magnetic properties are characterized. Organic paramagnets are treated as antioxidants, and their chemistry is considered. The role of stable radicals in the polymerization of vinyl monomers and in studying biological systems is emphasized.

7.1 CLASSIFICATION AND IDENTIFICATION

7.1.1 *FREE RADICALS AND ORGANIC PARAMAGNETS*

Free radicals are the chemical particles with unpaired electrons on the boundary orbitals, possessing paramagnetism and high reactivity. They can be neutral or charged (ionradicals), have one or more unpaired electrons (polyradicals), and be short lived (particles of seconds) or long lived (about several years) at 298 K, solids, liquids, or gas.

The stability of chemically pure samples of long-lived radicals is determined by the speed of disappearance of paramagnet particles in consequence of their recombination or disproportion, but without distinct determination of standard thermodynamic conditions conclusions about the stability of radicals do not make physical sense.

Chemical particles with distinct localized unpaired electrons (free valence) are related to belong to short-lived free radicals, for example,

$H_2N^{\bullet}$, $H_3C^{\bullet}$,$HO^{\bullet}$, 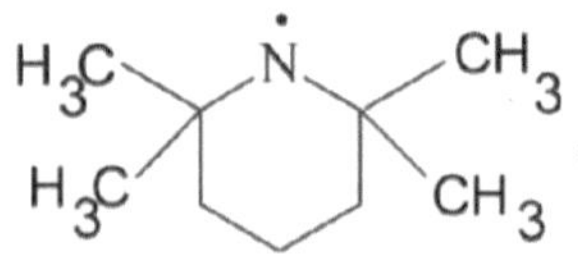 , $H_3Si^{\bullet}$, $(H_3C)_3Sn^{\bullet}$, $H_5C_6^{\bullet}$.

They are formed in the processes of photolysis, radiolysis, thermolysis, combustion, and explosion. With the help of impulse photolysis short-lived radicals are also generated, while their concentrations *in statu nascendi* can achieve 100%.

For the stabilization of radicals they often use the technology of low temperatures, applying in the capacity of cooling agents such as liquid He, H_2, Ne, N_2, and Ar with boiling points, accordingly: 4.2, 20.4, 27.3, 77.4, and 87.5 K. In vitrified inert solvents at low temperatures they are stabilized, passing into the so-called frozen state.

In different solid natural and synthetic materials, free radicals generate with the influence of ultraviolet, ultrasound, and radiation of high energies. Annealing of paramagnetic centers and other methods are used to simplify structural-kinetic studies of the paramagnetic particles conserved in the solid matrix.

Various physical and chemical methods of influence on substance are used to generate short-lived radicals; these include impulse photolysis and radiolysis, thermolysis and solvolysis, sound chemical and electrochemical methods, for example, electrolysis of carboxylic acid salts.

$$H_3C\text{-}CO_2^- \xrightarrow{-e} H_3C^{\bullet} + CO_2$$

The energy of luminous quantum which is absorbed by substance in the process of photolysis, apparently should be equal or exceed the energy of dissociation of split chemical bonds of original molecules. Thus, for example, in the synthesis of phenylmercuric bromide diphenylmercury (2 g) is irradiated by light of mercury lamp in bromoform (15 mL) for 15 h. Fallen out crystals (95%) after recrystallization from dichloroethane have the melting point of 551 K.

$$H_9(C_6H_5)_2 \xrightarrow{UF} C_6H_5^{\bullet} + C_6H_5Hg^{\bullet}$$

$$C_6H_5Hg^{\bullet} + HCBr_3 \xrightarrow{\Delta} C_6H_5HgBr + HBr_2C^{\bullet}$$

Easiness of thermic decomposition of molecules into radicals depends on the energy of dissociation of the appropriate chemical bonds (Table 7.1).

TABLE 7.1 Energy of Dissociation of Some Molecules by Free Radicals

Connection type	D (kJ mol^{-1})	Connection type	D (kJ mol^{-1})
H—OH	502.1	H_3C—CH_3	352.3
O—O	494.5	O_2N—CH_3	167.3
H—$OC(CH_3)_3$	460.2	ON—Cl	154.8
H—CH_3	426.8	ON—OCH_3	152.3
H—OCH_3	418.4	O_2N—Cl	125.5

For the generation of free radicals, thermolysis of the following organic peroxides is used: benzoylperoxide, benzoylhydroperoxide, *t*-butylhydroperoxide, cumylhidroperoxide, *tert*-butylperoxide, and cumylperoxide. *tert*-Butylperoxide, for example, is safe enough in usage and easily generates radicals in solutions under mild temperatures (half-value period at 398 K is about 12 h).

$$(H_3C)_3C\text{-}0\text{-}0\text{-}C(CH_3)_3 \xrightarrow{\Delta} 2\ (H_3C)_3C\text{-}O^{\bullet}.$$

tert-Butoxyl radicals perform rather effectively the separation of a hydrogen atom from hydrocarbon molecules, because the HO-bond in *t*-butanol is stronger than the HC-bond in hydrocarbons (Table 7.1).

$$(H_3C)_3C\text{-}0^{\bullet} + R\text{-}H \xrightarrow{\Delta} (H_3C)_3C\text{-}OH + R^{\bullet}$$

The so-called azobisnitrils, hydrogen peroxide, acetylcyclohexylsulphonyl peroxide, tri-*tert*-butylperoxyvinylsilicon, arylnitrosoamides, and triasenes are also applied.

The classical example of radicals' generation in solutions is the Gomberg's method.

$$(C_6H_5)_3CCl + Ag \longrightarrow AgCl + (C_6H_5)_3C^{\bullet}$$

Gomberg erroneously thought that the substance he had educed out of the solution was the stable free radical. In reality, he had got only the product of recombination–molecular dimer, which in the process of solution reversibly dissociated into free radicals (for benzene $K = 2 \times 10^{-4}$ mol dm^{-3} at 293 K).

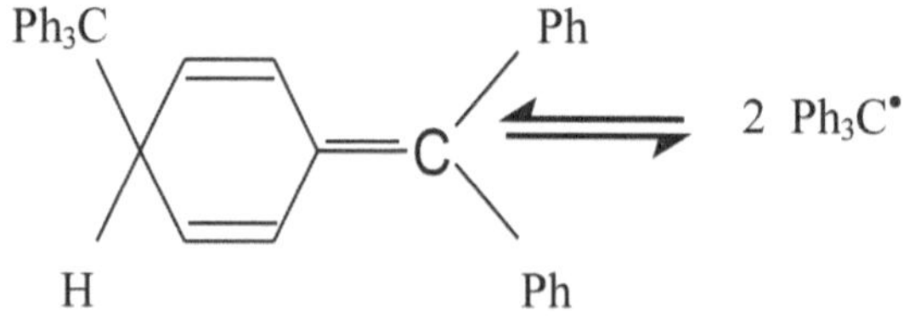

Specific and nonspecific radical solvation in solutions mostly changes electronic structure geometry, and reactivity; that is why radical solvation complexes of radicals can significantly be different in the properties from the conforming chemically pure paramagnets. Some radicals can exist exceptionally in solutions, being in thermodynamic balance with molecules. For example, in the process of solution of the Fremi salt darkviolet paramagnet solution of hydrating anionradicals $ON^\bullet(SO_3^-)_2$ are formed in water. The properties of this solution practically do not change in the air at 298 K for many hours. The stability of radical solvate complexes is determined by their chemical structure and physicochemical properties of solvents. Trityl and its numerous analogues belong to the group of organic solvated radicals existing in solutions in thermodynamic equilibrium with their dimeric molecules. For complicated conjugated systems, mesomeric and steric peculiarities of paramagnetic particles influence the balance condition in solutions between radicals and their dimers, except solvation factors.

Facts on the dissociation of triarylmethyls dimers in benzene solutions given below confirm this (Table 7.2):

TABLE 7. 2 Dissociation Degrees of Some Solvational Radicals

Solvational radical	Degree of dissociation (%)	Solvational radical	Degree of dissociation (%)
$(C_6H_5)_3C^\bullet$	2	$(CH_3)_3C-(n-C_6H_5C_6H_4)_2C^\bullet$	7 4
$n-C_6H_5C_6H_4(C_6H_5)_2^\bullet$	1 5	$(C_6H_5)_2C=CH(C_6H_5)_2C^\bullet$	8 0
$(\beta-C_{10}H_7)_3C^\bullet$	2 4	$(n-C_6H_5C_6H_4)_3C^\bullet$	1 0 0
$\alpha-C_{10}H_7(C_6H_5)_2C^\bullet$	6 0	$(C_6H_5)_3C(C_6H_5)_2C^\bullet$	100

In a number of cases for the radical generation in water, catalytic decomposition of ascorbic acid or hydrogen peroxide with Fe^{2+} salts (the Fenton's reaction) is used:

$$H_2O_2 + Fe^{2+} \longrightarrow \boldsymbol{HO^-} + Fe^{3+} + HO^\bullet$$

An example of molecular-induced radical generation is the spontaneous polymerization of chemically pure styrol.

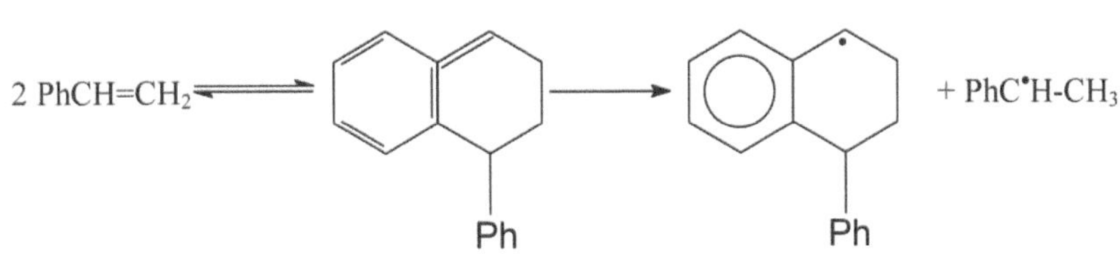

At normal thermodynamic conditions, it is impossible to get chemically pure substances, consisting of short-lived radicals and their stability and reactivity are mostly evaluated with the help of quantum chemical calculations on the basis of spectroscopic facts.

The basis for contemporary theoretical investigations of the radical is magnetic resonance spectra and quantum chemical calculations of the so-called magnetic resonance parameters. The useful information is drawn from the comparison of experimental and calculated measures of electron paramagnetic resonance (EPR) spectra. For the convenience of using the quantum chemical methods of calculation all free radicals are conditionally subdivided into two groups, disregarding their chemical nature.

In π-electronic radicals, the unpaired electron is mostly localized on $2p$ or π-orbitals and accordingly the atomic nucleus are in the nodal plane of these orbitals. Alkyl, allyl, and benzyl radicals, and also ion radicals of aromatic hydrocarbons, cyclooctatetraene, and divinyl, and similar systems belong to π-electronic ones.

It should be underlined that the problem of ion radical particle structure cannot be presented so simplified, because ions, solvated ions, and chemically pure salts of ionradicals are extremely different in their properties. Considerations in connection with the approximation convenience of real geometry of alkyl radicals to trigonal π-electronic systems do not mean, of course, the well-defined solution of the problem. There is nothing extraordinary in this, because every natural regulatory is based on observations. That is why the laws, rules, and theories are characterized by some uncertainty peculiar to them, and there are no absolutely exact scientific conclusions at all.

The so-called benzyl paradox serves as the illustration of imperfection of quantum chemical approaches to radical studying: calculations of distribution of π-electronic spin density in benzyl radical with the use of the Chukkel, McLachlan, and Parizer–Parra–Popl methods with and without regard for configurational interaction do not correspond with experiment.

The attempts to determine the constants on examples of other π-electronic free radicals do not most often give satisfactory results.

It would be relevant here to recall Leonardo da Vinci who was convinced that the sciences which were not born out of experiment, the basis of the knowledge are useless and full of fallacies, and if the investigators run into conflicts or conflicting facts, the solution of the problem is better to lay aside for the future.

In σ-electronic radicals, unpaired electron is preeminently localized on σ-orbital, and atom with free valency keeps electronic configuration of the original molecule. Phenyl, vinyl, and formyl radicals, and also carboxyl and pyridyl ionradicals belong to σ-electronic free radicals, accordingly.

$H_5\dot{C}_6$, $H_2C\dot{C}H$, $H\dot{C}O$, $C\dot{O}_2^-$ $C_5H_5\dot{N}$ (planar configuration)

Numerous chemical particles with pyramidal paramagnet center belong to the same group of radicals, for example:

$F_3\dot{C}$, $H_3\dot{S}i$, $H_3\dot{G}e$, $H_3\dot{S}n$ (nonplanar configuration)

Calculations of magnetic resonance parameters of σ-electronic radicals stimulated the development and perfection of semi-empirical quantum chemical investigating methods of short-lived paramagnet particles. Nevertheless, the application of semi-empirical methods for the evaluation of constants of isotropic and main meanings of tensor of anisotropic ultrafine interaction (UFI) even in the case of simple σ-electronic radicals often leads to conflicting results.

Apparently, when using semi-empirical calculation methods in valence approximation, the necessity in this kind of speculative classification of free radicals is no longer relevant.

It should be stressed that the impossibility of predicting the substance properties proceeding from the structure of its separate chemical particles, for example, radicals. Besides, in preparative practice substance is mostly important, but not its separate particles.

Paramagnet substances belong to long-lived free radicals.[1] The chemical particles of these substances possess strongly delocalized unpaired electrons and for the greater part steric screened reactions centers.

Chemically pure samples of long-lived radicals of the triarylmethyl row are stable in air at 298 K, brightly colored crystal or amorphous bodies containing unpaired electrons close to 6×10^{23} spin mol^{-1}, for example, 2,2,2,6,6,6-hexamethoxytriphenylmethyl (melting point 409 K) and the others. Brightgreen 1,3-bis(diphenylene)-2-(*n*-isopropylphenyl) allyl (melting point 462 K) and the so-called inert radicals: and other – high-melting substances of orange-red color possess a unique thermo-and chemical stability

$$(C_6Cl_5)_2\dot{C}Cl,\ (C_6Cl_5)_3\dot{C},\ (C_6Cl_5)_2\dot{C}C_6Cl_4OH$$

The aroxyl radicals, being stable in air, are paramagnet intermediates of reactions of single electronic excitation of the according phenol derivatives. Only single examples of chemically pure substances of this group are described in literature, for example, galvinoxyl (melting point 426 K) and indophenoxyl (melting point 409 K)

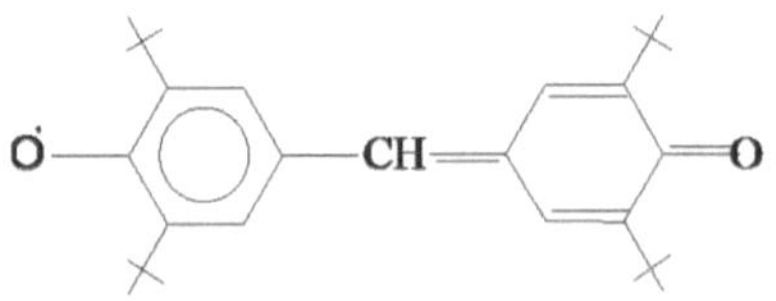

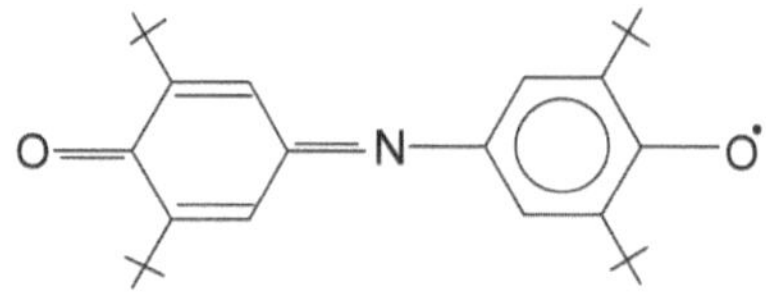

where + is *tert*-butyl group.

The most part of solvated and chemically pure aroxyl radicals only conditionally can be called longlived, because they react with oxygen quickly and that is why operations with such substances are conducted in artificial inert atmosphere or using vacuum techniques.

Stable in the air at 298 K chemically pure paramagnets of azil line for the greater part are obtained with the help of oxidation of the appropriate secondary amines. Thus, for example, crystal darkblue 1,3,6,8-tetra-*tert*-butyl-9-carbazyl (melting point 418 K), 2,4,8,10-tetra-*tert*-butyl-6,6-diphenyldibenzdioxazastannocyl (melting point 489 K) and cherry-red substituted phenazyl radicals are synthesized as follows:

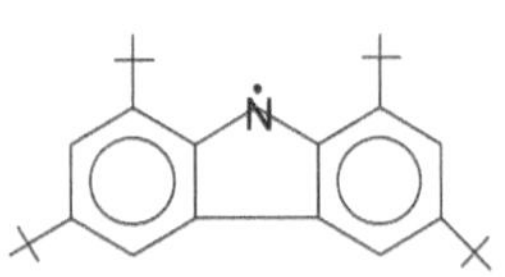

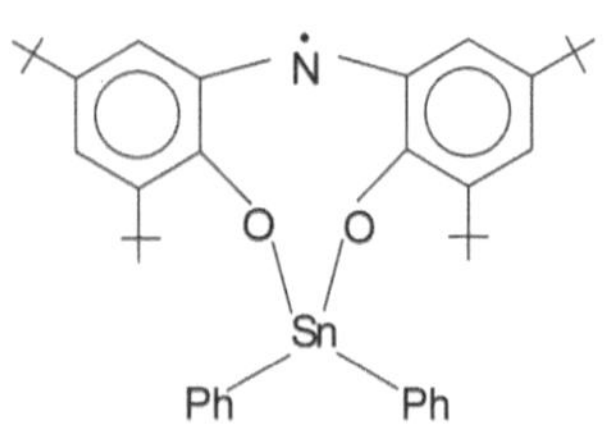

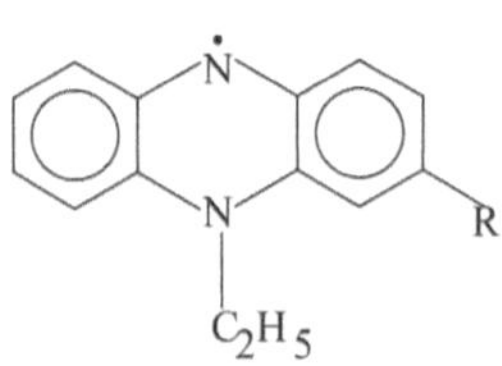

where R=CN, SO_3H.

Long-lived hydrazyl radicals represent intensively colored crystalline solids stable in the air. The darkviolet 1,1-diphenyl-2-picrylhydrazyl (melting point 411 K) is a typical representative of this class of substances.

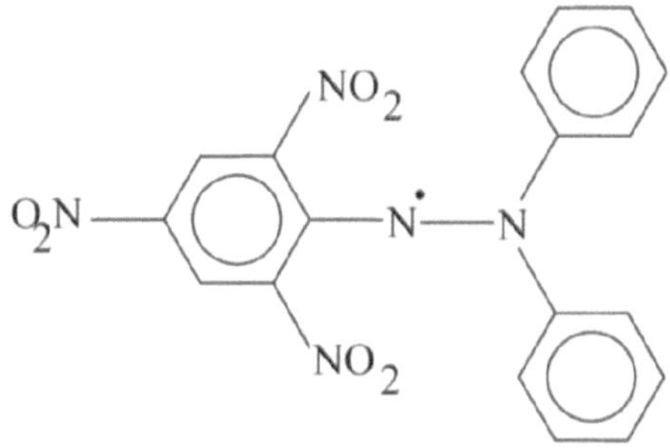

The expressed tendency to special and non-special solvation of this radicals group makes definite difficulties in the process of receiving chemically pure samples of these substances with stable physicochemical characteristics.

Verdazyl radicals (tetrazyls) belong to the number of the most stable organic paramagnets. The half-lives of some chemically pure samples in the air at 298 K are built up for years. The characteristic representative of radicals of this group of substances is 1,3,5-triphenylverdazyl (dark green needles with mp416 K)

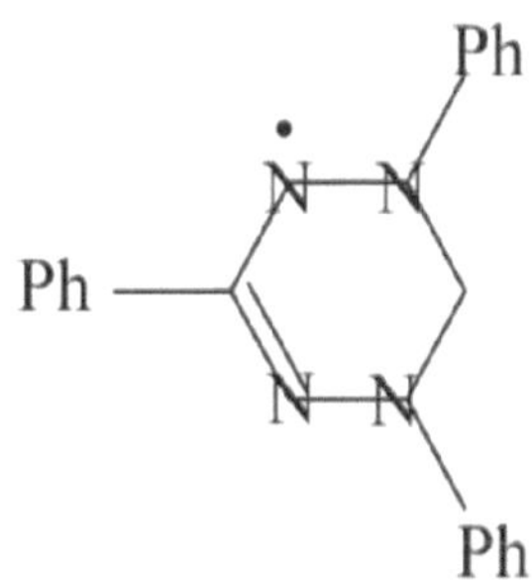

Classical nitroxyl radicals with strongly delocalized unpaired electrons[2] are traditionally called nitric oxide or nitroxide because of their formal similarity with the trivial oxidesof tertiary amines

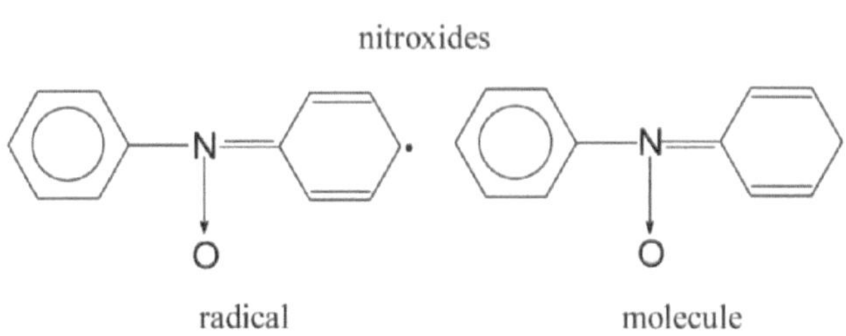

The first stable organic radical of this class called "perphyrexyd" is brick-red crystals, stable in air and soluble in water developing ruby-red solutions. The general condition of this radical because of the strong spin-charged delocalization can be presented by the superposition of several valence tautomeric structures.

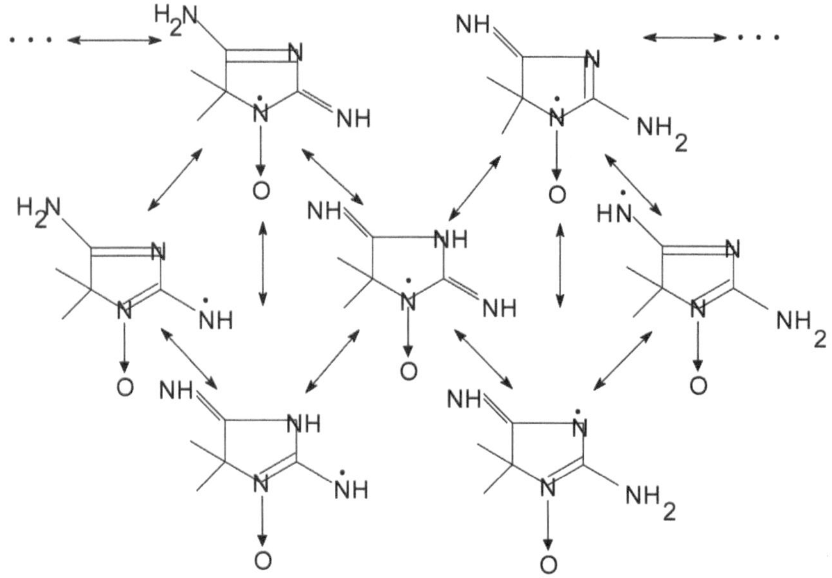

The evident delocalization of the unpaired electron (thermodynamic factor) and steric screening of reaction centers (kinetic factor) explain the stability of many nitroxyl radical of aromatic, aliphatic–aromatic, and heterocyclic row/line, for example, 4,4-dimethoxydiphenylnitroxide (melting point 434 K), 2,6-dimethoxyphenyl-*tert*-butylnitroxide (melting point 375 K), and imidazole nitroxide nitroxyl radical of Ulman (mp373 K)

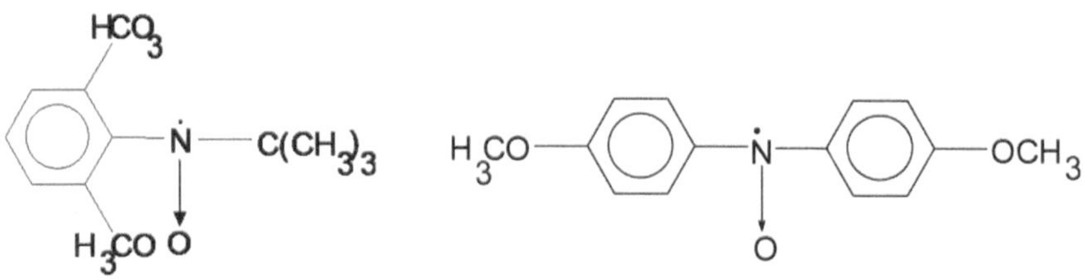

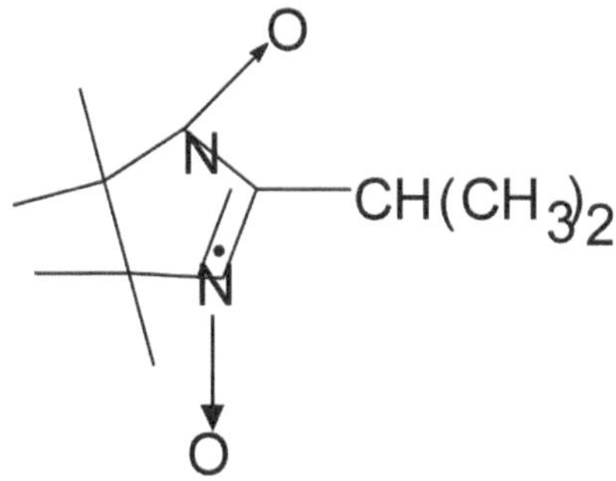

Principally new type of nitroxyl paramagnets are the so-called imminoxyl radicals, discovered in the USSR.[3–5] In spite of the presence of a strongly localized paramagnet center, imminoxyl radicals are stable in air and are easily obtained in chemically pure state, for example, darkred 2,2,6,6-tetramethylpiperidine-4-oxyl (melting point 311 K), brightorange 2,2,6,6-tetramethyl-4-oxylperdine-1-oxyl (melting point 309 K), yellow 2,2,5,5-tetramethyl-3-carboxypyrroline (melting point 484 K), and nitroxydnitroxyl derivatives of imidazolin, accordingly.[6]

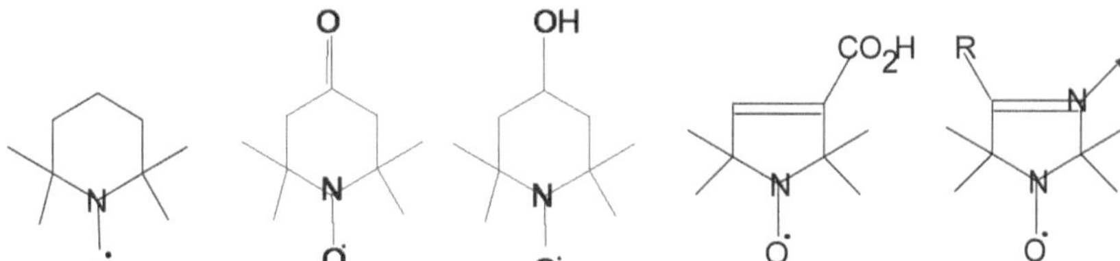

The synthesis of chemically pure radicals is carried out with different methods of preparative organic chemistry, including single electronic reduction–oxidation reactions without touching upon paramagnet center. Thus, for example, spin-marked analytic reagents, high-molecular multiradicals, and biologically active paramagnets are obtained.

Radicals identification is produced with the help of magnometry colometry, volumetry, infrared (IR), and the Raman spectroscopy, mass spectrometry, chromatomass spectrometry, and chromatographic and kinetic methods. In some cases, short-lived radicals are transferred into stable connections (spin clamp method), which later are analyzed by appropriate physicochemical methods. Specific methods of qualitative and quantitative radical analyses are mirrors method, magnetic conversion of *ortho-para*-hydrogen, flash photolysis, microwave spectroscopy, and especially methods of EPR and NMR (nuclear magnetic resonance).

Taking into consideration facts and observations, gathered for the last 30 years of formation and development of radical chemistry, we can come to a conclusion that it seems to be the most appropriate: there are no methodical limits for the registration of any paramagnet particles existing in gas, liquids, or solids. But not nearly all free radicals can be isolated in the form of chemically pure condensed phase; stable phase is enough in the normal thermodynamic conditions. In other words, at present it is easy to reregistrate any free radical, but it is not always possible to obtain chemically pure paramagnet in the form of stable enough product. For example, stable in solution-free radical 2,2,6-trimethyl-4-oxopiperidine-1-oxyl in separation out is transferred into diamagnetic mixture, consisting of heterocyclic nitroxide and the appropriate to it hydroxylamine.

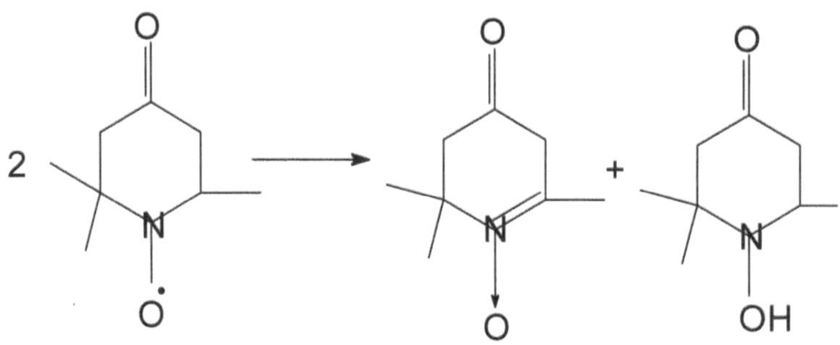

Reactivity of radicals is determined chiefly by the presence of the free valency in particles, due to which they can undergo reaction of radical substitution, addition, decomposition, isomerization, and disproportionation; every time the unpaired electron takes directly part in the forming of new chemical bonds, for example:

$$Cl^{\cdot} + CH_3CH_2CH_3 \longrightarrow HCl + CH_3C^{\cdot}HCH_3,$$
$$Br^{\cdot} + CH_2{=}CH_2 \longrightarrow C^{\cdot}H_2CH_2Br,$$
$$CH_3CH_2CH_2C^{\cdot}H_2 \longrightarrow CH_2{=}CH_2 + CH_3C^{\cdot}H_2,$$
$$C_6H_5C(CH_3)_2C^{\cdot}H_2 \longrightarrow C_6H_5CH_2C^{\cdot}(CH_3)_2,$$
$$C_6H_5C^{\cdot}H_2 + C_6H_5C^{\cdot}H_2 \longrightarrow C_6H_5CH_2CH_2C_6H_5,$$
$$CH_3C^{\cdot}H_2 + CH_3C^{\cdot}H_2 \longrightarrow CH_2{=}CH_2 + CH_3CH_3.$$

The outstanding contribution in the theory of building and reactivity of short-lived free radicals was made by the school of academician N. N.

Semenov, who discovered and proved chain reactions with degenerated propagation. For example, spontaneous oxidation of hydrocarbons in liquid phase occurs according to the following scheme:

RH + O2 $\rightarrow$	R• + HO2•	Initiation
2RH + O2 $\rightarrow$	2R• + H2O2	
R• + O2 $\rightarrow$	RO2•	Propagation
RO2 + RH $\rightarrow$	RO2H + R•	
RO2H $\rightarrow$	RO• + HO•	Branching
RO2• + RO2• $\rightarrow$		Chain termination
R• + R• $\rightarrow$		
R• + RO2• $\rightarrow$		

Reactions of new type radicals were discovered and investigated by one of the authors in the beginning of 60s. During such reactions, free valency of radicals does not participate in the forming of new covalent bonds (non-radical reactions of radicals), for example,

$$HOC_9H_{17}NO^{\cdot} + RCOCl \xrightarrow[-\,HCl]{B:} RCO_2C_9H_{17}NO^{\cdot}$$

These unusual reactions have become the basis for the formation of modern synthetic chemistry of free radicals, which has got later the widespread development in this country and abroad.

At more than one-third of all world chemical products are produced at present using free radicals, including ethylbromide, acids, carboxyl (ic) compounds, phenol, carpolactam, polyvinylchloride, synthetic rubber, different kinds of motor fuel, explosives, and nitric fertilizers. Thermal cracking of hydrocarbons is also a free radical chain process, during which numerous reactions of free radicals are released. Radicals play an important role in reactions of oxidizing polymerization during drying of lacquers and paints, and also in the oxidation and light-oxidizing processes of destruction and radiation modification of polymeric materials.

Radicals (mainly imminoxyl ones) are used as effective inhibitors in polymerization reactions, thermo- and lightoxidation of different organic materials, for example, for the stability increase of acrylonitrile, vinylacetate, vinylidene chloride, styrol, oligoesteracrylatess, furfural, synthetic rubber (SR), natural rubber (NR), fats, oils, and carotene-containing feeds. Long-lived radicals are applied for the intensification of chemical processes, for the increase of selectivity of catalytic systems and improving the product quality in the manufacture of anaerobic hermetics, epoxy resins, polyolefins, and methacrylic acid. Stable paramagnets are used in biophysical and molecular biological research on their capacity as spin marks and sounds, in forensic diagnostics, analytical chemistry, for the increase of adhesion of polymeric coatings, in the process of making film photomaterials, in instrument making, in oil field geophysics, and defectoscopy of solids.

Radical formation in living systems is stimulated by radiolysis, photolysis, actions of oxidants, molecule homolysis of xenobiotics, and ions of variable valency metals. Getting older human organism is accumulating ageing pigments (lipofuscins and ceroids) as a result of the oxidation of membrane lipids and lipoproteids with the help of free radicals. The affection of lungs, organs of vision acceleration of age hardening, infarct myocardium, neoplasms and the number of other pathologies, which are induced by photochemical smog, exhausts, and ozone, are caused by initiation of free radical reactions in different organs and tissues. The pollution of the air with different waste is accompanied by the increase of radical contents, stabilized with sorption in dust particles.

Controlled by ferments, the intermediate formation of paramagnet states also takes place in the processes of normal vital activity, for example, in biosynthesis of prostaglandins, in transporting of electrons in mitophagocytizing in neutralizing bacteria by phagocytic cells, in microsomal hydroxylation of xenobiotics, and so forth.

Organic materials undergo oxidation transformation in air: the caloric value of liquid and solid fuels is reduced, the characteristics of lubricating oils and hydraulic liquids get worse, the strength of products made of polymers and plastics deteriorate, and pharmaceuticals, cosmetics, explosives, fat-containing feeds, and foodstuffs get out of order. For example, when butter gets rancid, toxic products of oxidizing reactions are accumulated in it: spirits, carbonyl connections, peroxides, hydroperoxides, oxiranes, and others.

Radicals stabilized by solid matrix occur in meteorites, natural minerals, and caustobrolites. Mineral salts, diamonds, and other crystal substances often include paramagnet admixtures, among them free radicals giving them different color shades.

"Short-lived" radicals: H, K, Na, Fe, Ti, CH, CN, and others, whose density achieves 3×10^{-24} g cm^{-3} diffused in interstellar space, are of great importance of astronautics and astrophysics. Research of comets indicates the presence of radicals OH, CH, NH, CN, C—C—C, CO^+, NH_2, N_2^+, NC^+ in them. Judging by the specties of aurora polaris, atmosphere of the earthy atom contains H, O, O^+, O_2^+, CH, N_2^+,and CN. The maximum concentration of oxygen is fixed at the height of 105 km above the sea level (10^{13} cm^{-3}). In the atmosphere of the sun and colder stars H, CH, PH, CaH, CP, B, HSi, NH, CN, H, and other free radicals are discovered. It is assumed that the varying color of Jupiter surface is due to the presence of stabilized radicals in its atmosphere.

It is relevant to remind that the chemistry of free radicals is based on two levels or approaches to the solution of problem. Synthesizing or investigating substances, the chemist works with directly tangible objects, liquids and steam, crystals, and melts, which are possible to manipulate – weigh, study, and put into chemical reactions. On the other hand, discussing the structure and reactivity of substances consideration is carried out on the level of invisible, submicroscopical objects, from which the investigated substances are made of. The confidence of the appropriateness of such approach is based on the belief that behavior and properties of condensed phase immediately reflect behavior and properties of discrete particles of the investigated liquids or solid substances.

It is extremely important not to mix these two levels of consideration, but it happens rather often in chemical discussions. One should remember that our ideas of properties and behavior of radicals and molecules are based on the level of physical observations, and this level is the highest criteria of truth of our knowledge.

7.1.2 *IDENTIFICATION AND PURIFICATION OF PARAMAGNETS*

Individual structure peculiarities of chemically pure paramagnets are connected with their physical properties, the sum of which is a characteristic sign of a substance, its "fingerprint."

In other words, molecular architectonics of paramagnet is connected with unique set of its physical properties, which, in principle, is the substance of passport. That is why experimental exposure of interrelationship between the structure and properties of an organic paramagnet makes sense only working with the maximum purified substance.

The strength of physical methods in the establishment of the substance structure is proved by numerous examples, showing how the similarity of molecular structures appears in the likeness of physical properties of chemically pure substances. In chemist's opinion, an organic substance is pure if it is built of similar molecular blocks or in the case of paramagnets – of similar free radical particles.

Of course, in practice the ideal situation, when separately taken radical is investigated in the absence of fields, interactions and collisions, is simply not accessible. This situation is also impossible in experiments with microscopic paramagnet crystal with the number of atom nucleus 10^{23} and immense number of electrons.

Working with chemically pure substance (absolutely pure substances exist only in imagination), we consciously simplify the real picture of interrelationship of chemical particles in condensed phase, for example, in liquid or "molecular" crystal. Even in the case of gaseous paramagnets, for example, hexafluorodimethysiminoxyl

```
   F       F
   |       |
F - C - N - C - F
   |   |   |
   F  O•   F
```

The properties of this violet gas cannot be explained, basing just on the structure of separate free radical particles without taking into consideration and quantum statistics laws.

Unfortunately, quantumchemical information about the structure and properties of separate free radical particles, received during the solution of appropriate tasks, does not give an opportunity to predict thermal stability of the chemically pure substance (condensed phase of homogeneous particles). Attempts of using the factors connected with the structure of separate particles, for prediction of stability of paramagnet substances, are doomed to failure.

One of these curious cases is the example of 4,8-diaza-4,8-dioxidadamantian, which in its chemically pure condition does not possess paramagnetic properties at all and only in solution at 373 K a superthin structure of spectrum EPR occurs, which can be explained by the presence of free dioxyl biradicals 4,8-diaza-4,8-dioxidadamatan.

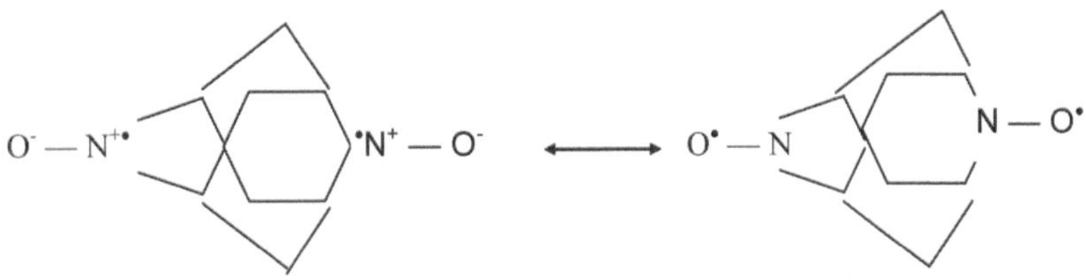

Registration of stable signal of EPR in solution does not mean, of course, that chemically pure paramagnet can be separated out of it, the particle structure of which corresponds to the structure of non-solvated free radicals. Moreover, it is not correct to affirm the getting a stable free radical, having in fact only stable enough paramagnet solution.

Thus, for example, it is extremely easy to generate in solution free radicals such as triphenylmethyl, diisopropylimminoxyl, diphenylazyl, and paramagnet anions Phremi, but it is impossible to get them in chemically pure condition. Thus, concentration in vacuum of comparatively stable in air solution of phenyl-*t*-butylnitroxyl results in receiving of diamagnetic mixture of products of radical particles disproportionation.

In general, the stability of separate paramagnet particle (in matrix or in vacuum) and paramagnet body, built of paramagnet particles, should never be equated. The first is connected with the changing the character of surface profile of potential energy of a separately taken radical, the second with analysis of the dependence of thermodynamic potentials of the concentration of radical particles.

The criterion of purity from the chemical point of view (molecular level of consideration) is chemical homogeneity of a sample. To get true facts about properties of free radicals, first of all, it is necessary to find a way of removing the admixtures from a sample. Only after good purification we can say that the measuring data and observations received during the investigation, for example, of magnetic properties, are not connected with artifacts, with admixture of outside paramagnets or ferromagnetics. A dirty substance often possesses physicochemical properties, which constitute something intermediate between properties of pure components of mixture. In the last case, the choice of the method of separating the mixture into chemically homogeneous fractions is an exceptionally important moment of purification. The effectiveness criterion of a chosen method is the control of changing the physicochemical properties of a substance during purification. During purification properties of preparation are significantly changed, reflecting the transition from mixture of substances into chemically homogeneous compound. When the following operation of purification does not result in the changing of physical properties of a material, one can consider that chemically pure substance is received and the following purification does not make any sense.

If it is assumed that two samples of a paramagnet, possessing identical physicochemical properties, should consist of identical free radical particles, this circumstance can be used for structure identification of free radicals.

At present all synthesized paramagnets are described in chemical literature by a set of physicochemical constants. The reliability of these data is guaranteed by multiple reproductions of experiments described in literature. As a result of such multiple independent checkings, wrong or adulterated results are rejected and changed for more safe. In the process of producing new paramagnets, their physicochemical properties are compared with constants of already described compounds of a similar structure. Practical example is mixture of paramagnet substances, received by oxidizing of condensation products of phorone and ammonia on which one of the authors managed to release chromatographic fraction of a crystal paramagnet with distinct melting point.

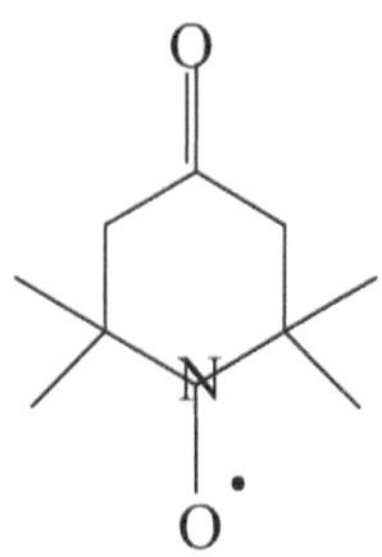

Thus, the first chemically pure stable radical able to react without touching upon paramagnetic center was received. After fixing its melting point, other physical properties of this compound were investigated, which were then being used as standard physicochemical constants.

In the considered case like in most of the others the melting point of a substance serves as one of the most important physical characteristics of solids. A chemically pure substance often possesses a higher melting point than any other mixtures of this substance, even if separately taken components of a mixture have the same melting points. On the contrary, the absence of melting point depression of mixed sample is considered as an indisputable argument of chemical identity of mixed substances. One should take into account that the melting points reflect passing of a substance from solid into liquid state in conditions of balance with vapor of a substance and air. Melting point is considered to be a point of the first appearance of liquid phase of a substance in the capillary or on the heating table under microscope is marked. Usually, for pure substances the temperature interval from the beginning up to the complete melting should not be more than 0.5–1.0°C. The results of determination can be influenced not only by admixtures in a substance but also by the purity of the inner surface of capillary or the outer plate surface board of the heating table. For example, in crude glass capillaries, melting points of pure organic preparations can fall in comparison with table values to 5–10°C because of the influence of alkaline products.

Transfer of thermal energy from the heating surface into condensed phase of the sample and dispersion of hidden heat of melting take definite time that is why too fast heating up can lead to the significant distortion of the determination results. Depending on the heating speed melting point of, for example, glycocholic acid can change between 132–178°C. Satisfactory results can be achieved, if in the process of determination of a melting point the heating speed of a sample does not exceed 1°C min^{-1}.

When keeping the determination technology extreme, melting point of the sample indicates most often the satisfactory level of its purification. Recrystallization of a substance from two different solvents is the effective technique of checking the purity of preparation. If this technique does not lead to the achievement of an extreme melting point, the substance should be subjected to chromatographing, sublimation, or molecular vacuum distillation. There are rare cases when mixtures melt as sharply as individual substances. Thus, for example, equimolecular adduct of oxotetraphenylpropanol (melting point 159°C) with tetraphenylpropenon (melting point 154°C) has sharp melting point 169°C). Something similar is observed with hydrocarbon adducts of the stable paramagnet 2,2,6,6-tetramethyl-4-oxopiperidine-1-oxyl (Table 7.3).

TABLE 7.3 Characteristics of 2,2,6,6-tetramethyl-4-oxopiperidine-1-oxyl Adducts

Included hydrocarbon	Melting point (°C)	Nitrogen content (%)
n-heptane	46.5	8.2
n-octane	54.0	7.9
n-nonane	59.5	7.8
n-decane	57.0	7.7
n-dodecane	60.0	7.7
n-hexadecane	63.0	6.7

These and other facts in the chemical literature show that preparations with highest melting points are not the purest ones.

Preparations of free radicals with high melting points are usually less stable than common organic substances and that is why they melt with decomposition. For receiving of reproducible results, a capillary with sample is recommended to be put into a heating block (a bath or a camera) already heated almost up to a melting point or decomposition of an investigated substance. For finding such a temperature one should make estimating definitions of a melting point of a sample during its fast heating. In case of sample decomposition, it first agglomerates and then frothes and carbonizes. All these phenomena should be characterized with the appropriate temperature intervals.

In some cases, working with complicated organic substances inclined to forming several crystal modifications, we may observe sequence of phase transitions, which is expressed in appearance of multiple melting points. Classical examples of triple melting are different triglycerides (trilaurin: 15.35 and 46°C; tristearin: 54.65 and 71°C).

The most precise method of a melting point definition is the analysis of a cooling curve built according to time and temperature coordinates. This method allows to fix the most reliable certain melting point, the purity level of preparation, and even admixture percentage if it is present. Data for construction of graph are received registering the melting point in the process of cooling. Most often, the passing of a liquid state into solid state is accompanied with phenomena of supercooling and freezing of a melt, which can be easily removed. According to Raoul's law, lowering of the freezing point in half period of melt freezing should be twice as much than in the beginning of determination. Knowing temperature in the beginning and in half period of melt freezing, one can evaluate the exact melting point of a substance summing up the temperature difference value with the initial temperature. The theory of the method is very simple and basically does not differ from the theory of cryoscopic evaluation of the substance molecular mass.

When determining admixture percentage simple function is usually used:

$$n = \frac{m}{100K} \Delta T,$$

where n is number of admixtures moles, m is substance weight, T is melting point depression, and K is molar depression of a freezing point in 100 g of a solvent.

If the chemical nature of admixture is known, one can estimate admixture percentage not only in molecular percent but also in weight percent. The purity level of the preparation and exactness of its evaluation does not depend on thermometer calibration, because determination is carried out on a horizontal plot of the freezing curve. The method is convenient at work with samples of organic paramagnets, because the results do not depend on the heating speed of preparation; its radical purity is registered immediately without repeated crystallizations of the product. It is quite obvious that for receiving freezing (or cooling) curves greater amounts of substances than usually working with capillaries or heating tables are needed.

7.1.3 *NOMENCLATURE OF ORGANIC PARAMAGNETS*[1,2,6]

There is a rather widespread opinion that all organic compounds can be unambiguously described with the help of formulae of the chemical structure that is why there is no need for their systematic names. However, without precise and distinct terms, it is impossible to describe and explain observed phenomena simply and clearly. "We must be grateful to God that He created the World in such a way, that all simple – is true, and all compicated is false" said P. L. Kapitsa.

A structural formula is only a conditional symbol of a rough molecular model, which is necessary but insufficient for systematization of chemical knowledge and automatized search of the information we need. Finally, fundamental chemistry makes sense in receiving and systematization of objective knowledge of a substance, and discovery of a scientist acquires real value when it becomes the common property of all the others. The condition for this is the proper use of the scientific language. If an idea is not expressed distinctly and clearly it can be understood, evaluated, and accepted only with great difficulty. Arguments lose their power if they are expressed indefinitely or formulated vaguely. When working with paramagnets, it is especially important to choose precise expressions and terms and to use chemical language and symbols correctly. The nonobservance of these recommendations leads to curious faults and causes additional problems working with literature. To illustrate the above mentioned facts, one can regard the example of a substance (dioxide diazadamantane), which was mistakenly classified by some authors as stable free biradicals,

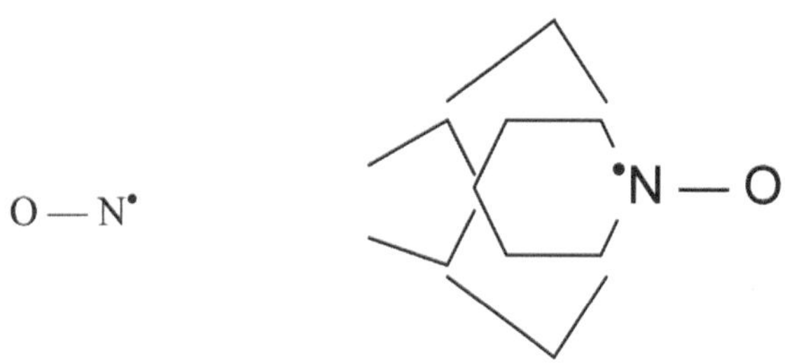

though the EPR spectrum, appropriate to the structure given above, occurs only in solutions at 100°C.

Other example of a diamagnetic substance, to which the structure of a stable paramagnet groundlessly attributed, is the reduced product of adamsite.

It is amazing that on the eve of 21st century many chemists continue to use universally in publications such routine terms as benzol and pyrrol and others, though the suffix «ol» indicates the presence of hydroxyl group in a molecule.

The use of the terms nitroxide, azotoxide, and azotoxide radical for the designation of free radicals also seems archaic, because their names should be ended in a characteristic suffix "yl," for example, hydroxyl, methyl, hydrazyl, imminoxyl, nitroxyl, phenoxyl, and so forth. In spite of the evident historically formed expedience, there are still rough nomenclature distortions in publications producing additional problems when working with chemical literature.

The example of such semantic indefiniteness in the modern system of chemical designations is iminogroup, which is generally marked as NH. The term «imin» was borrowed from the German language in the beginning of the 20th century and had firmly been implanted in the Russian systematism of saturated heterocyclic compounds. Compare, for example, methylenimine with saturated heterocyclic compounds of ethylenimine (dimethylenimine), propylenimine (trimethylenimine), and butylenimine (tetramethylenimine):

$$CH_2NH, (CH_2)_2NH, (CH_2)_3NH, (CH_2)_4NH$$

At the same time under the influence of French and Italian authors, imino group in compounds with open chain lost sense phonetic sounding and got indefinite sense meaning compare, for example, iminodiacetic acid with methylcarbylamine and dimethylamine:

$$HO_2CCH_2NHCH_2CO_2H, H_3CNC, (H_3C)_2NH.$$

After founding of the special international nomenclature committee (Paris, 1889) and summing up its work at the chemical congress (Geneva, 1892), main rules of composing names of organic compounds were formulated: the nomenclature is based on principle of hydrogen atoms substitution; characteristic radical (common to all the group) is the base for the name, substitutes are expressed by affixes, modifying the root name.

Thus, the Geneva principles have given chemists a sum of rules, according to which depending on circumstances names of any organic substances could be composed. One should note that the assumptions accepted in Geneva about composing the "official" names of organic substances, logically following from their chemical structure formulae, have many drawbacks, so they are often broken in practice. Nomenclature divisions of aromatic and heterocyclic compounds, among which later the most long lived organic paramagnets were discovered, were especially unfinished.

The Beilstein's reference book had a definite influence on the nomenclature of organic compounds, especially its IV edition (1913) in which all the substances were subdivided into three types: acyclic, carbocyclic, and heterocyclic compounds. Functional characteristic was the basis for more detailed classification. Unfortunately, but the great prestige of this reference book promoted the deviation from the strict principles of the single official classification.

The Sheltsner's reference book, rather popular in its time brought even more disorganization in the field of chemical nomenclature, but the majority of chemists were still aiming at such a terminology in which the genetic connection of the functional derivatives with the initial chemical functions automatically traced. This tendency has found its expression in the rules of organic compounds nomenclature of International Unity of Pure and Applied Chemistry (IUPAC). According to these rules the exact form of words with the help of affixes should be brought in conformity with the peculiarities of particular language by special regional committees on chemical nomenclature.

It is relevant to mention that after the legal rights of the authors for original trademarks and names of the chemical products have got to be registered and protected by law in all legal states.

Rules of organic compounds nomenclature of IUPAC recommend to use endings in the names of radicals, for example, hydroxyl, hydroperoxyl, methyl, azyl, iminoxyl, nitroxyl, thiyl, selenyl, and so forth.

Side substitutes in radicals are designated with figures with the start of counting from the atom with free valency, for example, 2-phenylethyl or 2,2,6,6-tetramethyl-4-hydroxypiperidine-1-oxyl. Positions of multiple connections are also designated with figures, if it is necessary, for example, butane-2-yl, 2,2-dimthyl-butene-2-yl, and so forth.

In contrast to free radicals, their remnants combined into a molecule can exist without characteristic ending, for example, ethylene, trimethylene, ethylenilydene, and tetramethylene. It is important that in the new nomenclature the enumeration order of affixes or remnants (bond radicals) remains arbitrary, and nomenclature ortholoxy as applied to free radicals in most cases is unjustified.

Thus, one can state that there are still no satisfactory rules of systematization and composition of organic free radicals names.

Among objective reasons of the crisis situation in the field of nomenclature of organic paramagnets one can distinguish at least two of them. First, discrediting of the so-called radical theory, the first knock on which was made by the Duma's work (1834) and then an absolute rejection of the notion of radicals as freely existing particles (Shorlemmer, 1863). After formation of a "type theory" serious chemists attitude toward free radicals became mainly ironical. Secondly, conflicts between chemistry of free radicals and dogmas of classical theory of organic compounds chemical structure, which were already outlined in the first decade of the 20th century. Butlerov's structural theory states that the radicals bound into a molecule keep their chemical structure in the reaction products during any chemical reactions, while experiments with stable radicals have indicated the possibility of structural rearrangements, for example:

$$2\ (C6H5)3C\bullet \longrightarrow (C6H5)3C{-}C6H4CH(C6H5)2$$

The structure theory implied that the only formula of a chemical structure for each organic compound, which gives the idea of its properties, but experiments with stable radicals brought to the conclusion that properties of a substance can be explained only with the help of a combination of several formulae:

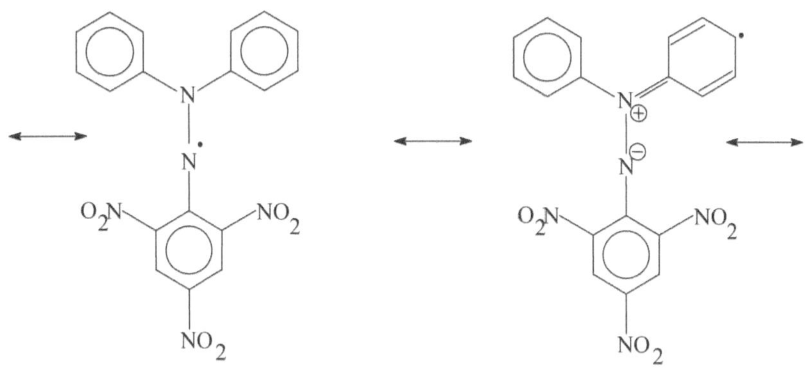

Chemical structure theory asserted that properties of compounds are strongly depended on succession of the bonds between atoms in the molecules, and free radical can easily be described with formulae with different successions of simple and multiple bonds, because electrons are delocalized in such compounds

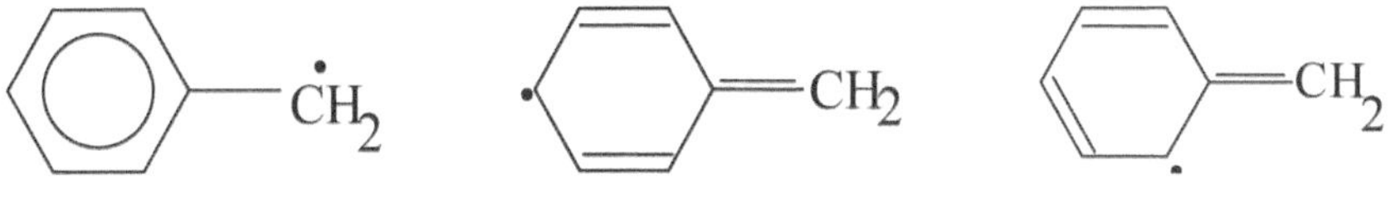

The idea of constant tetravalence of the carbon atom was the basis for classical chemical structure theory, and chemistry of free radicals showed a lot of examples of trivalent carbon compounds existence, for example, triphenylmethyl and its derivatives: $(C_6H_5)_3C^{\bullet}$; $(C_6H_5C_6H_4)_3C^{\bullet}$; $(C_6Cl_5)_3C^{\bullet}$.

It should be underlined that for all these structural formulae and postulates of classical chemical structure theory were and still are the basis for official nomenclature and systematism of organic chemistry. Progress in the chemistry of free radicals has become a powerful factor of activating of fundamental researches in the field of the structure and reactivity of organic compounds; it promoted the perfecting of experimental techniques and technology of physical instrument making, but did not influence much on the improving of the situation in the nomenclature field and systematization of free radicals.

To illustrate problem of organic paramagnet systemization pointed above we may consider the reasons provoking to subdivide the most wide class of nitroxyl radicals into paramagnetic nitroxides with strongly delocalized unpaired electron, for example:

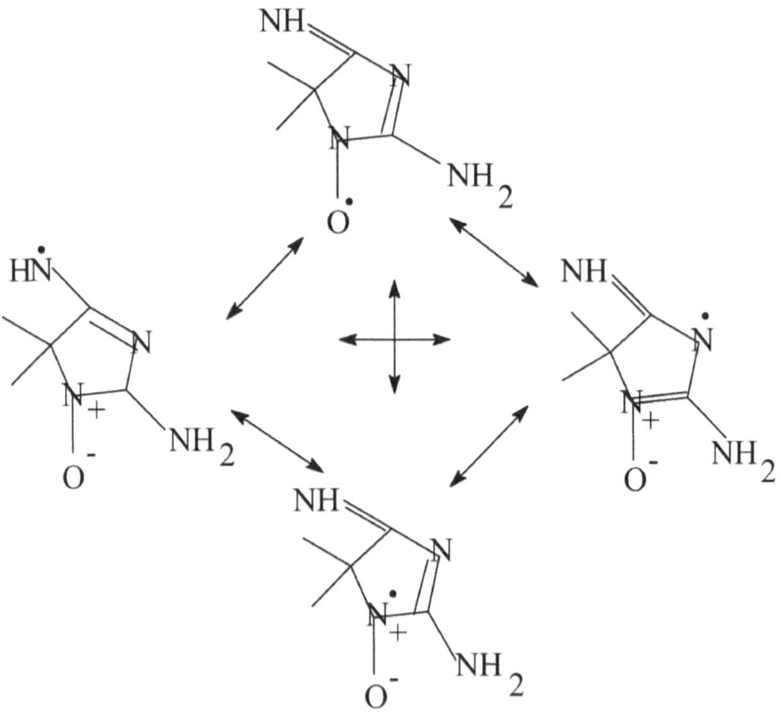

and iminoxyls with distinctly localized paramagnetic center, for example,

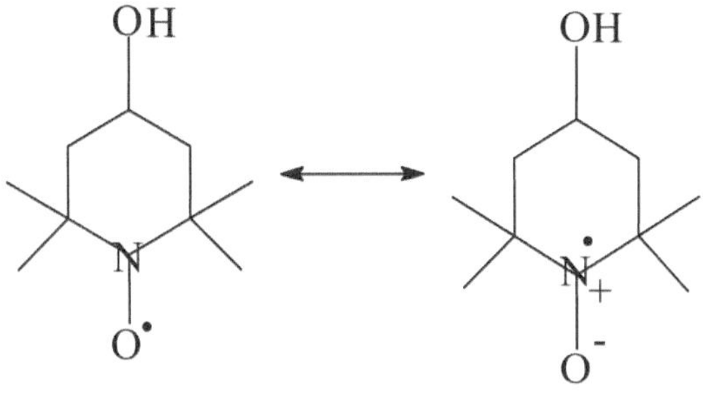

Due to the system of conjugated multiple bonds classical paramagnetic nitroxides are inclined to radical reactions with transfer of the reaction center to the attack place, for example, to the aromatic nuclear:

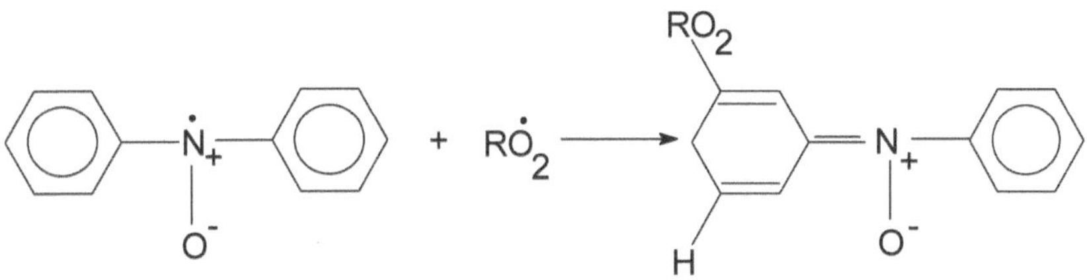

On the contrary, radical reactions of iminoxyl radicals take place exceptionally regioselectively on the localized paramagnetic center, for example:

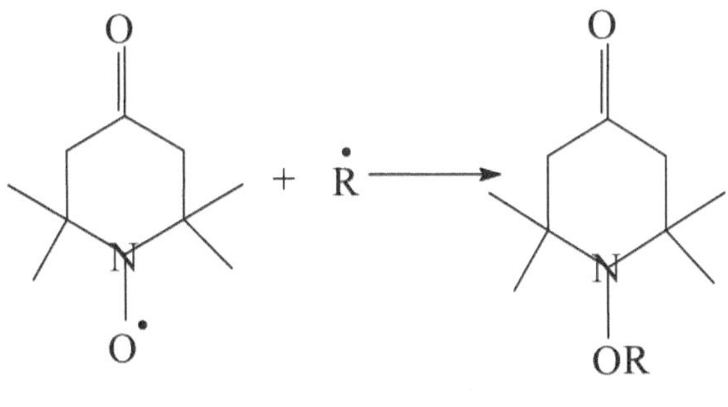

As for the transfer of radical reaction center to carbonic atoms, in the case of free iminoxyds it is absolutely forbidden because of the insignificant contribution of a particle of valence structures of nitroxyd type to the main state of a particle.

Widely spread classification of conjugated radicals, based on the formal structural sign of "key" atom is evidently not satisfactory enough, because of the delocalization of unpaired electrons cloud different in their origin radicals, for example, aroxyls, classical nitroxyls, aromatic azyls, and verdazyls, can be formally united into one and the same group of compounds of trivalent carbon.

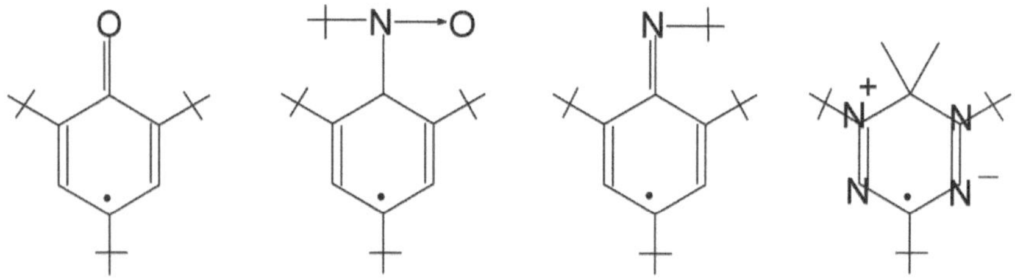

7.2 MAGNETISM AND ORGANIC PARAMAGNETS

7.2.1 GENERAL STATEMENTS

When a magnetic field H is imposed, all substances show a macroscopic magnetic moment M. The value M relates to the imposed field H with a coefficient of proportionality χ (the magnetic susceptibility):

$$M = \chi H$$

In diamagnetic substances with completely filled orbitals, the induced moment is oriented *against* the external field; their magnetic susceptibility is negative and temperatureindependent. In paramagnetic substances with half-filled orbitals, the induced moment vector under the influence of the imposed magnetic field is directed *parallel* to the latter. For noninteracting (independent) spins, the value of the magnetic moment is inversely proportional to temperature and their susceptibility can be approximated by Curie's expression:

$$\chi = C/T,$$

where C is Curie's constant, T is absolute temperature.

The value of magnetic susceptibility is usually recalculated to the effective magnetic moment μ_{eff} defined as:

$$\mu_{eff} = [(3k/N_a)\,\chi T]^{0,5} = \mu_s g\,[S\,(S+1)]^{0,5},$$

where k is Boltzmann's constant, N_a Avogadro's number, $\mu_Б$ Bohr's magneton, and S is spin.

For the case of interacting spins, numerous deviations from Curie's law are known. As a first approximation, such behavior is described by the Curie–Weiss law (Fig. 7.1):

$$\chi = C/(T-\theta)$$

Here, the "characteristic temperature" θ is determined by the crystal field and may be either positive, corresponding to ferromagnetic interactions (with parallel spin orientation), or negative, corresponding to antiferromagnetic interactions (with antiparallel spin orientation).

The interradical interactions of uncoupled electrons are classified into two types, namely: dipole–dipole ones and exchange ones; the latter is determined by the overlap of the wave functions of uncoupled electrons and quickly decreases with increasing distance. The exchange interaction averages both the dipolar interaction between uncoupled electrons and the intraradical superfine interaction of uncoupled electrons with atomic nuclei. When there are a couple of electrons on neighboring centers with pronounced overlapping wave functions, some interaction between the spins S_1 and S_2 arises. It leads to the formation of singlet and triplet states. According to Heitler-London's description of chemical bonds, this interaction is expressed by the Hamiltonian:

$$\hat{H} = -2J\hat{S}_1\hat{S}_2 \tag{7.1}$$

Extension of eq 7.1 to a multielectron system is described by Heisenberg's exchange Hamiltonian as:

$$\hat{H} = -\sum_{i,j} J_{i,j}\hat{S}_i\hat{S}_j, \tag{7.2}$$

where $J_{i,j}$ is the exchange integral between atoms i and j having total spins S_i and S_j.

The exchange integral *J* characterizes the exchange interaction degree and is expressed in energy units. Negative *J* values correspond to interactions of the antiferromagnetic type (the state of the lowest energy with the antiparallel spin orientation, the ground state being a spin singlet). A positive exchange integral is associated with the ferromagnetic interaction (the ground state is a spin triplet).[1]

In 1963 McConnell[2] formulated an idea of the possible presence of particles with high positive and negative atomic π-spin densities. In a crystal, such compounds can be packed in parallel to each other into stacks to form conditions for strong exchange interactions between the atoms with a positive spin density and the atoms with a negative spin density in the neighboring radicals. Ferromagnetic exchange interaction expressed by Heisenberg's exchange integral

$$H^{AB} = -\sum_{i,j} J_{ij}^{AB} S_i^A S_j^B = -S^A S^B \sum_{i,j} J_{ij}^{AB} \rho_i^A \rho_j^B \quad (7.3)$$

is a consequence of the incomplete compensation of the antiferromagnetic coupled spins, where S^A and S^B are the total spins of radicals A and B; ρ_i^A and ρ_j^B are the π-spin densities on atoms i and j in radicals A and B, J_{ij}^{AB} the interradical exchange integral for i and j.

Buchachenko[3] stated that "it would be almost impossible to realize this way since it is impossible to construct such a crystal lattice of the radical 'to turn-on' the intermolecular exchange interaction among the atoms with opposite spin densities only and to 'turn-out' it among the atoms with spin densities of an identical sign." McConnell's model, nevertheless, was involved to interpret complex interradical interactions found in the crystals of stable organic paramagnets. Rather recently[4], direct experimental evidence on *bis*-phenylmethylenyl-[2,2]-*p*-cyclophanes has been obtained that ferromagnetic exchange can be reached within McConnell's model.

Pseudo-*o*-, pseudo-*m*-, and pseudo-*p*-*bis*-phenylmethylenyl-[2.2]-*p*-cyclophanes

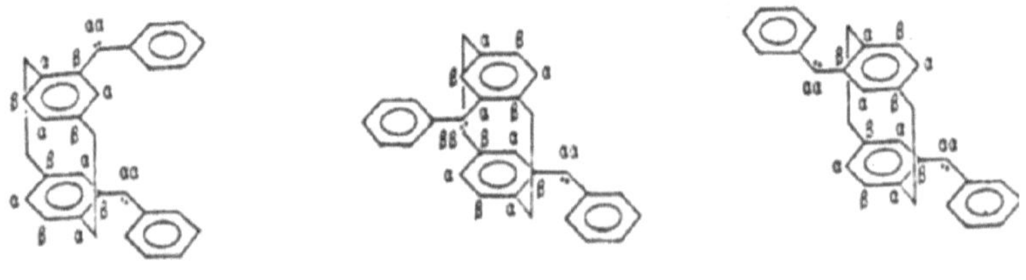

$S = 2$ $S = 0$ $S = 2$

respectively, were obtained through photolysis in a vitrified matrix at low temperatures. The spin–spin interaction between the two triplet diphenylcarben fragments built into the [2,2]-*p*-cyclophane frame, was explored by the EPR technique.

For the pseudo-*o*-dicarben, a quintet state has been revealed, and within a temperature range of 11–50 K the EPR signal intensity obeys the Curie law. When $T > 20$ K, another signal caused by a changed population of the triplet level was observed. Therefore, the pseudo-*o*-isomer is in its ground quintet state with $D = 0.0624$ and $E = 0.0190$ cm^{-1}, and the triplet state lies 63 cm^{-1} higher by energy.

Pseudo-*m*-*bis*-phenylmethylenyl-[2.2]-*n*-cyclophane gives no resonance signal at 11 K. But a triplet state with $D = 0.1840$ and $E = 0.0023$ cm^{-1} was recorded with increasing temperature. The pseudo-*m*-isomer is in its ground singlet state, and the value of singlet–triplet splitting is 98 cm^{-1}. At 15 K for the pseudo-*n*-isomer the quintet nature of the ground state with $D = 0.1215$ and $E = 0.085$ cm^{-1} has been established, but it is not stable chemically.

7.2.2 MAGNETIC INTERACTIONS IN STABLE ORGANIC PARAMAGNETS

As our work deals with stable radicals only, it seems expedient to analyze literature data, having limited ourselves to stable organic paramagnets. According to our goal, it is worthwhile focusing attention mainly on measurements of magnetic susceptibility and, in particular, on clarification of the dependence of the magnetic properties of substances on their chemical structure. One of the most studied stable aroxyls, the so-called galvinoxyl, possesses a highly delocalized uncoupled electron. The formula of galvinoxyl is:

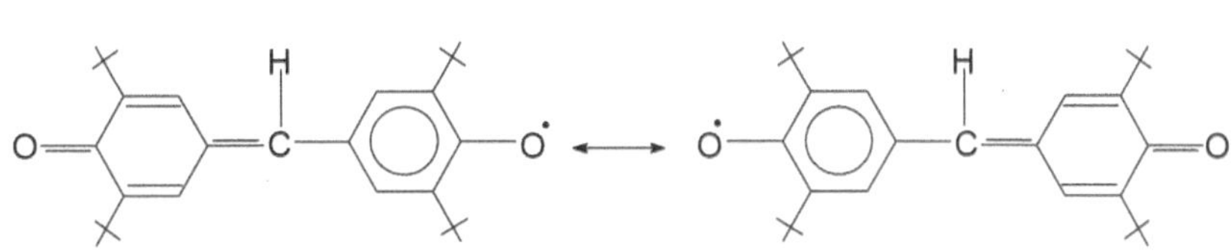

The crystals of galvinoxyl have monoclinic symmetry with the elementary cell parameters a = 23.78, b = 10.87, and c = 10.69 nm and the

angle of non-orthogonality β = 106.6o; a second-order symmetry axis; a 12o deviation from coplanarity, and a 134o angle formed by the C—C bonds at the central carbon atom.[5] The crystal structure of galvinoxyl allows the possibility of the formation of a magnetic linear chain structure extended along the c axis.

The temperature dependence of the paramagnetic susceptibility of galvinoxyl obeys the Curie–Weiss law with a positive Weiss constant θ = +19 K above 85 K, which allows one to assume ferromagnetic interactions between neighboring particles. However, at 85 K a phase transition is observed, upon which the paramagnetic susceptibility sharply decreases, and at 55 K its value corresponds to the content of free radicals 1.1%.[6]

It is interesting that galvinoxyl radicals form couples in a diluted crystal, which have a ground triplet state, and a thermally achievable excited singlet state lies $2J$ higher.[7] Therefore, a ferromagnetic interradical exchange interaction with $2J_F = 1.5 \pm 0.7$ meV is realized in every radical couple. In other words, within a temperature range 10–100 K a diluted galvinoxyl crystal shows no phase transition since it retains ferromagnetic interactions. On the contrary, antiferromagnetic-type interactions with $2J_{AF} = -45 \pm 2$ meV prevail in chemically pure galvinoxyl below 85 K. Apparently, the phase transition in this case is caused by radical dimerization.

This is also confirmed by data on the temperature dependence of the magnetic susceptibility of mixed galvinoxyl crystals. From magnetization curves, it follows that the spin multiplicity is almost proportional to the radical concentration in a mixed crystal. As calculations show,[8] the ferromagnetic intermolecular interactions in galvinoxyl can be explained by superposition of the effects of intraradical spin polarization and charge transfer between free radicals.

Hydrazyl and hydrazidyl radicals are inclined to the formation of various complexes with solvents. This circumstance slightly influences the value of the *g* factor but strongly changes the EPR linewidth. The discordance in the magnetic data of different researchers is probably caused by the presence of impurities in the samples studied, owing to experimental difficulties in purification of organic paramagnets.

The magnetic susceptibility of 1,3,5-triphenyl verdazyl[8]

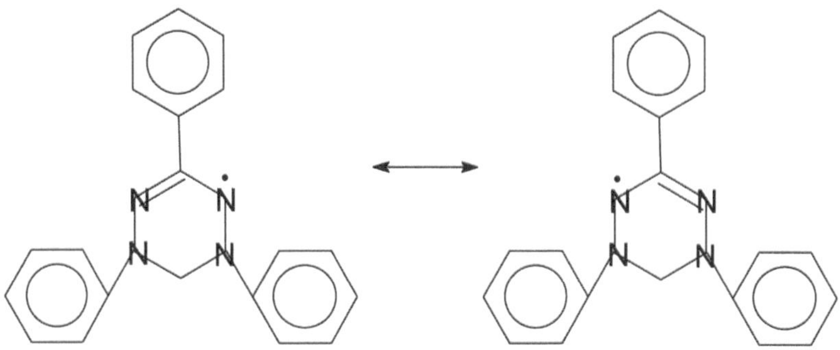

was measured in a temperature range of 1.6–300 K.

In the high-temperature range the magnetic susceptibility obeys the Curie–Weiss law with a negative Weiss constant $\theta = -8$ K. The susceptibility deviates from the Curie–Weiss law at lower temperatures and shows a wide maximum near 6.9 K.

The usage of Heisenberg's linear model with isotropic exchange interaction with $J/k = -5.4$ K above 6 K provides satisfactory agreement with experiment. The distant order of interactions caused by ferromagnetic-type interchain interactions arises at 1.7 K. The crystals of 1,3,5-triphenylverdazine have orthorhombic symmetry with the elementary cell parameters: $a = 18.467$, $b = 9.854$, and $c = 8.965$ nm. All the four nitrogen atoms and the substituent at position 3 are almost coplanar, the two other phenyl groups turned relative to the C—N bond by 23° and 13°, respectively. The radicals in a possible magnetic chain are shown[9] to be bound with each other by a second-order screw axis parallel to the *c* axis so that interchain ferromagnetic exchange interactions are formed between these antiferromagnetically ordered chains.

In this regard, verdazyl biradicals with strongly delocalized uncoupled electrons are of interest, namely: *n*-di-1,5-diphenyl-3-verdazyl benzene and *m*-di-1,5-diphenyl-3-verdazyl benzene:

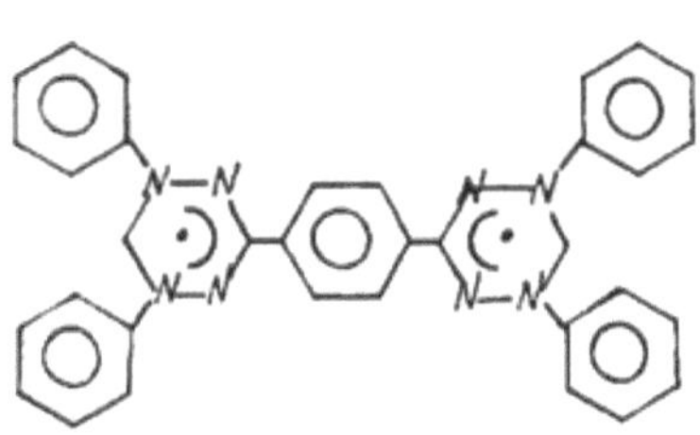
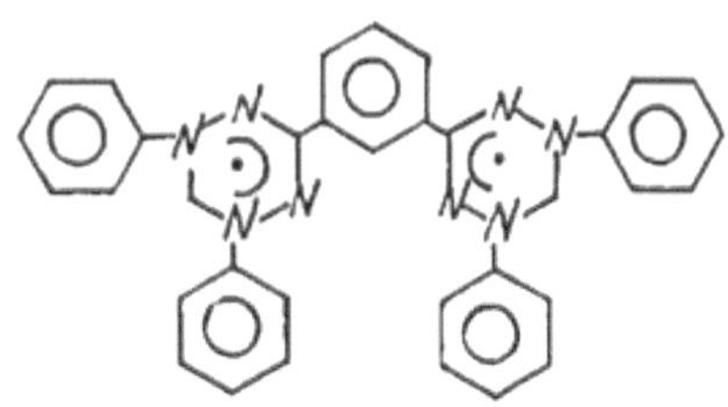

The susceptibility of the *n*-isomer obeys the Curie–Weiss law above 100 K with a Weiss constant $\theta = -100 \pm 20$ K and a Curie constant $C = 1.0 \pm 0.01$ K·emu mol^{-1}, and the χ versus T curve passes through a maximum at 19 ± 1 K when temperature reduces.

In the case of the *m*-isomer, the susceptibility follows the Curie–Weiss law over the whole temperature range studied 1.8–300 K ($C = 0.90 \pm 0.05$ K·emu mol^{-1} and $\theta = -12 \pm 3$ K). Both biradicals are supposed to exist in a ground triplet state ($J/k > 300$ K). The J'/k value of the exchange interaction between the triplets in *n-bis*-verdazyl was estimated from the location of the maximum, it was negative (−7 K).

Classical aromatic hydrocarbonic radicals are often classified as so-called π-electronic radicals wherein an uncoupled electron is delocalized over the whole aromatic bond system. In their majority, arylmethyl radicals in solution exist in thermodynamic equilibrium with their dimer.

Ballester's perchloro-triphenylmethyl radicals sharply differ from classical hydrocarbonic ones by properties: they are rather stable in the absence of light and completely monomeric in both solution and their solid state.

The perchloro-triphenylmethyls studied in ref 10 within the range 293–77 K obey the Curie–Weiss law (Table 7.4).

TABLE 7.4 Characteristic Temperature θ (K) of Some Perchloro-Triphenylmethyl Radicals: Ar,Ar^1, Ar^2 C$^{\bullet}$.

Ar	Ar^1	Ar^2	θ,	μ_{eff}
4H—C_6HCl_4	C_6Cl_5	C_6Cl_5	−4.8	1.76
4H—C_6HCl_4	4H—C_6HCl_4	4H—C_6HCl_4	+1.9	1.73
3H,5H—$C_6H_2Cl_3$	C_6Cl_5	C_6Cl_5	−10.4	1.76
3H,5H—$C_6H_2Cl_3$	3H,5H—$C_6H_2C_3$	3H,5H—$C_6H_2Cl_3$	−10.1	1.74
2H—C_6HCl_4	C_6Cl_5	C_6Cl_5	−12.0	1.71
2H—C_6HCl_4	2H—C_6HCl_4	C_6Cl_5	−3.3	1.69

The antiferromagnetic-type interactions found in the stable paramagnets of the trichloro-triphenylmethyl series are well described by McConnell's above model, being in agreement with the crystal structure and spin density values.

Unlike classical aromatic radicals, the NO group in the iminoxyl radicals takes no part in the formation of a conjugated bond system; the uncoupled electron in such radicals is therefore mainly localized on the nitrogen–oxygen bond. The rather reliable steric shielding of the uncoupled electron (due to the effects of the voluminous methyl groups and the σ-bond system interfering uncoupled electron delocalization) provides conditions for non-radical reactions to proceed in the row of functionalized radicals of this class. This allows synthesizing many chemically pure paramagnets of various chemical structures (Fig. 7.2).[11]

One can easily see that the majority of works is devoted to 2,2,6,6-tetramethyl-4-hydroxypiperidino-1-oxyl (TEMPOL) derivatives obtained by Rozantsev.[12] TEMPOL crystallizes in a monoclinic cell with the axis parameters $a = 0.705, b = 1.408$, and $c = 0.578$ nm; $\beta = 118°40$ and belongs to the spatial group C (Fig. 7.3).

Chains of the radicals bound to each other by hydrogen bonds are formed in a TEMPOL crystal (Fig. 7.3). It is supposed that the strongest exchange interactions of radicals are oriented along the Z axis through the oxygen atoms. The direction along the *a* axis through the hydrogen bond could probably be the interaction next in contribution. Proceeding from structural reasons, weaker magnetic interactions can be expected between the *ac* planes.

Rozantsev and Karimov[13] investigated the magnetic susceptibility of chemically pure TEMPOL by the EPR method for the first time in 1966. They showed the EPR signal strength to deviate from Curie's law and to exhibit a wide and smooth maximum near 6 K. A wide maximum on the thermal capacity curve was found at 5 K. In the high-temperature range, the susceptibility obeys the Curie–Weiss law with a negative Weiss constant $\theta = -6$ K. At lower temperatures, the χ versus T curve deviates from the Curie–Weiss law and has a flat maximum at 6 K.

Such behavior of paramagnets is well described by Heisenberg's one-dimensional model with isotropic antiferromagnetic interactions. For Heisenberg's linear system with $S = 1/2$ the magnetic susceptibility should have a flat maximum determined by $\chi_{max}/(N_a g^2 \mu^2_B/J) \approx 0.07346$ at $kT_{max}/J \approx 1.282$.

The value of the exchange J/k parameter is estimatedas -5 K. Therefore, independent studies of the magnetic susceptibility of TEMPOL evidence strong exchange inteactions experienced by radicals in one direction, which results in the near order of interactions and the formation of linear antiferromagnetic chains near 6 K.

As nonzero interaction always exists between the chains in one-dimensional magnetic systems, it could be expected that below some critical temperature it would get rather expressed to cause transition to a distant order of interactions. In the case of TEMPOL, the distant order caused by interchain interactions arises at $T_N = 0.34$ K, the interchain to intrachain interaction ratio (J/J') estimated as 0.003, $J/k = 0.013$ K.

An alternating linear chain arises if $\gamma < 1$. The case $\gamma = 0$ corresponds to a simple dimer where paired interactions act only.

The stable di-2,2,6,6-tetramethyl-1-oxyl-4-piperidyl sulfite biradical exemplifies the alternating chain (Fig. 7.4). In the high-temperature range, its magnetic susceptibility is described by the Curie–Weiss law with $\theta = -9$ K. When temperature falls, the susceptibility of this biradical deviates from the Curie–Weiss law near 25 K, and then sharply drops down to 2 K.

As is seen from Figure 4, the susceptibility maximum of this biradical is less flat than it would be for a regular spin chain with $\gamma = 1$. The use of the model of an alternating spin chain with $J/k = 9.6$ K and $\gamma = 0.55$ provides satisfactory agreement with experiment, and the maximum exchange between interacting spins $J/k = 9.6$ K is a result of structural exchange interactions in the crystal.

The adduct of copper hexafluoro acetylacetonate with the iminoxyl radical of 2,2,6,6-tetramethyl-4-hydroxypiperidyl-1-oxyl is another example of the alternating linear chain.[14]

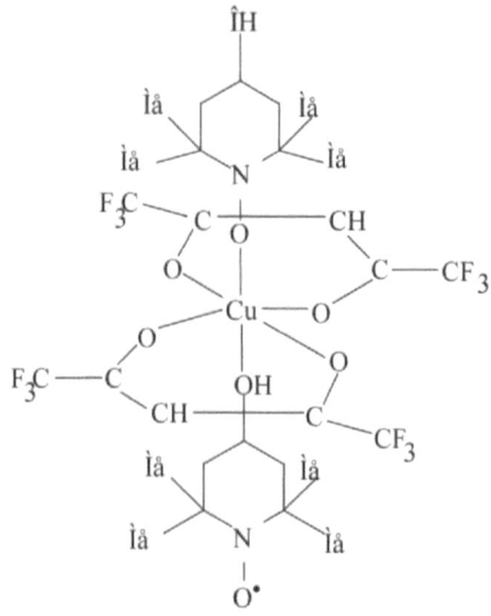

In experiments, a strong (19 K) ferromagnetic interaction between the copper ion and iminoxyl was revealed, which followed from the data obtained at temperatures above 4.2 K. At temperatures below 1 K, the magnetic susceptibility sharply increased, and a flat maximum, characteristic of antiferromagnetic linear chains, was found at ~ 80 meV. Analysis of these data was carried out in the assumption that the substance consists of chains with spins $S = 1$ and weak ($2J = -78$ meV) antiferromagnetic interaction between spins. Therefore, the alternation in this case arises because of alternation of the strong ferromagnetic and weak antiferromagnetic interactions.

For the silicon-organic iminoxyl polyradicals

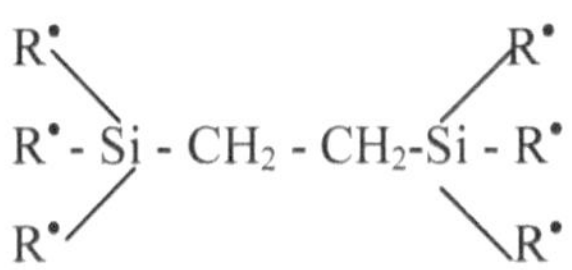

I

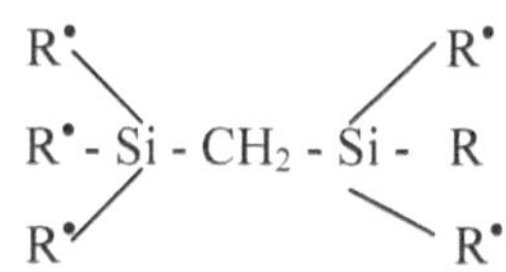

II

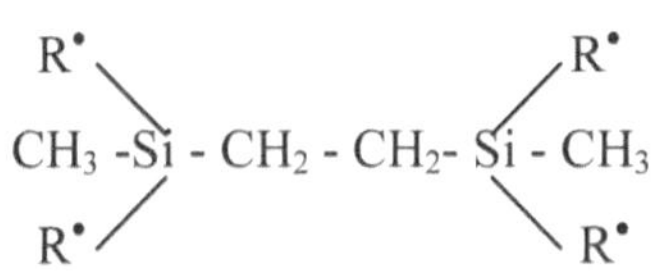

III

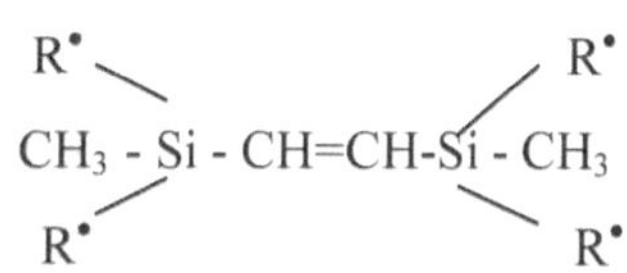

IV

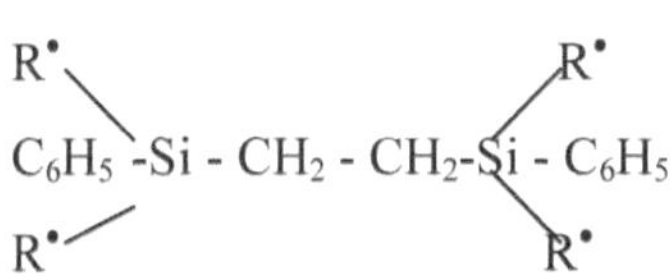

V

($R^{\bullet} \equiv$ 2,2,6,6-tetramethyl-1-oxyl-4-piperidyl fragment) paired spin–spin interactions are characteristic. All the studied paramagnets (I–V) exhibit low-temperature deviations of the course of their magnetic susceptibility from Curie's law $\chi = \text{const}/T$ (Fig. 7.5a), which are due to the existence of correlation between uncoupled electrons. For example, the susceptibility of tetraradical V passes through a maximum at 8 K and decreases by

10 times when temperature falls down to 2 K (Fig. 7.5b). Such a course of susceptibility is well described by a model offered for paired exchange interactions of uncoupled electrons:

$$\chi = CT^{-1}\,[3 + \exp\,(J/kT)]^{-1}$$

If the ground state of such a couple is a singlet and the thermally excited state is a triplet (or a triplet magnetic exciton), it mainly contributes to the magnetic susceptibility. Excitons get energy for their excitation from thermal energy. Therefore, when kT becomes less than the exchange value J between electrons, the number of triplet states sharply falls and, hence, the susceptibility sharply decreases. From analysis of the course of susceptibility, the exchange parameters of strongly bound spins (Table 7.5) were estimated.

TABLE 7.5 Exchange Interaction Parameters of Silicon-Organic Polyradicals

Radical	I	II	III	IV	V
J/k, K	2.2±0.5	3.2±0.5	4.6±0.5	5.2±0.5	14.6±0.5

It is interesting that for tetraradical V the exchange interaction parameter J'/k between spin couples is 0.1 K.

The crystal structure of organic radicals, because of the asymmetry of the majority of chemical particles, as a rule, allows one to resolve topological linear chains of most strongly interacting spins. A study of the structure of, for example, radical V (Fig. 7.6) has shown that the nitrogen atoms of one radical heterocycle form a chain of paramagnetic centers with a link length about 6 nm parallel to the axis, and the nitrogen atoms of the other heterocycle form another spin chain parallel to the first one with the length of an elementary link of 6.6 nm. Besides, each paramagnetic center in the chain, thus, has two neighboring spins from other chains at distances of 6.4 and 6.6 nm, respectively.

In ref 15, the paired intermolecular interaction of the basic spin system in 1,4-*bis*-2,2,6,6-tetramethyl-1-oxyl-4-piperidyl-butane crystals was reported. This interaction is distinctly seen on the temperature dependence curve within a range of 10–300 K as the presence of a characteristic maximum near 40 K:

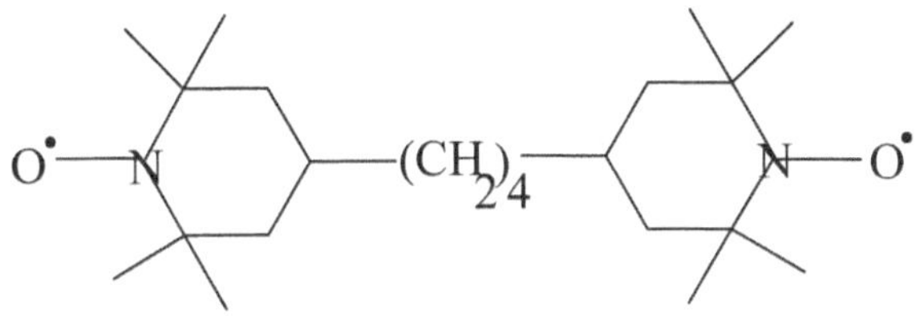

The monoclinic crystals of this biradical have the elementary crystal cell parameters $a = 11.754$, $b = 10.980$, and $c = 8.693$ nm with the $P_1 2/b$ spatial group (Fig. 7.7).

This structure features the existence of two systems of pairs of =NO$^{\bullet}$ radical fragments which are mirror symmetric about the *ab* plane. For the mirror symmetric couples, the angle between the lines connecting the centers of the iminoxyl fragments is 50°. Inside each couple, the oxygen atoms are at a distance of 0.351 nm, and the nitrogen atoms are at a distance of 0.485 nm. The short distance between the NO fragments in a couple and the relative location of the C—N—C planes promote direct electronic exchange in these couples (Fig. 7.8).[16,17]

Really, the temperature course of the paramagnetic susceptibility in the crystals is well described within the model of antiferromagnetic paired exchange with a constant $J = -33.5$ K. The intramolecular exchange interactions J' transferred through the —$(CH_2)_4$— bonds appear less than the hyperfine coupling constant, which corresponds to $J'J' < 2 \times \cdot 10^{-3}$ K, that is, $J/J' > 10^4$.

The tanolic ester of octanoic acid

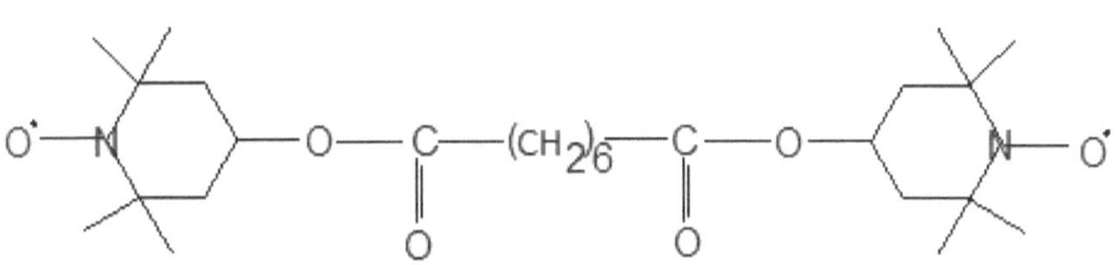

obeys the Curie–Weiss law with a positive constant $\theta = +1$ K in a temperature range 1.9–300 K. All the magnetic interactions of interest are rather weak and manifest themselves at temperatures below 1 K only. Apparently, a magnetic transition at $T = 0.38 \pm 0.01$ K proceeds in the system due to ferromagnetic ordering. Neutronography has established that the crystals of this paramagnetic are layered: the neighboring particles inside each layer are bound ferromagnetically with $J_1 = +1.1$ K and $J_2 = 0.07$ K, but the layers are connected among themselves by weaker antiferromagnetic interactions with $J' = -0.015$ K. It is believed[18] that the substance behaves as a metamagnetic with two-dimensional ferromagnetic ordering. For 2,2,6,6-tetramethyl-4-oxopiperidyl-1-oxyl azine (TEMPAD), a maximum at 16.5 K is found on the curve of the temperature course of paramagnetic susceptibility, and its change under the Curie–Weiss law ($\theta = -15$ K) is observed within 77–273 K. In the case of diluted TEMPAD crystals, two values of the Weiss constant have been found: about −10 K in the high-temperature range and about −1 K in the low-temperature one (Fig. 7.9).[19]

The magnetic behavior of TEMPAD was interpreted within the theory of magnetic triplet paired transitions.[46–48] Nobody can exclude the existence of strong intermolecular exchange interaction along the *a* axis (J_1, $J_1/k = -12.8$ K), weak intramolecular interaction (J_2, $J_2/k \sim 2{\cdot}10^{-2}$ K), and interlayer interaction with $J_1'/k \sim 1$ K.[20]

In 2,2,6,6-tetramethyl-4-oxypiperidyl-1-oxyl phosphite (TEMPOP)

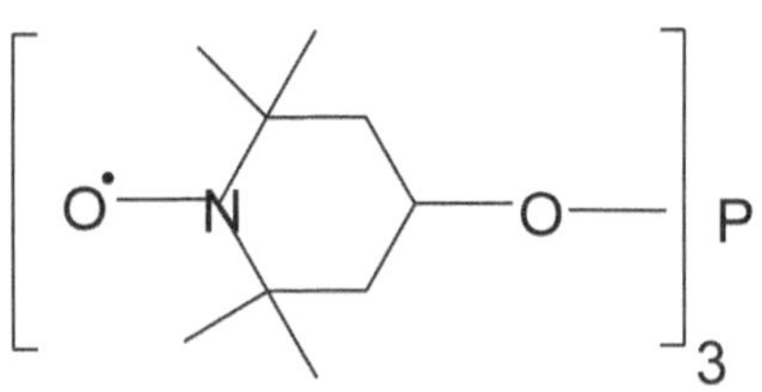

no near order of interactions was found, though at very low temperatures the course of inverse paramagnetic susceptibility deviated from the Curie–Weiss law (Fig. 7.10). Theoretically, the effective magnetic moment of three noninteracting spins with $g = 2.00$ should be equal to 3.00 μ_B; in experiment (at high temperatures), a value of 3.01_B was obtained.

Interesting studies on nitronyl nitroxyls (or nitroxide nitroxyls) NIT(R) ≡ 2R-4,4,5.5-tetramethyl-4,5-dihydro-1H-imidazolyl-1-oxyl-3-oxides

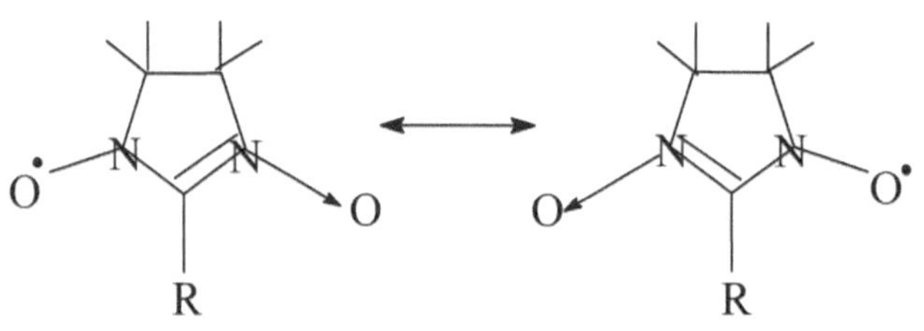

with metal ions are presented in refs 21–32. Owing to the conjugation of a nitroxyl group with a nitroxide one, the exchange interaction between one oxygen atom and a metal ion can be transferred to another oxygen atom without attenuation. Therefore, these radicals with a delocalized uncoupled electron are capable of forming not only mononuclear complexes with metals[21–23] but also magnetic chains of various natures.[24–27]

There exist a metal-containing compound $Cu(hfac)_2(NIT)Me$, where **hfac** is a hexafluoroacetylacetonate ion $[CF_3C(O)CHC(O)CF_3]^-$, behaves as a one-dimensional ferromagnetic with an interaction constant of 25.7 cm^{-1}. The effective magnetic moment equal to 2.8 μ_B at 300 K monotonously increases with decreasing temperature and gets 4.9 μ_B at 4 K. This means that the effective spin of the system increases almost up to three.

Replacement of a copper ion by Ni (II) or Mn (II) in the $Ni(hfac)_2(NIT)R$ and $Mn(hfac)_2(NIT)R$ complexes with R = Me, Et, i-Pr, n-Pr, and Ph leads to stronger antiferromagnetic-type interactions with $J = -424$ cm^{-1} for nickel and $J = -230$–330 cm^{-1} for manganese derivatives. When exploring $Mn(hfac)_2(NIT)iPr$ monocrystals, noticeable anisotropy was discovered. As follows from the temperature dependence of susceptibility along the easy magnetization axis (this direction in the crystal coincides with the spin direction orientation), phase transition to a ferromagnetic state occurs at 7.6 K.

In $Mn(hfac)_2(NIT)Et$, $Mn(hfac)_2(NIT)nPr$, and $Ni(hfac)_2(NIT)Me$ ferromagnetic ordering occurs at 8.1, 8.6, and 5.3 K, respectively. Calculations confirm the dipole–dipole nature of the magnetic interactions in these compounds. Higher temperatures of magnetic phase transition were found for $[Mn(F_5benz)_2]_2NITEt$ and $[Me(F_5benz)_2]_2NITMe$ (where F_5benz is pentafluorobenzoate): 20.5 and 24 K, respectively. But there is no unambiguous confidence which type (ferrimagnetic or weak ferromagnetic) this ordering belongs.

The binuclear complex $[CuCl_2(NITpPy)_2]_2$, where NITpPy ≡ 2-(4-piridyl)-4,4,5,5-tetramethylimidizolinyl-3-oxide-l-oxyl, was studied. Potentially, the radical NITpFy could be a tridentate ligand. The copper ions are shown to be coordinated with two nitrogen atoms in the pyridine rings and three chlorine atoms, two of which are bridge ones. The NO groups of radicals belonging to different mononuclear fragments are rather close to each other. On the basis of magnetic susceptibility data and EPR spectra recorded at 4.2 K, it is supposed that six spins $S = 1/2$ are bound by antiferromagnetic exchange interaction. The interaction between copper and the radical through the pyridine cycle's nitrogen is preferable, which allows the authors to consider NITpFy as promising ligands in the synthesis of metal-containing magnetic materials.

7.1.3 ORGANIC LOW-MOLECULAR-WEIGHT AND HIGH-MOLECULAR-WEIGHT MAGNETS

The interest to low-molecular magnets[33–40] and high-spin compounds[41–57] is associated with the hope of obtaining compounds possessing spontaneous magnetization below their critical temperature. Though the critical temperatures reached are quite low, it is possible to state with confidence that the understanding of the necessary conditions for the design of a high-temperature organic magnetic material has become clearer than several years ago.

Wide-range studies on ion radical salts like $D^{\cdot +}A^{\cdot -}D^{\cdot +}A^{\cdot -}$, where D is a cation (donor) and A is an anion (acceptor), were carried out by Miller et al.[33–40] Decamethylferrocene $Fe(II)(C_5Me_5)_2$ is often used as a donor and flat 7,7,8,8-tetracyan-*p*-quinodimethane (TCNQ) and tetracyanoethylene (TCNE) serve as acceptors

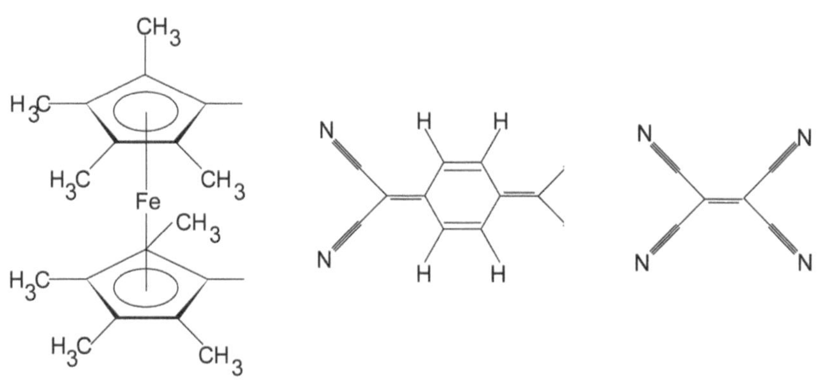

TCNQ TCNE

The complex $[Fe(C_5Me_5)_2]^{+}[TCNE]^{\cdot +}$ is characterized by a positive Curie–Weiss constant, $\theta = +30$ K, Curie's temperature (T_c) equal to 4.3 K; in a zero magnetic field, spontaneous magnetization is observed for a polycrystalline sample ($M \sim 2000$ emu·Gs mol^{-1}).[35] The saturation magnetization is 16300 emu·Gsmol^{-1} in oriented monocrystals. This result is in good agreement with the theoretical magnetic saturation moment at ferromagnetic spin alignment of the donor and acceptor and is by 36% higher than for metal iron (per 1 g atom). At 2 K, hysteresis with a coercive force of 1000 Gs is observed, corresponding to the values for magnetically hard materials. Above 16 K, the magnetic properties are described by Heisenberg's one-dimensional model with ferromagnetic interaction ($J = +27.4$ K). At temperatures near T_c, three-dimensional ordering prevails.[36]

The compound $[Fe(C_5Me_5)_2]^{+}[TCNQ]^{-}$ shows metamagnetic signs with a Néel temperature $T_N = 2.55$ K and a critical field of ~ 1600 Gs. As a rule, metamagnetics are substances with strong anisotropy and, in the presence of concurrent interactions therein, first-order transition to a phase with a total magnetic moment can be observed.[1] For example, for the salt $[TCNQ]^{-}$, the magnetization in fields with $H < 1600$ Gs is characteristic of a antiferromagnetic, while when $H > 1600$ Gs, a hump-like increase of magnetization occurs up to the saturation value, which is characteristic of the ferromagnetic state.[37]

Of the 2,5-disubstituted TCNQ salts of decamethylferripinium $[Fe(C_5Me_5)_2]^{+}[TCNQR_2]^{+}$ (R = Cl, Br, J, Me, OMe, OPh),[39] $[Fe(C_5Me_5)_2]^{+}[TCNQI_2]^{-}$ possesses the highest effective moment $\mu_{eff} = 3.96\mu_B$. Above 60 K, the magnetic susceptibility obeys the Curie–Weiss law ($\theta = +9.5$ K) and the substance is a one-dimensional ferromagnetic. This feature, in combination with that $[TCNQI_2]^{-}$ exhibits stronger interchain antiferromagnetic interactions in comparison with $[Fe(C_5Me_5)_2]^{+}[TCNQ]^{-}$, provides no three-dimensional ferromagnetic ground state at temperatures above 2.5 K.

The complex[40]

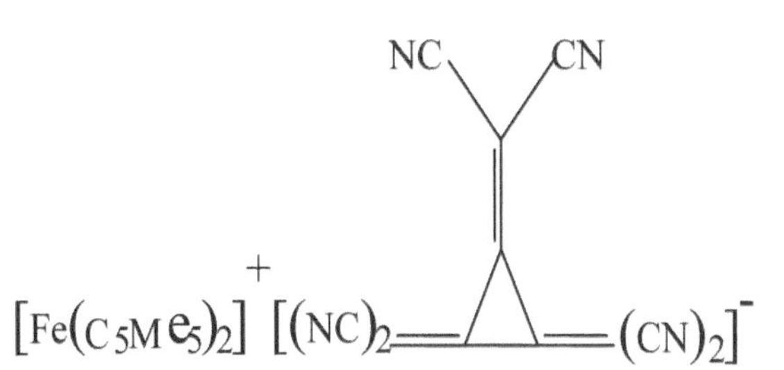

can exist in two polymorphic modifications, namely, monoclinic and triclinic, both obeying the Curie–Weiss law with $\theta = -3.4$ K, $\mu_{eff} = 2.98\ \mu_B$, and $\theta = -3.4$ K, $\mu_{eff} = 3.10\ \mu_B$, respectively. Below 40 K, in a magnetic field of 30 Gs, the monoclinic compound shows the Bonner–Fisher type of one-dimensional antiferromagnetic interaction, that is, it has a typical flat maximum about 4 K. This is attributed to antiferromagnetic interaction along the cation chains with an exchange parameter of $J/k = -2.75$ K. To explain the magnetism of ion radical salts, the model of configuration interaction of the virtual triplet excited state with the ground state offered by McConnell[57]is applied. For example, in the case of donor–acceptor pair, $D^{\cdot+}A^{\cdot-}$, it is supposed that the wave function of the ground state has the maximum "impurity" to the wave function of the lower virtual excited state with charge transfer. This state can arise due to direct virtual charge transfer ($D^{\cdot+}$ $+A^{\cdot-}$) → (D^{2+} $+A^{2-}$), reverse charge transfer ($D^{\cdot+}$ $+A^{\cdot-}$→ D° +A°) or disproportionation (2$D^{\cdot+}$ D^{2+} + D°).

If any of the states with charge transfer (either donor D or acceptor A, but not both) is triplet, the ground ferromagnetic state of the $D^{\cdot+}A^{\cdot-}$ pair will be stabilized. Therefore, for ferromagnetism manifestation, an organic radical should possess a degenerated and partially filled valent orbital. An essential contribution of the lower virtual excited state with charge transfer to the ground state of the system is necessary, and the structure of the radical ion should be highly symmetric, without any structural or electronic dislocations breaking the symmetry and eliminating the degeneration.[33,38]

A whole series of high-spin polycarbenes has been so far synthesized by means of photolysis of the corresponding polydiazo compounds.[41–47]Attempts to get high-spin macromolecules by iodine oxidation of 1,3,5-triaminobenzene have failed. Breslow et al.[57–59] have succeeded to synthesize stable organic triplet systems with C_3 and higher symmetry on the basis of hexaminotriphenylene and hexaaminobenzene derivatives.

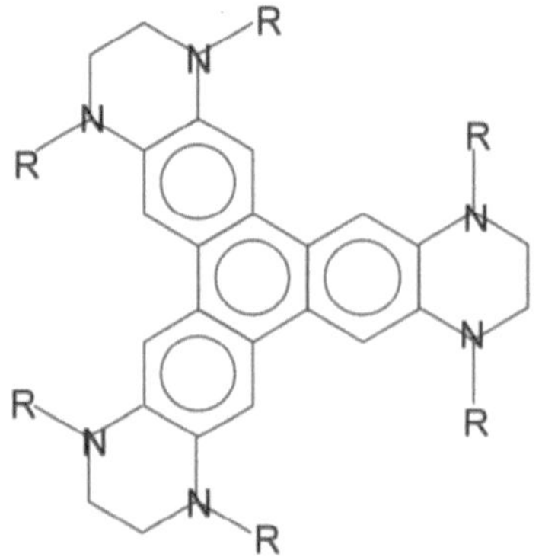

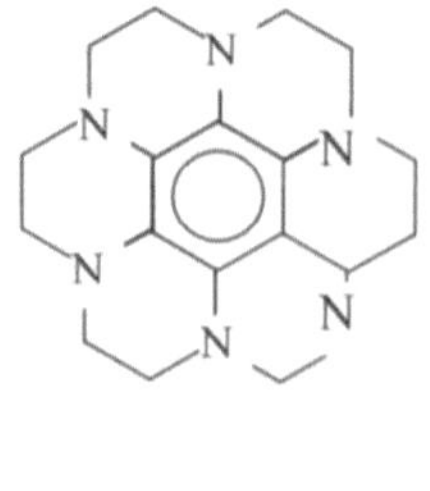

where R = C_2H_5, $C_6H_5CH_2$, CF_3CH_2,

Studies on the material obtained by spontaneous polymerization of diacetylene monomer containing stable iminoxyl fragments of butadiyn-*bis*-2,2,6,6-tetramethyl-1-oxyl-4-oxi-4-piperidyl (BIPO) is a highly mysterious story.

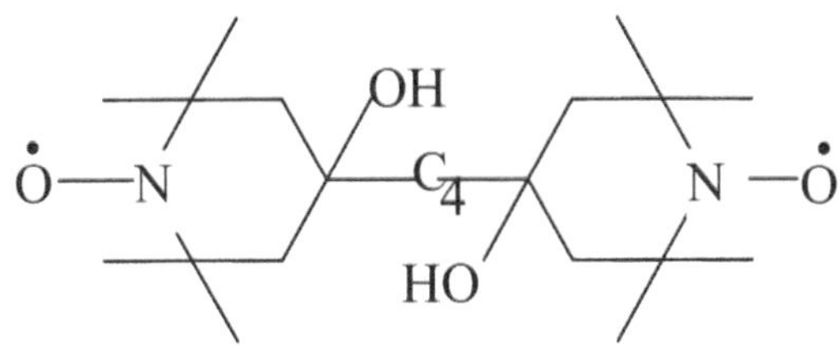

The magnetic susceptibility of BIPO[60] obeys the Curie–Weiss law with $\theta = -1.8$ K. The effective magnetic moment equal to 2.45 μ_B at high temperatures corresponds to two independent spins of $S = 1/2$ per monomer unit. The exchange constant derived from analysis of the EPR line has appeared to be $J \sim 0.165$ K (0.115 cm^{-1}), and an estimation in the approach of molecular field has given a value of $J \sim 0.155$ K (0.108 cm^{-1}).

The thermal or photochemical polymerization of BIPO leads to the formation of black powder whose insignificant fraction (0.1%) shows ferromagnetic properties, its magnetization reaches above 1 $Gs \cdot g^{-1}$. It is noted that ferromagnetism holds up to abnormally high temperatures (up to 200–300°C) and the paramagnetic centers thus die during polymerization (in some cases no more than 10% of their initial quantity remains).

Contrary to the earlier published analysis of these intriguing data in a subsequent work,[60] it was noted that the products of thermal decomposition of BIPO showed neither signs of three-dimensional ferromagnetism nor magnetic interaction. Detailed static and dynamic magnetic data indicate the existence of weak intradimeric ferromagnetic (triplet) interaction with J ~ 10 K only.

Obviously, while solving this problem, the degree of reliability of obtained results will strongly depend on the chemical purity of the materials studied. It is possible to state without exaggeration that natural sciences progress is associated with obtaining and studying chemically pure materials.

Unfortunately, even superficial analysis of the available publications convinces us that the majority of experimental works in this field is associated with studying of structurally disordered "dirty" systems like spin glasses[61] with no coordinated magnetic interactions between chemical particles. The relative simplicity of obtaining "dirty systems" provokes the avalanche-like spreading of "impressive results," various fantastic models and theories, having nothing in common with true science.

It would be thoughtless to consider that chemical purity is sufficient to achieve success in the basic research of high-spin nonmetallic systems. Precision-measuring equipment and a methodology, including automated X-ray diffraction analysis and modern magnetometry with the usage of superconducting quantum interferometers (squids), without being limited to EPR equipment and high-temperature magnetic measurements, are, undoubtedly, other necessary conditions.

Only the successful development of the basic research of the magnetic properties of pure systems and their constituent chemical particles can provide real breakthrough in the technology of the design of materials of a new generation suitable for manufacturing competitive organic ferromagnets, antiferromagnets, and ferrimagnets, including metamagnets and speromagnets.

In 1990, Emsley[62] published a paper under an intriguing title where he reported about the synthesis of a stable iminoxyl radical (nitroxide nitroxyl triradical) with its properties of a "molecular" organic magnet. In other words, the discovery of a metal-free "organic magneton" with cooperatively ordered electronic interactions at the level of a discrete chemical particle was claimed:

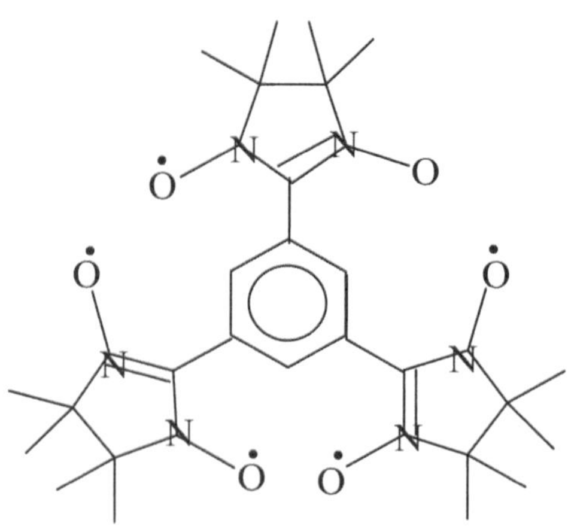

Dulog and Kim[63] have found that some blue powder obtained by them possesses a high value of magnetic susceptibility and can strengthen or weaken the intensity of the applied magnetic field like metal magnets. Provided that the remarkable properties of blue trinitroxyl are not a trivial consequence of metallic pollution, the new material will be able to find applications when designing magnetic registering devices of a new generation, magnetoplanes, and other equipment.

Using the principle of orienting effect of the intraradical electrostatic field of nitroxide groups, it could be possible to design high-molecular-weight magnetic materials with magneto-ordered organic domains (magnetons) like blue nitroxide nitroxyl triradical of Stuttgart's chemists as monomeric links therein.

Stable paramagnets have found practical applications as additives to polarized proton targets in the experimental physics of high energy. The method of reaching ultralow temperatures by ^3He dissolution in ^4He opens new opportunities in the technology of polarized targets. For example, the high polarization obtained by a usual dynamic method in a strong and uniform magnetic field (25 kOe) can be kept for a long time after the termination of the dynamic polarization "pumping" if the working substance of the target is cooled rather quickly down to a temperature about 0.1 ÷ 0.01 K. Then, the intensity of magnetic field can be lowered down to ~5 kOe. This opens new prospects of the use of such targets in physical experiments.

In the existing polarized proton targets operating at temperatures as low as 0.5 K, the main working substances are butyl alcohol and ethylene glycol as frozen balls. However, these substances are not technological for their usage in cryostats with ^3He dissolution in ^4He.

A substance, solid at room temperature, rather rich with protons, and containing radicals stable at room temperature as paramagnetic additives, would be most convenient. Therefore, polyethylene used as either a 200 mμ film or powder with ~200 mμ grains was selected as the working substance on the basis of recommendations from references. Stable iminoxyl radicals were taken as a paramagnetic additive. To introduce such a radical into polyethylene, the necessary amount of the radical and polyethylene was placed into a tight glass ampoule, heated up to 80°C, and maintained at this temperature for 8–10 h. To study proton polarization in polyethylene, preliminary experiments were conducted at a temperature of 1.3 K in magnetic fields of 13 and 27 kOe. The stable volatile radical 2,2,6,6-tetramethyl-piperidine-1-oxyl appeared most promising.

At an optimum concentration of the radical in a magnetic field of 13 kOe in a polyethylene film, a polarization of 5–7% has been reached, which corresponds to an amplification polarization factor of $E = 50 \div 70$. The period T_1 of proton spin–lattice relaxation was 2.5 min. Transfer into a magnetic field of 27 kOe led to no increase in the polarization amplification factor, only T_1 increased. However, it was revealed that the use of polyethylene of lower density and higher purity as powder with 200 mμ grains led to an increase in E up to $70 \div 100$. Thus, a 14–20% polarization was achieved in the field of 27 kOe at $T = 1.3$ K.

In experiments at ultralow temperatures, polyethylene powder was exposed to preliminary annealing in vacuum (10^{-5} mmHg) at 80°C during 5–6 days. Then, the ampoule with this powder was filled with pure gaseous helium and the powder was saturated by the vapors of the radical in a helium atmosphere. Such an annealing procedure does not influence the rate of spin–lattice relaxation considerably, but increases the final polarization almost twice. The sample prepared in this amount (150 mg) was introduced, at room temperature, into a glass camera of ^{3}He dissolution in ^{4}He located in a running wave microwave cell. The cell was placed into a superconducting solenoid and cooled down to a temperature of 1.3 K. The sample was in direct contact with the solution whose minimum temperature was 0.04–0.05 K.

For polarization pumping, a microwave generator of the OV-13 type ($\lambda = 4$ mm) with a power about 70 mW was used; bringing 1.5–2.0 mW into the cryostat was enough to increase the solution temperature up to 0.1 K.

Proton polarization up to 50% was reached in polyethylene samples as powder with an optimum concentration of the radical of 10^{-19}spins per polymer gram. The period of polarization pumping was about 3.5 h. After switching the microwave field off, the temperature of the ^{3}He–^{4}He system decreased down to 0.05 K. At this temperature in a magnetic field of 27 kOe, almost no polarization decay was observed. In a magnetic field lowered down to 5 kOe, the relaxation time was not less than 30 h and only with the magnetic field decreased down to 1.5 kOe it reduced to 1.5 h.

The EPR spectrum of tetramethylpiperidine-1-oxyl introduced into polyethylene in the above quantity at room temperature represents a well-resolved triplet with a distance between the hyperfine coupling components of 15 kOe. At temperatures below 1 K, the EPR spectrum is transformed to a line with its halfwidth of 80 Oe with an ill-defined superfine structure. To simplify the EPR structure, an attempt to replace the radical with ^{14}N by that with ^{15}N in the same concentration was made. However, such replacement gave no increase in the maximum polarization.

The experiments conducted have shown that now there is a real possibility to create a "frozen" polarized proton target in a magnetic field about 5 kOe.

Now, in geophysics and astronautics, nuclear precession magnetometers possessing a number of essential advantages are widely adopted. Their high sensitivity (to 0.01 gamma) and the accuracy of measurements, the absoluteness of indications, and the independence of temperature, pressure, and sensor orientation are advantages of nuclear magnetometers.

Cyclic operation is a feature of precession magnetometers. The process of measurement consists of two consecutive processes, namely: polarization of the working substance of the magnetometer sensor during which nuclear magnetization is established, and measurements of the signal frequency of the nuclear induction determining the absolute value of the field measured.

The use of the phenomenon of dynamic polarization of atomic nuclei allows one to overlap the stages of polarization and measurement, and to essentially increase the speed of the magnetometer. Nuclear generators based on the phenomenon of dynamic polarization of atomic nuclei, allowing continuous monitoring of changes in the magnetic field were designed.

Earlier, only Fremy's diamagnetic salt dissociating into paramagnetic anions in aqueous solution was used as the working substance of the sensors of nuclear magnetometers based on the phenomenon of dynamic nuclei polarization.

Saturation of any hyperfine coupling line in the EPR spectrum of Fremy's salt solution leads to a significant increase in the nuclear magnetization of the solvent. This is the effect of dynamic polarization. The hydrolytic instability is a demerit of Fremy's salt. Even in distilled water, the radical anion of this salt is hydrolyzed to diamagnetic products during several dozen minutes, the process of degradation having autocatalytic character. A paramagnetic solution stabilized with an additive of potassium carbonate preserves about a month provided that its temperature would not exceed 40°C. Stable paramagnets of the iminoxyl class have indisputable advantages over Fremy's salt.

Rozantsev and Stepanov[64] proposed 2,2,6,6-tetramethyl-4-oxipiperidine-1-oxyl and 2,2,6,6-tetramethyl-4-oxopiperidine-1-oxyl well soluble in many proton-containing solvents and possessing a resolved hyperfine coupling in their EPR spectra within a wide range of magnetic fields as working substances for nuclear precession magnetometers in 1965.

In weak magnetic fields, the bond between the electronic and nuclear spins of nitrogen is not broken off, and the set of energy levels is characterized by a total spin number S taking on two values (1/2 and 3/2) and a magnetic quantum number m^s taking on values $2S + 1$. The set of energy levels of iminoxyl radicals in weak magnetic fields is presented in Figure 7.11.

The rule of selection $\Delta S = 0 \pm 1$ and $\Delta m_s = \pm 1$ exists for π-electronic transitions. Saturation of one of such transitions by a strong radio-frequency field oriented perpendicular to the constant field, significantly changes the electronic magnetization M_z of the solution, which, in turn, results in an increase in the proton magnetization m_z of solvent molecules according to the expression:

$$m_z - m_o = fg(M_o - M_z),$$

where m_z, m_o, M_o, and M_z are the nuclear and electronic magnetizations of the solution in a stationary mode and thermal equilibrium at saturation of the transition in the system of electronic energy levels, respectively, f is a coefficient determining the contribution of electronic–nuclear interaction into the mechanism of nuclear relaxation of the solvent, g, a coefficient depending on the nature of the electronic–nuclear interaction.

Studying of iminoxyl radicals has shown that only saturation of the 1–6 and 4–5 transitions (Fig. 7.11) leads to a considerable dynamic polarization of the nuclei of the solvent. By the value of proton dynamic polarization in solutions of iminoxyl paramagnets, the latter ones are on a par with Fremy's salt applied earlier. They were more stable in water and organic solvents and did not change their initial characteristics within half a year. In the course of research, solutions of iminoxyl paramagnets were repeatedly heated up to 90°C. It is obvious that organic paramagnets represented essentially new working substances for nuclear precession magnetometers of a new generation. The main advantage of these new working substances consists in the possibility to choose a solvent for them to operate in any climatic conditions, with a high content of protons and a long period of proton relaxation. Little wider electronic transitions in comparison with an aqueous solution of Fremy's salt are easily eliminated by the usage of deuterated samples of iminoxyl radicals.[65] Numerous attempts of the authors to register the mentioned idea as a USSR author's invention certificate were steadily refused.

After these materials had been published, a group of French engineers designed a precession nuclear magnetometer with a deuterated solution of 2,2,6,6-tetramethyl-4-oxopiperidine-1-oxyl in dimethoxyethane as its working substance.

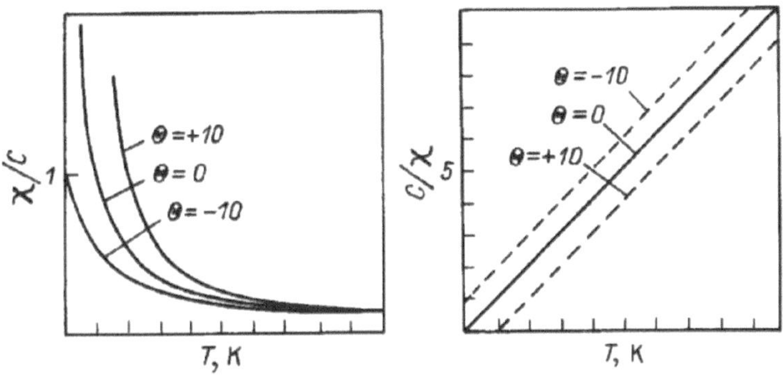

FIGURE 7.1 Temperature dependence of magnetic susceptibility (left) and that of inverse magnetic susceptibility (right) for noninteracting ferromagnetically coupled spins and antiferromagnetically coupled spins, respectively.

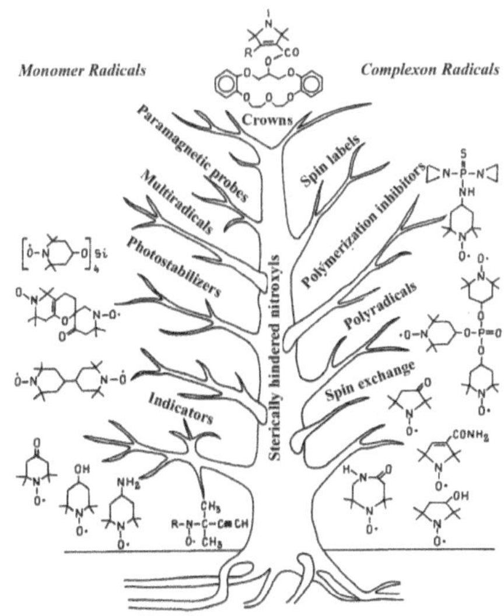

FIGURE 7.2 Genealogic tree of stable iminoxyl (nitroxyl) radicals, whose synthesis and application were promoted by the discovery of non-radical reactions of radicals.

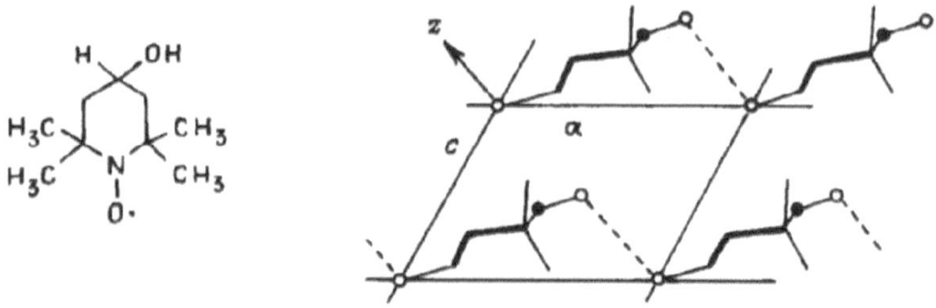

FIGURE 7.3 Projection of the 2,2,6,6-tetramethyl-4-hydroxypiperidyl-1-oxyl structure onto the *ac* plane.

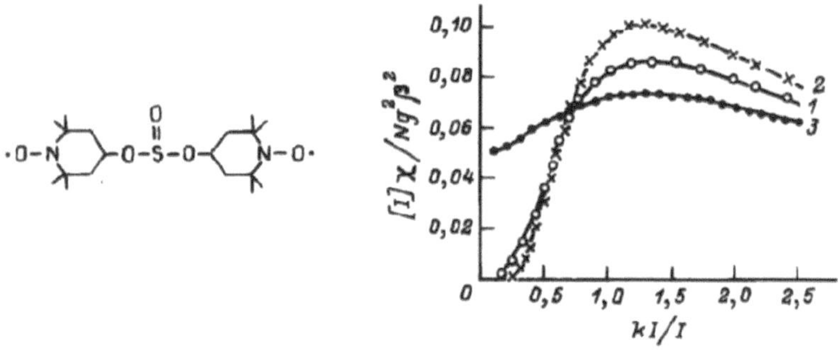

FIGURE 7.4 *1* A low-temperature fragment of the curve of magnetic susceptibility of the biradical, *2* calculated curve for Heisenberg's paired interaction with *J*/*k* = 9.6 K, *3* calculated curve for a regular spin chain with *J*/*k* = 9.6 K.

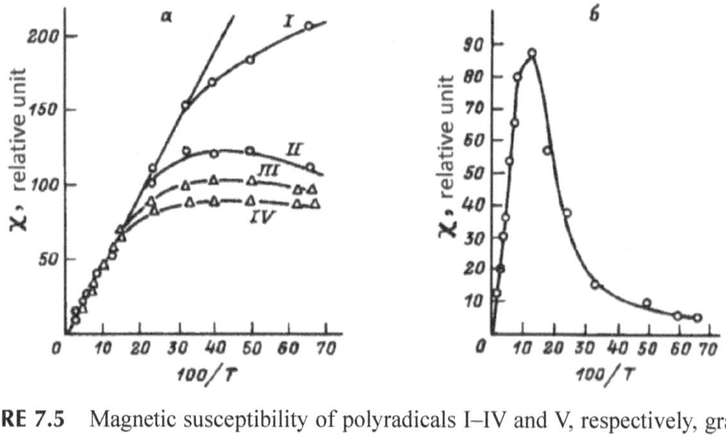

FIGURE 7.5 Magnetic susceptibility of polyradicals I–IV and V, respectively, graphs *a* and *b*.

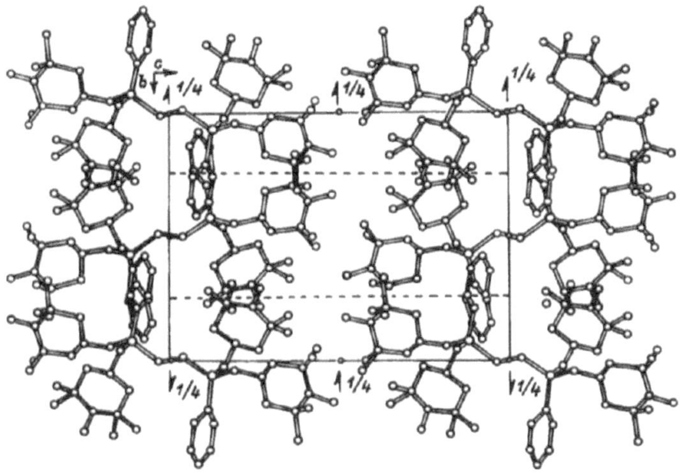

FIGURE 7.6 Projection of the tetraradical V structure onto plane (010).

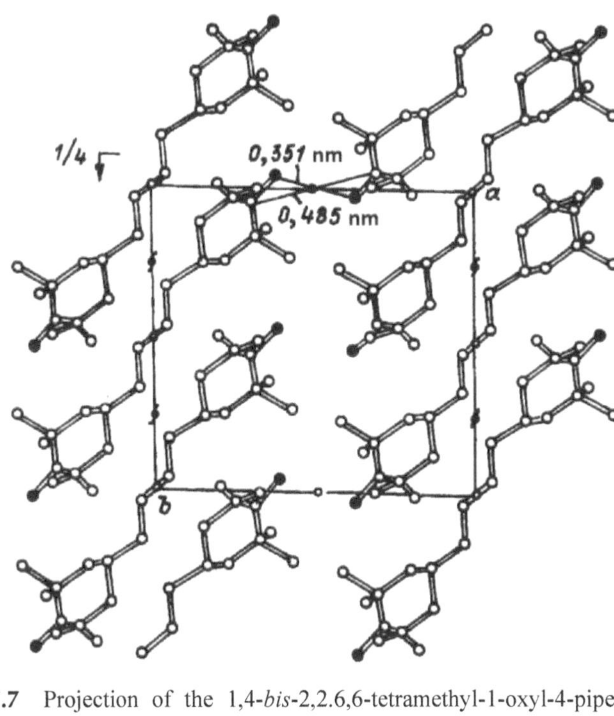

FIGURE 7.7 Projection of the 1,4-*bis*-2,2.6,6-tetramethyl-1-oxyl-4-piperidyl butane biradical structure.

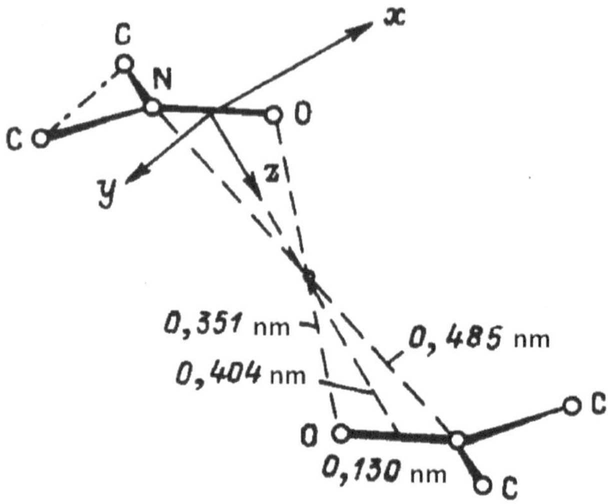

FIGURE 7.8 A scheme of the mutual arrangement of radical fragment =NO• pairs bound with strong exchange.

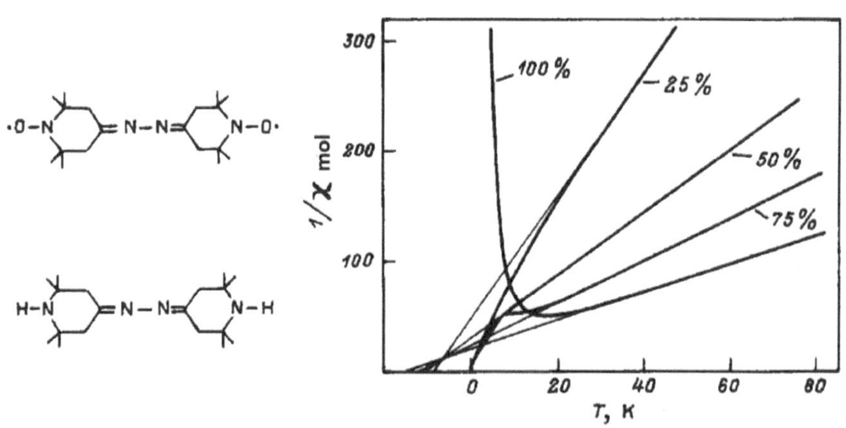

FIGURE 7.9 Temperature dependence of the inverse paramagnetic susceptibility of 2,2,6,6-tetramethyl-4-oxopiperidine-1-oxyl azine crystals in a matrix of triacetonamine azine.

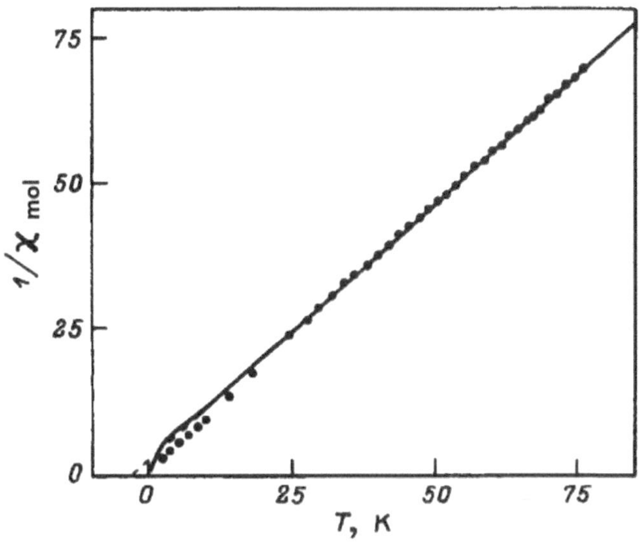

FIGURE 7.10 Temperature dependence of the inverse paramagnetic susceptibility of 2,2,6,6-tetramethyl-4-hydroxypiperidyl-1-oxyl phosphite (θ = −3.5 K). The solid line is calculated for a three-spin cluster with $J = -2.7$ K.

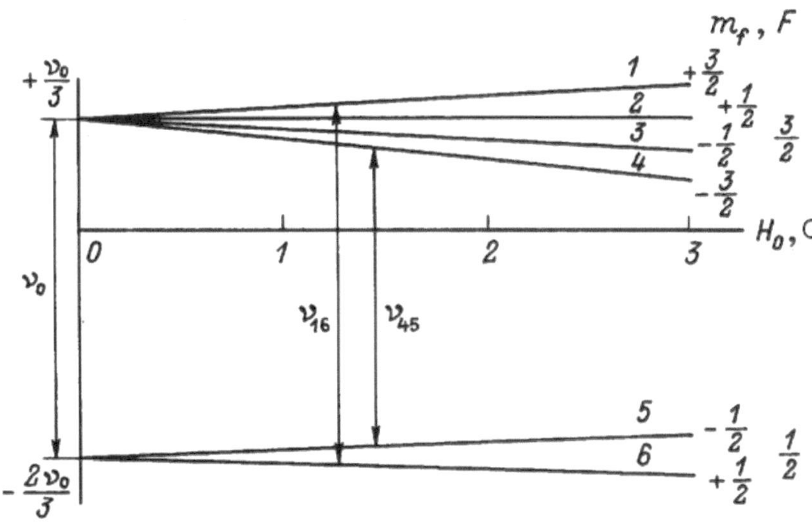

FIGURE 7.11 A scheme of the energy levels of iminoxyl radicals in weak magnetic fields.

7.3 A BRIEF DESCRIPTION OF THE CHEMISTRY OF PARAMAGNETIC SUBSTANCES AS ANTIOXIDANTS

The method of inhibiting the thermal oxidative degradation of polymers with additives of paramagnetic materials, which was proposed half a century ago,[1,66] was the beginning of a new scientific and practical lead, namely, the chemistry of antioxidants, associated with stabilization of organic materials. The stabilization efficiency was due to some peculiarities of the structure and some unique properties of the paramagnetic products of their oxidation:

$$AH + RO\dot{2} \longrightarrow RO_2H + A,$$

where AH is an antioxidant molecule and A, a paramagnetic particle.

It has been experimentally proven that the intermediate paramagnetic particles repeatedly die and regenerate during the oxidation of organic materials, leading to kinetic chain termination according to the scheme:

$$\dot{A} + \dot{R} \rightarrow AR$$

$$AR + RO_2 \rightarrow RO_2R + A$$

$$\dot{A} + \dot{R} \rightarrow AR, \text{ and so forth.}$$

As early as in developing methods of the preparative chemistry of free iminoxyl radicals it was shown that sterically hindered amines (SHA), being precursors of nitroxyl radicals, could be used as efficient methods for inhibiting the photodegradation and thermal degradation of high-molecular-weight compounds.

Features of the preparative chemistry of most effective inhibitors of the photodestruction and thermooxidative destruction of many types of polymers are discussed in more detail in Refs 67 and 68.

Piperidole-based stabilizers are used for the heat and light stabilization of polyolefins, polyvinyl chloride, polyvinylidene chloride, polyamides (PAs), and polyurethanes (they are much more effective as light stabilizers than the commonly known Tinuvin P).

The esters of hindered piperidoles (I) constitute a large group of light and heat stabilizers:

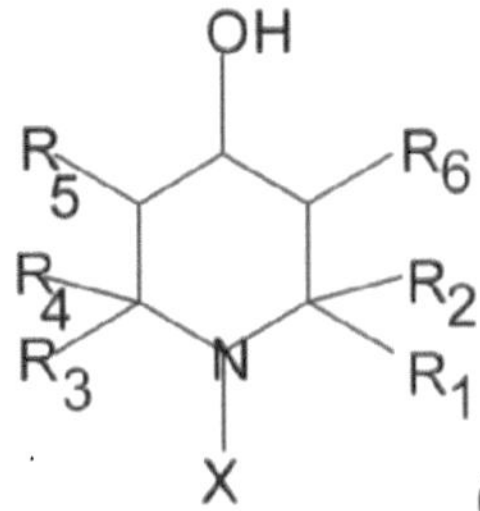

(I)

where R_1,R_2,R_5, and R_6= H or alkyl; R_3 and R_4= alkyl C_1—C_8, $(CH_2)_n$ with n = 4, 5 or $(CH_2CMe_2)N$; X=H, OH, aralkyl, alkenyl, epoxypropyl, amyl, alkoxycarbonylethyl, and so forth.

The inhibitors based on 4-aminopiperidines are effective for the light stabilization of polyurethanes, polyesters, polyimides, cellulose esters, polyvinyl chloride, and other polymers, for example, substituted guanidines like:

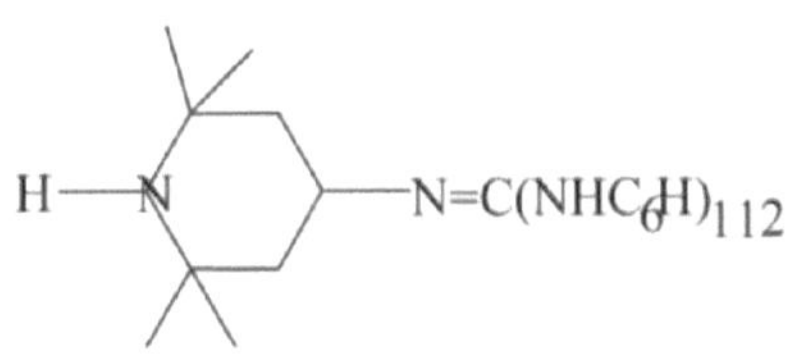

Amidothiophosphates show good light stabilization properties for polyurethanes.

The greatest number of studies on polymer stabilizers based on 2,2,6,6-tetramethylpiperidine is devoted to triacetonamine ketals:

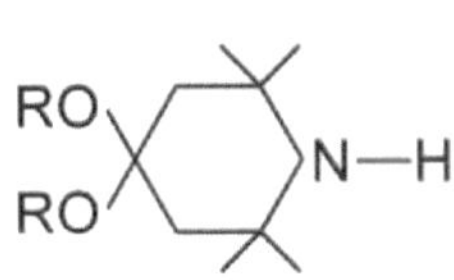

A series of stabilizers based on (I) (R + R = CHCOOH, R = H, and R = CH_2CH_2OH and $CH_2CH_2NH_2$) have been synthesized but the relative complexity of their production process makes them difficult to use. The light stabilizers like:

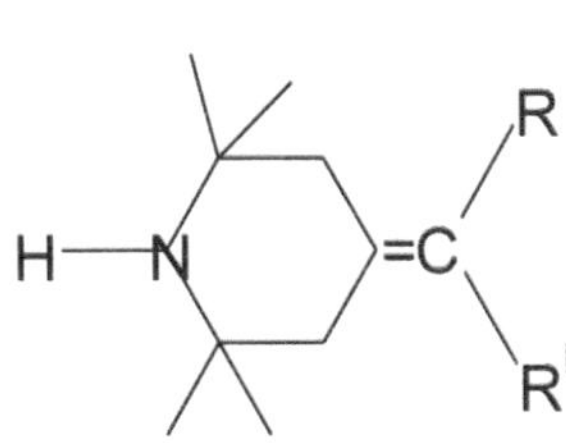

where Rand R' are cyano, aryl, acyl, or carbamoyl groups, are proposed for polyolefins.

Hydantoins and their derivatives are relatively available stabilizers, in particular, the salts of

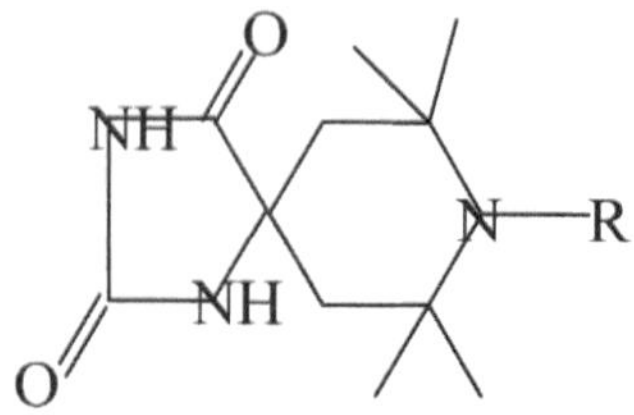

They are suggested for use to improve the heat resistance of polyethylene terephthalate and its copolymers. Hydantoins are prepared by interaction (I) with potassium cyanide and ammonium carbonate in nearly quantitative yield. The compounds containing several inhibitory centers possess the highest efficiency, for example, the compound:

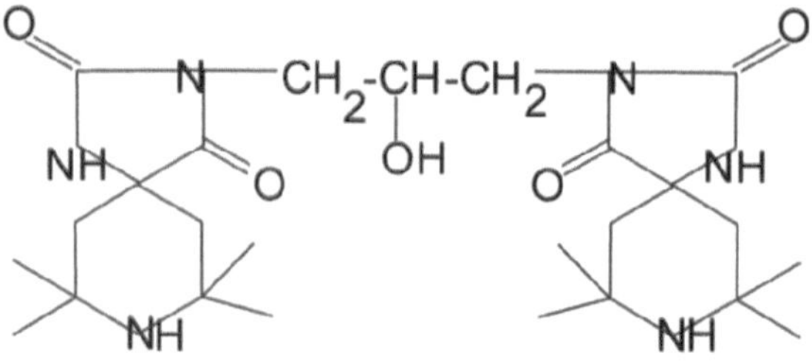

Some heterocyclic compounds close to hydantoin by structure:

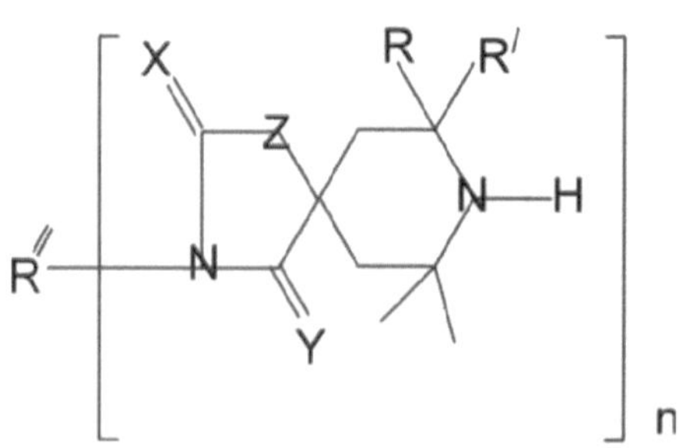

where X = Oor S; Y = Z=O or NH; n = 1–2; and R" = alkyl, benzyl (n = 1), alkylene, and phenylene (n = 2), are used for the light stabilization of polyethylene, polypropylene, polystyrene (PS), polyurethanes, and other polymers. In this case, the exposure time until embrittlement for polypropylene without any additives, with an additive of Tinuvin P, and with an additive of the above compound is 40, 80, and 400–780 h, respectively.

Sulfur-containing compounds obtained by a reaction like

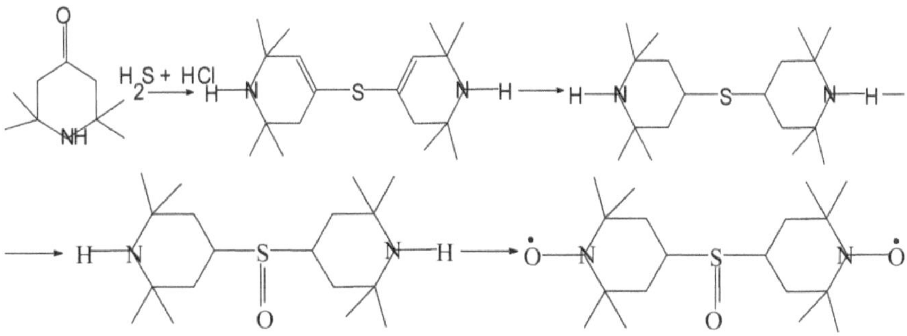

are light and heat stabilizers as well.

The derivatives of benzimidazole, benzoxazole, and benzothiazole are light stabilizers, for example, with the general formula:

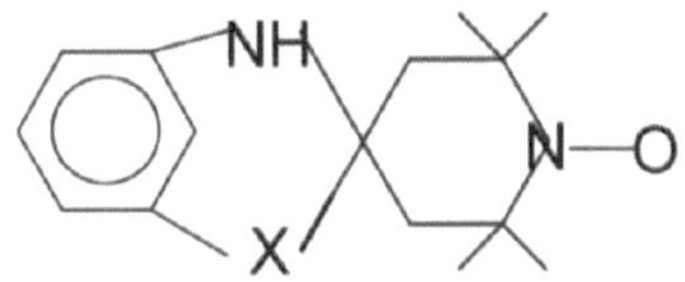

Below is a scheme of the synthesis of light stabilizers with conjugated double bonds capable of joining dienophiles:

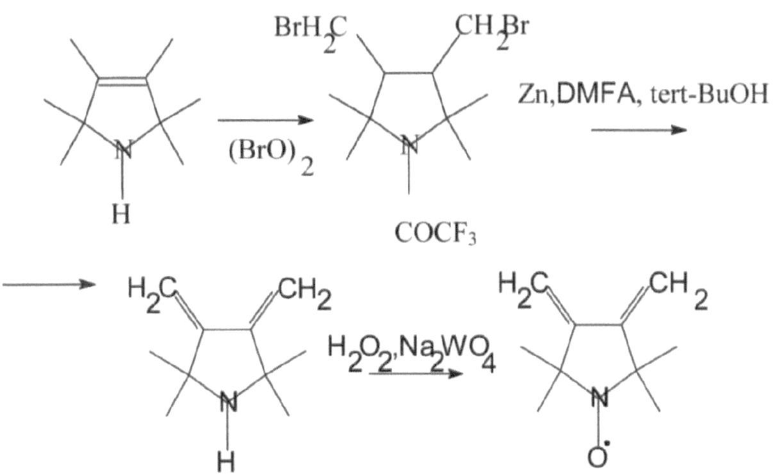

Studies of the antioxidant and light-protective properties of more than 20 compounds (tyroxyl radicals and SHA of the 2,2,6,6-tetramethyl-piperidine series) have led to the following conclusions:

1. biradicals are more active than monoradicals;
2. the antioxidant activity of a biradical decreases as the distance between the two piperidinoxyl moieties increases;
3. as the number of paramagnetic centers increases, the antioxidant activity of the compound increases in the following order:

$$Ph_3SiR < Ph_2SiR_2 < PhSiR_3 < SiR_4 < Ph_2POR < PhP(OR)_2 < R(OR)_3,$$

where R=

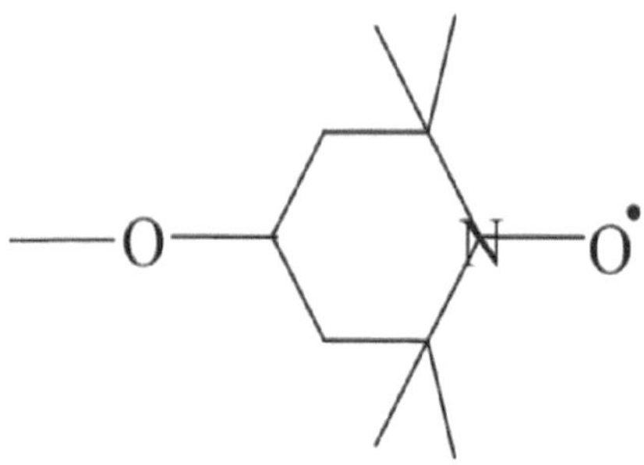

4. the stabilizing properties of radicals are enhanced in a mixture with Topanol® CA, 2,4,6-tri-*tert*-butylphenol or a light stabilizer of screening effect, for example,

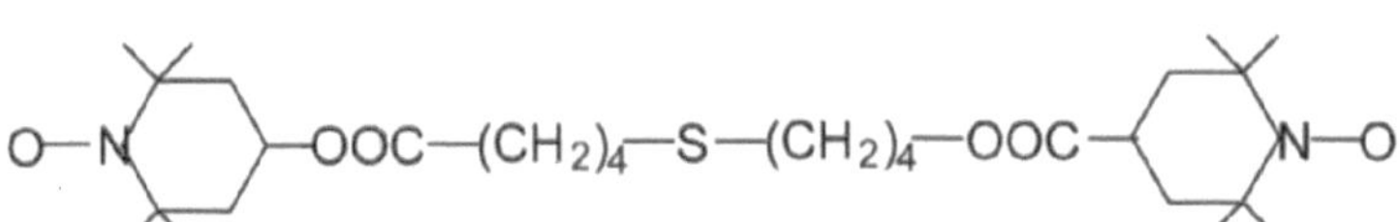

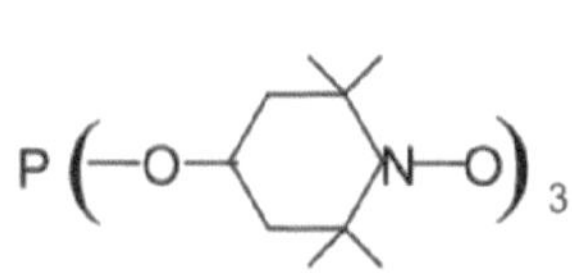

5. the shielded mono-, bi-, and polypiperidines are weak antioxidants but effective light stabilizers; and the incompletely shielded 2,6-dimethylpiperidine exhibits no light protective properties.

As antioxidants for rubber compositions, stabilizers are used based on sterically hindered hydrogenated quinoline derivatives, in particular, 1,2-dihydro-6-ethoxy-2,2,4-trimethylquinolin (ethoxyquin). The latter is also an effective stabilizer of polyolefins and particularly polyethylene.

These compounds are readily prepared from the corresponding aromatic amines and acetone. To protect rubbers, oils, and fats from light and thermal oxidation, the oxy-, *tert*-butyl, and methoxy derivatives of 2,2,4-trimethyl-1-1,2,3,4-tetrahydroquinoline and similar decahydroquinoline are proposed. For the light stabilization of plastics, paper, cotton, and cellulose esters, indoline derivatives are used:

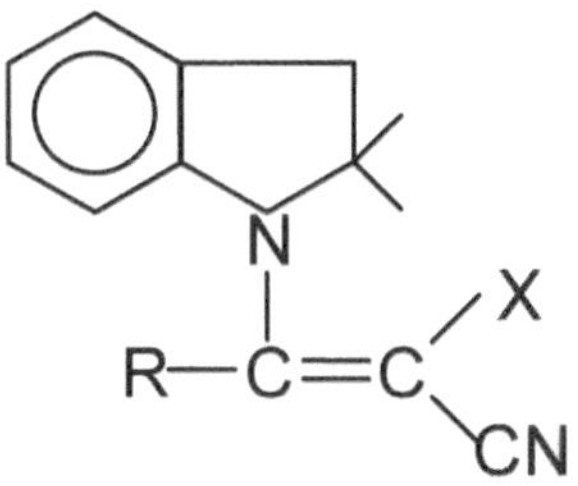

To stabilize polymers against heat and light destruction, sterically hindered heterocyclic amines are proposed, having a structural similarity with 2,2,6,6-tetramethylpiperidine and forming stable nitroxyl radicals by oxidation.

To protect such carbon-chain polymers as PS, polyvinyl chloride, polyvinyl acetate, polyethylene, and propylene against photooxidation, substituted piperazinedione

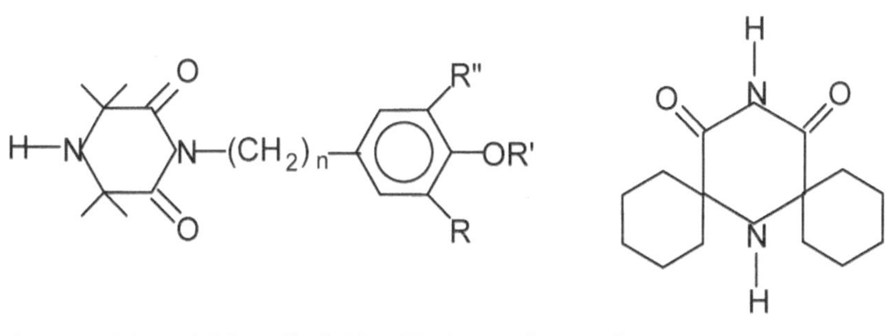

where R, R', and R"—alkyl, X = H, O, can be used.

In most cases, the light-shielding properties of stable nitroxyl radicals are higher than those of the corresponding amines. However, they are used as light stabilizers limitedly due to their low stability and coloration.

The nitroxyl radicals of 2,2,6,6-tetramethylpiperidine can be used not only for inhibiting polymerization (as mentioned above) but also as effective heat stabilizers of polymers and for protection of other substances against oxidation. For example, the paramagnets of this series effectively inhibit the oxidation of chlorinated hydrocarbons, the polymerization of low-boiling monomers during their distillation, and due to the increased volatility of radicals, inhibition occurs not only in the cube but also on the trays of the column, which prevents contamination of the industrial equipment with the ballast polymer.

7.3.1 MECHANISM OF THE STABILIZING EFFECT OF STERICALLY HINDERED AMINES

Hydrocarbon polymers constitute one of the most important classes of polymers requiring stabilization against light and heat. The photooxidative destruction of carbon-chain polymers can be represented by the following scheme:

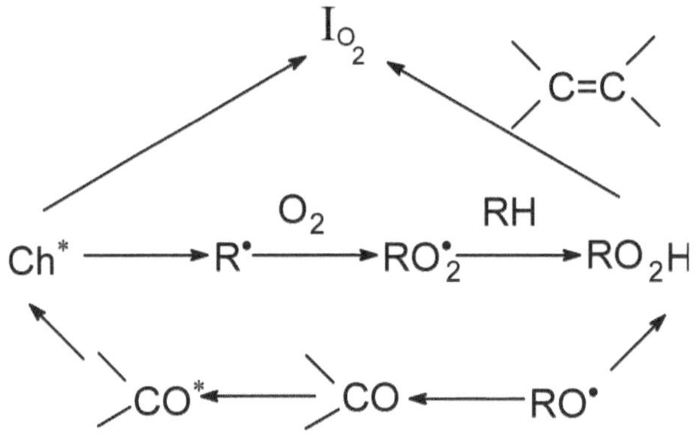

where Ch is a chromophore group or impurities; Io_2 an oxygen molecule in the singlet state; RHa polymer, C=C< unsaturated groups, >C=O carbonyl groups, R•, RO_2,•and RO• are alkyl, peroxyl, and alkoxyl radicals, respectively, *denotes the excited state of the group.

This scheme differs from that of thermal oxidation by some specificity of initiation only, which occurs under the action of light and includes processes involving not only hydroperoxyl and chromophore groups (or impurities) but also carbonyl groups.

In accordance with these representations, light stabilizers must perform one or more of the following functions:

1. to shield the polymer from the action of light, thus reducing the amount of light absorbed by the chromophore groups or impurities (and, therefore, the rate of initiation);

2. to quench the electron-excited states of the chromophore groups or impurities, as well as IO_2, which should lead to a decrease in the rate of initiation;
3. to catalyze the radical-free decomposition of hydroperoxides, reducing their concentration, which should reduce the rate of degenerate branching;
4. to interact with free radicals, terminating oxidation chains.

To date, main efforts in the development of light stabilizers have been directed to the preparation of compounds acting by the first three mechanisms. The fourth one seemed low promising, as it aims "to eliminate the consequences" of light, rather than direct prevention of the chemical changes occurring when light is absorbed. Common antioxidants (phenols and aromatic amines) react almost exclusively with peroxyl radicals, that is, when not only a macromolecule has been disintegrated but also a chemical change of the primary (alkyl) radical has occurred and is has been converted into a peroxyl one. These stabilizers, in addition, strongly absorb light and decompose in their excited state, acting as initiating groups Ch. Therefore, at the light stabilizers of polymers, they are applied only in mixtures with ultraviolet (UV) absorbers, protecting them from light. The first way of stabilization has a natural limit due to the extinction coefficients of organic compounds not exceeding 104–105 mol/(L × s). Therefore, those UV absorbers, which already have such parameters in the UV region and exhibit high light resistance, have approached their theoretical limit as light stabilizers.

The second way is considered ineffective because too high concentrations of quenchers are required in solid polymer at a uniform distribution of stabilizers, which cannot be actually used in practice.

Best results can be expected in the development of additives acting by the mechanisms of decomposition of hydroperoxides and interaction with free radicals, in spite of the fundamental difficulties associated with the polymer being a low-mobile medium. Such additives may also be effective, which do not directly act as stabilizers but are rapidly converted to a stabilizer in the oxidative degradation of the polymer.

Without going into further detail, we note the most important features of the action of sterically hindered amines:

- the mechanism of SHA action differs from that of most other light stabilizers, because it is not directly related to the quenching of excited states, shielding, hydroperoxide decomposition, or reacting with free radicals;

- the light-shielding effect of SHA is due to the products of their transformation, that is, stable nitroxyl radicals, hydroxylamines, and hydroxylamine esters.

Based on the data obtained for model systems, three possible reactions of SHA are usually considered:

$$>NR \xrightarrow{RO_2} >NO^{\bullet}$$

$$>NR \xrightarrow{I_{O_2}} >NO^{\bullet}$$

$$>NR \xrightarrow{ROOH} >NO^{\bullet}$$

The contribution of each process depends on the nature of the polymer, sample preparation conditions, and the properties of additives.

7.3.2 MECHANISM OF ACTION OF NITROXYL RADICALS, HYDROXYLAMINES, AND HYDROXYLAMINE ETHERS

The antioxidative action of nitroxyls depends on their ability to react with alkyl and hydroxyl radicals:

$$>NO^{\bullet} + R^{\bullet} \rightarrow >NOR$$

$$>NO^{\bullet} + OH^{\bullet} \rightarrow >N(O)OH$$

By their reactivity to alkyl radicals, nitroxyls are significantly superior to inhibitors of other classes, but usually are inferior to molecular oxygen. Nitroxyl radicals interact with hydroxyl radicals in aqueous solution at a rate constant of 10^9 L/(mol × c). This reaction can be very important, since the hydroperoxide decomposition process to form hydroxyl radicals is most important for initiation at the photooxidation of many polymers.

Compared with other antioxidants, such as sterically hindered phenols and aromatic amines, nitroxyls have three most important advantages, namely:

1. they enter into an addition reaction with active radicals while phenols and amines enter into a substitution reaction:

$$InH + RO_2^{\bullet} \rightarrow In^{\bullet} + ROOH,$$

where InH is phenol or amine; in this case, hydroperoxide is formed, involved in degenerated branching reactions.

2. the hydroxylamine esters formed by recombination of nitroxyls with alkyl radicals do not absorb light with $\lambda > 300$ nm, and therefore, are photochemically inactive; the reaction products of phenols and amines strongly absorb light and are oxidation sensitizers;
3. the reaction products of nitroxyl radicals, that is, hydroxylamine esters, are inhibitors, and, by their interaction with peroxyl radicals, nitroxyls can be regenerated:

$$>NOR+RO_2 \rightarrow NO^{\bullet}+ROOR$$

As a result, oxidation chains are broken many times at one inhibitor molecule. At the same time, nitroxyl radicals have some disadvantages typical for antioxidants. For example, at elevated temperatures or by light, they can react with the polymer or the products of its oxidation as destruction initiators:

$$>NO^{\bullet} + RH \rightsquigarrow NOH + R^{\bullet}$$

$$>NO^{\bullet} + ROOH \rightarrow NOH + RO_2$$

$$>NO^{\bullet} + RH \rightarrow >NOH + R^{\bullet}$$

The latter reaction, while playing a certain role, does not significantly decrease the efficiency of the light-protection action of nitroxyls, because it proceeds only under the influence of light, corresponding to the short-wavelength absorption band of nitroxyls. In this range, nitroxyls absorb much weaker than phenols and aromatic amines. Moreover, the hydroxylamines formed during phototransformation are extremely effective antioxidants, whereas the phototransformation of common antioxidants produces sensitizers, which, in turn, initiate polymer photooxidation. Hydroxylamines are effective inhibitors, which, unlike nitroxyls, react with peroxyl radicals:

$$>NOH+RO_2^{\bullet} \rightarrow >NO^{\bullet} + ROOH$$

The rate constant of this reaction is close to the corresponding constants of most active phenols. An advantage of hydroxylamines compared with phenols is the formation of a stable nitroxyl radical (which is also an inhibitor) as a result of oxidation chain termination. A demerit of hydroxylamines (like phenols) is due to the formation of hydroperoxide at chain termination and a possible reaction with it

$$>NOH + ROOH \rightarrow RO^{\bullet} + H_2O + >N\text{-}O^{\bullet}$$

Hydroxylamine esters can react with peroxyl radicals:

$$>NOR + RO_2 \rightarrow >NO^{\bullet} + \text{-C=C-} + ROOH$$

The nitroxyl radical is again regenerated.

In the presence of sufficiently effective acceptors of alkyl radicals (e.g., oxygen), the formation rate of nitroxyl radicals is significantly higher than that of hydroxylamine in vacuum. The higher the degree of electron delocalization in the radicalR• and the higher the solvent polarity, the higher the decay rate constant of the hydroxylamine ester.

Analysis of the reaction rate constants in model system indicates that at room temperature the main reaction leading to the regeneration of nitroxyl radicals is interaction with peroxide radicals, and, at elevated temperatures (beginning from about 80°C), it is basically the decomposition reaction of the hydroxylamine ester, especially, at low initiation rates. Therefore, in analysis of the mechanism of SHA action it is necessary to consider the following reactions:

$$> NH \xrightarrow{RO_2\ Io_2\ \ or\ \ ROOH} NOH^{\bullet}$$

$$>NO^{\bullet} + R \rightarrow >NOR$$

$$>NO^{\bullet} + RH \xrightarrow{h\nu, t} >NOH + R^{\bullet}$$

$$>NOH + RO_2 \rightarrow >NO^{\bullet} + R$$

$$||$$

$$>NOR + RO_2 \rightarrow >NO^{\bullet} + \text{-C=C-} + ROOH$$

$$>NOR \xrightarrow{O_2} >NO^{\bullet} + RO_2$$

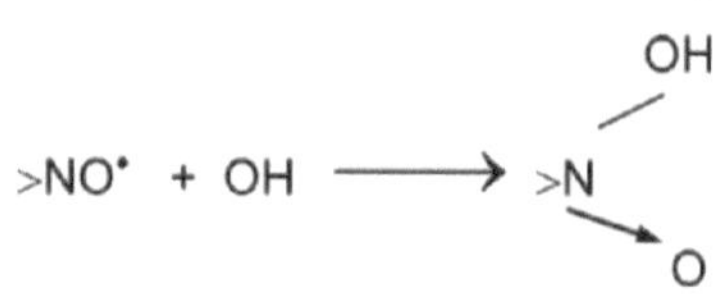

$$>NO^{\bullet} >NO^{\bullet} + \text{molecular products.}$$

This explains the complexity of not only quantitative but also qualitative description of destruction of the stabilized polymer, especially in the presence of additives. However, this scheme makes it possible (in some cases) to produce correct predictions of the effect of the stabilizer structure or the composition of a mixture of stabilizers on the light resistance of the polymer. Of great practical and theoretical interest is the dependence of the stabilization effect on the concentration of the stabilizer.

When the polymers stabilized with nitroxyl radicals are thermally oxidized, as noted previously, critical phenomena are possible, leading, in particular, to an increased relative efficiency of nitroxyls at concentrations exceeding the critical value. Analysis of the oxidation rate constants of the polymer and the interaction of hydroxylamine esters with peroxyl radicals in model systems allows one to obtain a value less than 0.02 mol L^{-1} for the critical concentration of the hydroxylamine ester or for the initial concentration of nitroxyl in photooxidation processes, which corresponds to 0.5% of the nitroxyl radical (i.e., the commonly used concentrations).

Really, the critical concentration should be significantly lower since nitroxyl radicals, like other additives, are distributed in the amorphous phase of the polymer, whose percentage, for example, is 40% for polypropylene. Meanwhile, the induction period of polymer photooxidation in the presence of SHA or nitroxyl radicals at concentrations ≤ 0.5 is proportional to the concentration, and changes with its further increase, relatively weak, reaching a constant value at concentrations $\geq 1\%$, which is probably due to the limited solubility of SHA and nitroxyls in polymers. The presence of a straight line fragment on the dependence of the induction period on the concentration of SHA or nitroxyl radical indicates no critical phenomena at photooxidation of the polymers stabilized in such a way. Apparently, this is due to some peculiarities of the photooxidation process and to the presence of photosensitizer functions in nitroxyls.

7.4 STABLE RADICALS (PARAMAGNETS) IN THE POLYMERIZATION KINETICS OF VINYL MONOMERS[69,70]

7.4.1 HYDROCARBON RADICALS

The American chemist Moses Gomberg, who synthesized hexaphenylethane, thought it as a purely triphenylmethyl radical. Only in 1909 it was found that Gomberg's radical was a dimer being in thermodynamic equilibrium with the true triphenylmethyl radical. For example, only about 2% of the dimer dissociates into free radicals in benzene solution at room temperature. Interestingly, a very stable hydrocarbon with melting point 234°C was found among the dimers of triphenylmethyl:

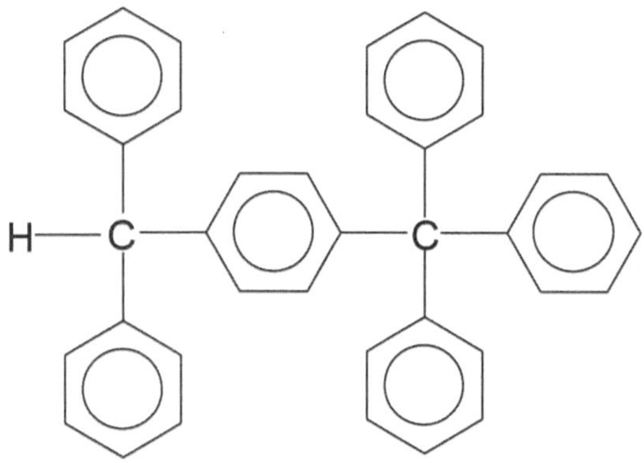

The triphenylmethyl radical inhibits the thermal polymerization of styrene. In this case, the measured duration of the induction period of polymerization is significantly shorter than that calculated by the inhibitor consumption. The dimer dissolved in styrene serves the source of triphenylmethyl radicals, 77 of its molecules being required to terminate a reaction chain.

In the presence of benzyl and diphenylmethyl radicals, which are more reactive than triphenylmethyl, the polymerization of acrylonitrile and methyl methacrylate has no induction period and its rate is comparable with that of an initiated reaction.

7.4.2 HYDRAZYLS

A large number of studies are devoted to the influence of 1,1-diphenyl-2-picrylhydrazyl (DPPH) on the polymerization rate of vinyl monomers. The high stability and relative availability of DPPH have allowed its use to study the photolysis and radiolysis of various compounds, cellular effects, and the dissociation of initiators. DPPH is an effective inhibitor of the polymerization of styrene and vinyl acetate. With increasing concentration of DPPH, the duration of the induction period (its relative value) reduces, which is associated with additional initiation due to the paramagnetic material. About 10% of the initial amount of DPPH is spent on chain origin. At the same time, it should be noted that the amount of hydrazyl spent to initiate the reaction depends on the nature of the monomer. By the end of the induction period, the stationary polymerization rate decreases. Despite the bifunctionality of DPPH, many researchers consider it possible to use this paramagnetic material to determine the kinetic characteristics of polymerization, such as the chain initiation rate, the rate constants of chain growth and termination. When DPPH interacts with a growing chain, a complex mixture of products is formed: diphenyl, diphenylamine, 1,4-cyanisopropylphenyl-1-phenyl-2-picrylhydrazine, and 1,1-diphenyl-2-picrylhydrazine. If the reaction proceeds in the presence of oxygen, peroxyl radicals reacting with DPPH further complicates its mechanism. The possibility of interaction of a DPPH particle with multiple free radicals was first observed experimentally in a study of the dissociation of α,α′-azo-*iso*butyric acid dinitrile (AAD). The death of dimethylcyanomethyl radicals with the participation of DPPH is accompanied by the regeneration of the latter:

$$CH_3\dot{C}CN + DPPH \rightarrow CH_2{=}C(C_3H)\ CN + H\text{-}DPPH$$

DPPH reacts ionically with mineral and organic acids:

$$DPPH + L^- \rightarrow + [DPPH\text{ -}L]^-$$

$$[DPPH\text{ - }L]^- + H^+ \rightarrow DPPH\text{ - }H + 0.5\ L_2$$

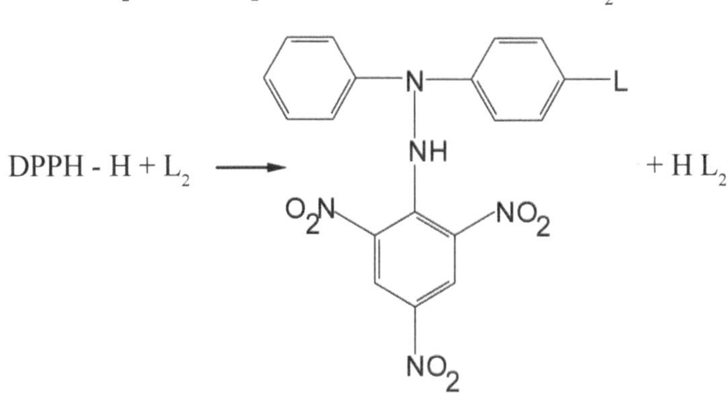

where L is an anion.

In this regard, the inhibition of polymerization processes with DPPH cannot be evidence of the radical nature of the reaction.

In the polymerization of vinyl acetate in the presence of DPPH additives, the induction period duration is not equal to the time of complete disappearance of the stable radical in the reaction system. Similar results were also obtained at the DPPH-inhibited thermal polymerization of styrene. The formation rate of reactive radicals was seven times higher than the value calculated from the ratio of the induction period to the initial inhibitor concentration. This difference in the initiation rates is due not only to the stoichiometry of inhibition but also to some features of the mechanism of thermal initiation and the possibility of DPPH degradation by UV radiation.

The reliability of the results obtained when DPPH is used to study the decomposition kinetics of various compounds and to determine their initiating activity is provided only when the concentration ratio of the paramagnetic material and initiator is not more than 0.5. Therefore, when using DPPH as a counter of active radicals, one needs to know and to take into account all side chemical reactions.

The reduced stationary rate settled by the end of the induction period with an increased concentration of the stable radical is most often associated with the presence of nitro groups. For example, at the polymerization of styrene in the presence of 1,1-diphenyl-2- (2', 6'-dinitrophenyl)-hydrazyl, additional initiation is absent and the stationary rate reduction is much smaller than in the case of DPPH. The inhibitory activity of hydrazyls containing no nitro groups, at the polymerization of vinyl monomers, depends on not only the structure of the paramagnetic substance but also the presence of oxygen in the system. The only exception is 1,1-diphenyl-2 (2', 4', 6'-tricarbometoxyphenyl)hydrazyl, which can be used for quantitative measurements of the rate initiation, since oxygen does not affect its inhibitory effect.

7.4.3 VERDAZYLS

In studies of the polymerization of styrene, methyl methacrylate and other esters of acrylic and methacrylic acid, the high efficiency of the inhibiting action of 1,3,5-triphenylverdazyl and the possibility of its use for determining the chain initiation rate have been found. The products of recombination of verdazyls with active radicals are capable of homolytic dissociation to form a stable radical. However, in the polymerization of some acrylic monomers (e.g., oligourethanacrylate) no verdazyl regeneration is observed and the inhibition efficiency markedly reduces as temperature increases. Unlike the monoradical of (2,4,6-triphenyl-3,4-dehydrotetrazine-1-verdazyl), the participation of phenylene-*bis*-verdazyl

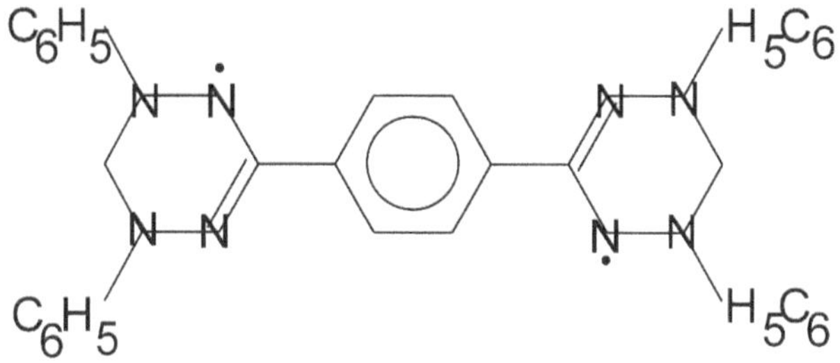

in the reaction leads to a drastically reduced polymerization rate.

Triphenylverdazyl reacts with acrylonitrile by the double bond with an activation energy of 45 kJ mol^{-1}, which indicates the relatively high initiating activity of verdazyl compared with many other paramagnetic materials.

The accuracy of determining the initiator decay rate using triphenylverdazyl depends on the inhibitor–initiator concentration ratio. This ratio should not exceed 2.

The ability of verdazyl radicals to relatively rapid oxidation by peroxide compounds at relatively low temperatures (10–20°C) allows determining polymerization initiation characteristics in their presence.

7.4.4 AROXYLS

In this class of radicals, only galvinoxyl is used as a counter of active radicals during polymerization:

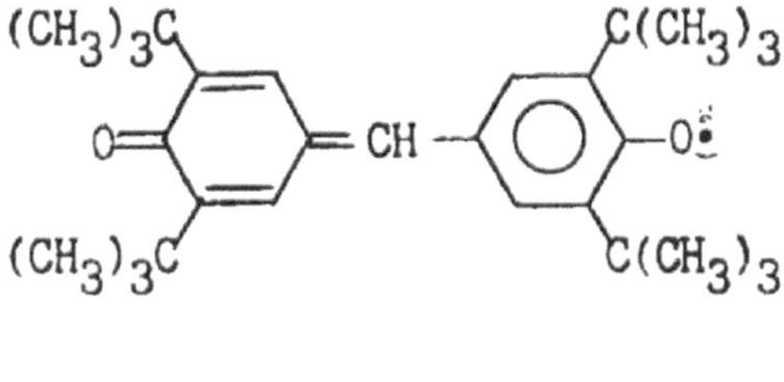

The decomposition reactions of azocumene, some peroxides, AAD, and the mechanism of the thermal polymerization of deuterated pentafluorostyrene have been studied with its help. Galvinoxyl interacts with dimethylcyanomethyl radicals mainly by recombination. At the same time, when using it for accurate kinetic measurements, one must take into account the conditions of the reactions studied, since they determine the photochemical and thermal stability of galvinoxyl.

7.4.5 *NITROXYLS*

Until the early 1960s, only one classic nitroxyl (Banfield's radical, or the nitroxyl–nitroxide radical) was used as a polymerization inhibitor and free-radical acceptor:

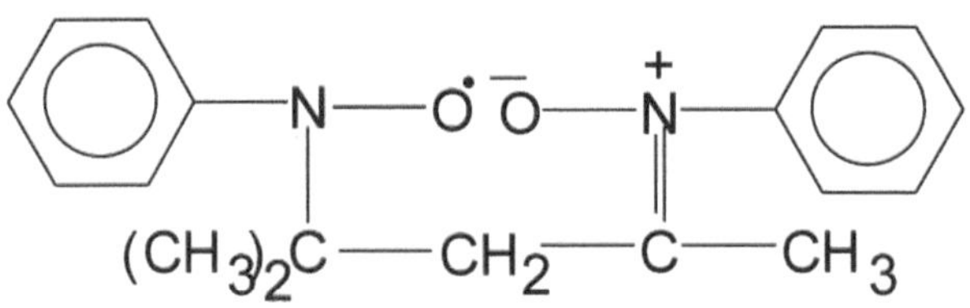

This radical can inhibit most vinyl monomers. Banfield's radical is low stable in air but stable in an inert atmosphere. The induction period of polymerization is usually shorter than the theoretically calculated value, which is due, according to many researchers, to the participation of this radical in the initiation reaction.

The effect of Banfield's radical on the polymerization rate depends on the monomer nature, and, when copolymerization, the monomer concentration ratio. This radical is not suitable to find quantitative characteristics of acrylonitrile polymerization since it reacts with the monomer itself. In the copolymerization of acrylonitrile with styrene, the stoichiometric inhibition ratio exceeds unity. At the copolymerization of acrylonitrile and methyl methacrylate, the contribution of Banfield's radical to chain origination is negligible. When styrene is copolymerized with maleic anhydride in acetonitrile, acetone and tetrahydrofuran solutions, the induction period of the reaction increases with decreasing monomer concentration.

As mentioned above, the successful development of the chemistry of stable radicals in the past 25 years has led to the discovery of a new class of highly stable non-aromatic nitroxyls (iminoxyls) with localized free radical centers and nitroxides with delocalized ones.

The classical aromatic nitroxyls react with $RO_2^{\bullet}$, for example, at 60°C, and the rate constant of the reaction of 4,4-dimethoxydiphenyl-nitroxyl with peroxyl radicals is 6×10^5 $dm^3/(mol \times s)$. The product of interaction of arnitroxyl with peroxyl radicals is most likely a quinoid compound

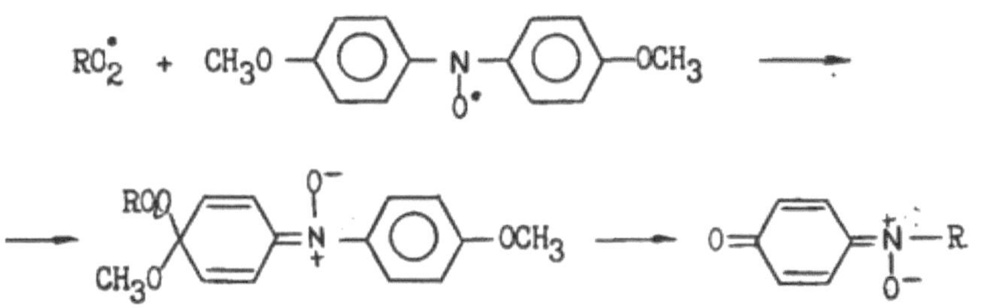

This scheme is well consistent with data on the accumulation of quino-nitroxide compounds in the reaction products. In contrast to the classical aromatic nitroxyls, non-aromatic nitroxyls recombine with alkyl radicals much easier than with peroxyl ones. Iminoperoxyl-type peroxyl radicals are formed by oxidation of primary and secondary aliphatic amines:

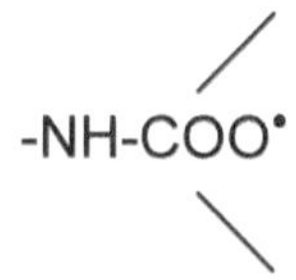

These radicals react with nonclassical nitroxyls to produce the corresponding hydroxylamine and the subsequent regeneration of the paramagnetic material:

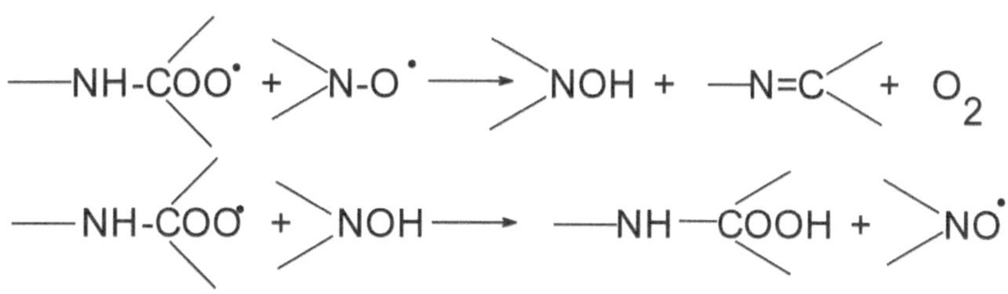

Similar results were obtained when inhibiting the oxidation of dimethylamino ethyl methacrylate. When the alkyl esters of methacrylic acid are oxidized, polyperoxyl radicals having no reduction properties are active intermediate chain centers. Chain termination occurs as a result of the interaction of nitroxyls with alkyls, and the inhibitor is not regenerated. When a tertiary amino group is introduced into methacrylate, the termination mechanism changes and nitroxyl is regenerated.

The ester of heterocyclic hydroxylamine is the primary product of the reaction of 2,2,6,6-tetramethyl-4-oxo-piperidine-1-oxyl with the dimethyl cyan methyl radical.

$$>NOR + ROO^{\bullet} \longrightarrow >NO^{\bullet} + \text{products}$$

The common opinion is that recombination is the only way of interaction of nitroxyls with active radicals. For example, the unique ability of nitroxyls to regenerate in their inhibition of the oxidation of olefins, polyolefins, and hydrocarbons is explained on this basis. However, some works are not consistent with these ideas.

For example, 2,2,6,6-tetramethyl-4-oxo-piperidine-1-oxyl regenerates approximately 10 times faster than if this regeneration was carried out by the above reaction. This is due to the fact that chain termination occurs not only through recombination but also through disproportionation leading to the formation of hydroxylamine.

The efficiency of nitroxyls as free radical acceptors has promoted their use to study polymerization initiation mechanism. It has been found that adding nitroxyl into the monomer–initiator system suppresses polymerization, but nitroxyl has time to react with a portion of the initiator radicals only, and at the use of peroxides (as initiators) it does not interact with the primary radicals at all. Therefore, the products of the reaction of primary radicals with monomer are formed in almost quantitative yield in this case. Under certain conditions, nitroxyl can catalyze peroxide decomposition.

In a study of the interaction of stable biradicals with primary radicals from AAD, it was found that the probability of recombination of the radical center depended on the remaining unpaired electron in the biradical. The first reaction (with a localized free-radical center) proceeds easier than the second one, which is explained in the framework of the adiabatic nature of the recombination reaction.

Under the conditions of the thermal polymerization of styrene and α-methyl styrene, a portion of the introduced stable radicals is consumed in reactions with the products initiating the process. Assuming thermal initiation by Mayo's mechanism, an equation was obtained for the rate of nitroxyl consumption in the monomer:

$$-\frac{d[X]}{dt} = 2k_{ДА}[M]^2 + 2k_M[M][X],$$

where $k_{ДА}$ is the rate constant of the reaction between iminoxyl with Diels–Alder's adduct; [M] the monomer concentration.

2,2,6,6-tetramethyl-4-hydroxypiperidine-1-oxyl was one of the first nitroxyls used as counters of active centers in the polymerization of tetrafluoroethylene. In this case, the polymerization induction period is directly proportional to the initial inhibitor concentration, and the reaction does not decelerates after its completion.

When the polymerization of styrene and methacrylic acid esters is inhibited with 2,2,6,6-tetramethyl-4-oxo-piperidine-1-oxyl, the recombination of active and stable radicals leads to the formation of the corresponding esters. The products do not affect the polymerization rate and the recovered radical (hydroxylamine) causes a strongly decreased reaction rate.

Most of the data obtained contradict the viewpoint of a simple termination stoichiometry. For example, a comparative evaluation of the inhibitory activity of iminoxyl and arnitroxyl radicals at the initiated polymerization of vinyl monomers shows that the stoichiometry of the reaction with macroradicals varies for different nitroxyl classes.

2,2,6,6-tetramethyl-4-oxo-piperidine-1-oxyl, under identical conditions, causes a longer polymerization induction period of oligourethanacrylate than 2,2,6,6-tetramethyl-4-hydroxypiperidine-1-oxyl does. This means that 2,4-*bis*-(2,2,6,6-tetramethyl-4-ureido-piperidine-1-oxyl)-toluylene biradical inhibits oligourethanacrylate polymerization more effectively than would be expected from a simple termination stoichiometry. When the copolymerization of styrene with acrylonitrile is inhibited with 2,2,6,6-tetramethyl-piperidine-1-oxyl, the stoichiometric inhibition ratio depends on the initiator nature and significantly exceeds unity when AAD is used.

At the polymerization of many known vinyl monomers in the presence of nitroxyl radicals, the nitroxyl consumption rate is lower than the initiation rate (AAD being the initiator). Earlier, it was associated with the inhibitor regeneration. However, the polymerization rate often becomes lower than the uninhibited process rate after the end of the induction period. Furthermore, the induction period duration relatively reduces with an increased inhibitor concentration (this is typically due to additional initiation). Consequently, the effective inhibition of polymerization by iminoxyls is accompanied by complex transformations. Esters do not affect the polymerization kinetics, and only the process rate decreases in the presence of hydroxylamines. Consequently, the regeneration can be explained by the occurrence of disproportionation reactions of iminoxyls with growing chains, resulting in the formation of nitroso compounds. They inhibit the polymerization, terminating two kinetic chains per inhibitor molecule.

The radical chain polymerization scheme of vinyl monomers initiated by azo compounds in the presence of iminoxyls can be represented as follows:

$$I \longrightarrow 2R_o^{\cdot} + N_2\uparrow (I)$$
$$R_o^{\cdot} + M \;\; R^{\cdot} (II)$$
$$R^{\cdot} + M \xrightarrow{k_u^o} R^{\cdot}(III)$$
$$X + M \xrightarrow{k_p} R^{\cdot} (IV)$$
$$R + M \xrightarrow{k_1} RX(V)$$
$$X + R \xrightarrow{k_2} XH + \text{inactive products (VI)}$$
$$X + M \xrightarrow{k_3} R'\text{-N=O} + HM \text{ (VII)}$$
$$R'\text{-N=O} + R^{\cdot} \longrightarrow + R'\text{-N-R (VIII)}$$

|

O˙

$$R'\text{-N-R} + R^{\cdot} \longrightarrow + R'\text{-N-R (IX)}$$

| |

O˙ OR

$$X + R^{\cdot} \longrightarrow X^{\cdot} + RH \text{ (X)}$$
$$R^{\cdot} + R^{\cdot} \longrightarrow \text{polymer (XI)}$$

Additional initiation by stable radicals is slow, secondary inhibition (retardation) is insignificant, and the fraction of quadratic termination in the total chain termination is small during the induction period. Therefore, without considering reactions (IX), (X), and (XI), the kinetic equation describing the consumption of iminoxyl reads:

$$-\frac{d[X]}{dt}=W_{u}(k_1+k_2+k_3)/(k_1+k_2+k_3) \qquad (1)$$

The stoichiometric ratio μ of inhibition in these cases is calculated according to the formula:

$$\mu=1+\frac{2k_3}{k_1+k_2+k_3} \qquad (2)$$

The time dependence of the inhibited polymerization rate obeys Bagdasarian-Bamford's equation:

$$-\frac{1}{\phi}+\ln\frac{1+\phi}{1-\phi}=(k_X+k_3)[R]_{\infty}t+const \qquad (3)$$

where $\phi=\frac{W_t}{W_\infty}$ is the ratio of the polymerization rate at any time to the stationary rate W_∞, $[R]_\infty$ is the stationary concentration of polymeric radicals; $k_X = k_1+k_2$.

Equations 1 and 2 are valid at relatively low temperatures (40–60°C). As temperature increases, the consumption of iminoxyls increases by suppressing thermal initiation and joining them to the monomer, and the experimentally determined value of $\mu_{оп}$ reduces. The contribution of these reactions can be taken into account if the inhibitor consumption rate in the monomer at an elevated temperature is calculated $\omega_o = (-d[X]/dt)_o$. Then, denoting the iminoxyl consumption rate in the inhibition reactions as $\omega = d[X]/dt$, we obtain:

$$W_и = \mu_{оп}(w_o+w). \qquad (4)$$

The true value of the stoichiometric inhibition coefficient μ is given by

$$\mu=\frac{\mu_{on}W_u}{W_u-\mu_{on}\varpi_o} \qquad (5)$$

The set of eqs 2, 3, and 5 makes it possible to evaluate the rate inhibition constants k_X, the inhibitor regeneration rate constants k_3, and the activation energy (*E*) of these reactions (Table 7.6). Comparing k_X, k_3, E_X, and E_3, one can conclude that the substituent in the iminoxyl radical has a major impact upon the pre-exponential (steric) factor rather than the activation energy.

TABLE 7.6 Kinetic Parameters of Inhibition of the Polymerization of Styrene and Methyl Methacrylate with Some Iminoxyl Radicals (60°C, AAD Being the Initiator)*

Radical	K_x 10^{-5}, $dm^3(mol\ s)^{-1}$	K_3 10^{-5}, $dm^3(mol\ s)^{-1}$	E_x, kJ mol^{-1}	E_3, kJ mol^{-1}
(A)	5.0, 6.4	1.9, 1.6	16, 12	26,23
(B)	7.0, 6.8	1.5, 1.2	13, 14	26, 29
(C)	2.0, 4.9	0.7, 1.6	12, 10	21, 22
(D)	2.5, 6.8	0.8, 1.2	13, 13	23, 27
(E)	3.5	0.7	11	25, –
(F)	3.0, 6.0	0.2	16, 17	36, –

*The first and second values are for styrene and methyl methacrylate, respectively.

**(A)– 2,2,6,6-tetramethyl-4-oxo-piperidine-1-oxyl ester of α-cyano-β,β′-diphenylacrylate acid,

(B)– 2,2,6,6-tetramethyl-4-salicylaliminopiperidin-1-oxyl,

(C)– 2,2,6,6-tetramethyl-4-palmetyliminopiperidin-1-oxyl,

(D)–2,2,6,6-tetramethyl-4-stearyliminopiperidin-1-oxyl,

(E)–2,2,6,6-tetramethyl-4-hydroxypiperidine-1-oxyl,

(F)–2,2,6,6-tetramethyl-4-benzoylpiperidin-1-oxyl.

The kinetics and mechanism of nitroxyl-inhibited polymerization significantly change at the transition from azo compound initiation to peroxide initiation. The mechanism of interaction of iminoxyls with peroxide compounds is determined by the chemical structure of the latter and the nature of the solvent. When the iminoxyl concentrations exceed the peroxide concentrations approximately twice, the induced decomposition of peroxide occurs in the monomers to form a heterocyclic nitroxide (nitron) and the corresponding acid.

When the concentration of peroxide is much higher than the nitroxyl paramagnetic concentration, the formation of byproducts is apparently suppressed and the induced decomposition accompanied by the formation of a donor–acceptor complex Q_1 becomes the basic process:

$$ROOR + X \rightleftharpoons Q_1 \rightleftharpoons 2RO^{\bullet} + X$$

Therefore, iminoxyl paramagnets are unsuitable for determining the reaction rate of chain initiation by peroxide initiators.

The significant differences in the reactivity of iminoxyls and aromatic nitroxyls lead to their different effects on the peroxide-initiated polymerization kinetics. While the role of iminoxyls reduces to accelerating the decay of peroxide (except, of course, inhibition), nitroxides (e.g. 4,4-dimethoxidiphenylnitroxyl, DMFN) react with benzoyl peroxide (BP) to form non-radical products. The reaction in the monomers proceeds by a more complex multistage mechanism:

$$BP{+}DMFN \quad 2C_6H_5COO{+}DMFN$$

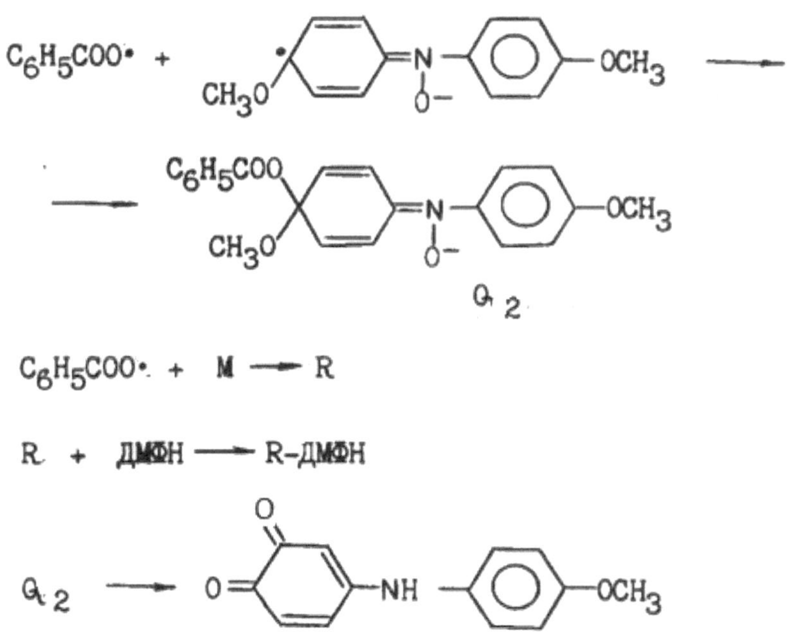

The formed aminoquinone (P) is the most likely polymerization inhibitor. As can be seen from the scheme, if two stable radicals are spent per PB molecule in an inert atmosphere, the stoichiometry changes due to arnitroxyl regeneration in the presence of air:

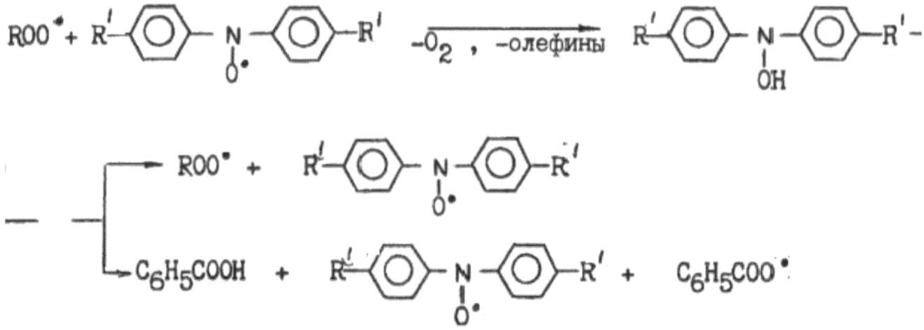

The change in stoichiometry leads to no new kinetic effects. At the end of the induction period due to the presence of aminoquinone, the polymerization proceeds at a constant rate W_∞ which is lower than the uninhibited reaction rate W.

The following equation corresponds to radical chain polymerization in the presence of aminoquinone in view of secondary inhibition:

$$-\frac{1}{\phi} - \ln(1-\phi_t) + (1-\phi_\infty^2)\ln\phi_t + \phi_\infty^2(\phi_t - \frac{1}{\phi_\infty^2}) = \frac{k_X W_\infty}{k_p[M]}t + const$$

where $\phi = W_t/W_\infty$, $\phi = W_\infty/W$.

The values of the inhibition constants k_X for various monomers vary within 10^4–10^5 dm^3/(mol × s), and the activation energy of the inhibition reaction is 23 kJ mol^{-1}.

The macromolecular paramagnetic nitroxyl is part of the polymer chain link and effectively inhibits the polymerization of vinyl monomers initiated by azo compounds, as well as the polymerization proceeding at the synthesis of unsaturated esters.

7.4.6 ANION RADICAL OF TETRACYANOQUINODIMETHANE

In contrast to neutral stable radicals, charged radicals usually initiate ionic polymerization. The initiating activity of anion radicals depends on the monomer–initiator difference in the electron affinity. Therefore, the anion radicals formed from compounds with very high electron affinity may be unable to initiate anionic polymerization.

With an excess electron, anion radicals often have donor properties and participate in a variety of charge-transfer reactions.

The effect of anion radicals on the polymerization of vinyl monomers, which proceeds by a radical mechanism, was studied in most detail on the example of tetracyanoquinodimethane salts, which dissociate in solution to form the tetracyanoquinodimethane anion radical ($TCNQ^{\overset{-}{\bullet}}$). When introduced into a solution of the lithium salt of tetracyanoquinodimethane (Li TCNQ), the polymerization initiated with benzoyl peroxide accelerates sharply, and the molecular weight of the resulting polymer decreases. The rate increase is observed only when the salt concentration exceeds the peroxide concentration not more than twice, and the highest rate of the process is achieved with an equal content of Li TCNQ and PB.

It has been found that in the polymerization of vinyl monomers, one electron transfer occurs between an anion radical and a peroxide molecule, with subsequent reactions between the formed neutral TCNQ and the benzoate anion R^-, and between the benzoate radical $R^\bullet$ and another TCNQ anion radical:

$$TCNQ^{\overset{-}{\bullet}} + ПБ \longrightarrow R^\bullet + R^- + TCNQ$$

$$TCNQ + R^- \longrightarrow Q^-$$

$$TCNQ^{\overset{-}{\bullet}} + R^\bullet \longrightarrow Q^-$$

These reactions proceed at a high rate just before the start of polymerization. Initiating radicals are formed by redox interaction of the intermediates with unreacted PB molecules:

$$BP + Q \longrightarrow R^\bullet + R^- + R_Q$$

Polymerization initiation under the action of the PB–Li TCNQ system proceeds several tens of times faster than by PB thermal decomposition.

If the initial concentration of Li TCNQ exceeds the peroxide concentration more than twice, free TCNQ anion radicals remain in the reaction medium, which inhibits the polymerization.

It has been found that the anion radical only interacts with those peroxides which have a sufficiently high electron-acceptor capacity (benzoyl and lauryl peroxides). Otherwise TCNQ only inhibits polymerization (e.g., of cumene peroxide). In the presence of lauryl peroxide, whose acceptability is lower than that of BP, the initiation rate reduces.

Polymerization initiation by the peroxide–TCNQ radical anion system is characterized by a low activation energy (~ 40 kJ mol^{-1}), which allows conducting the reaction at a high rate at relatively low temperatures.

The possibility of anion radicals to act as radical polymerization inhibitors was first demonstrated on the anion radical salts of 7,7,8,8-tetracyanoquinodimethane with dimethyl viologen. It has been found that TCNQ effectively inhibits the polymerization of styrene, the esters of acrylic and methacrylic acids, acrylonitrile, causing pronounced induction periods. TCNQ interacts not only with active radicals to inhibit the polymerization but also with monomer molecules, which leads to the emergence of new active centers, which immediately terminate at anion radicals. This additional initiation causes a relative reduction in the induction periods with increasing concentration of the inhibitor. The rate constant of additional initiation significantly depends on the monomer nature and decreases in the order: styrene > methylmethacrylate > methyl acrylate.

Radical anion complexes dissociate in solution differently, depending on the cation nature, concentration, and the temperature of the medium. Generally, TCNQ anion radicals exist in solution in the form of both free anions and ion pairs with different reactivity. Additional initiation is caused by the TCNQ anion radicals, which are present in solution as ion pairs with cations, while free radical anions are more active in the inhibiting reaction.

Changes in the composition of the reaction medium influence the degree of dissociation of the TCNQ complexes and, consequently, the contribution of additional initiation to the total kinetics of the process. As a result, the duration of the polymerization inhibition period by TCNQ anion radicals is determined not only by the ratio between the concentrations of the inhibitor and the rate of initiation but also by factors affecting the dissociation degree of the radical anion salts.

The dependence of the induction period duration of styrene polymerization on the TCNQ concentration is described by the equation:

$$\tau = [2k_{pac}f\alpha(1-\gamma)]^{-1}\ln\{1-\alpha(1-\gamma)\frac{[X]_o}{[I]_o}\}$$

Where $\alpha = k_и [M]/k_{рас} f$, $k_и$ is the rate constant of the additional initiation of polymerization by the ion pairs of TCNQ radical anion, $k_{рас}$ the rate constant of the initiator decomposition,

f is the initiation efficiency,

$[I]_o$ and $[X]_o$ are the initial concentrations of the initiator and the radical anion salt.

In view of the reaction of additional initiation, the time dependence of the inhibited polymerization rate is described by the equation:

$$-2\ln(\phi_t - \theta) - (\frac{1}{\theta} - 1)\ln(1 - \phi_t) + (\frac{1}{\theta} + 1)\ln(1 - \phi_t) + 0.5$$
$$(\frac{1}{\theta^2} - 1)\ln[\frac{\phi_t - \theta}{\phi_t - \theta}] = \frac{k_X W_\infty}{k_p[M]}(\frac{1}{\theta} - \theta)t + const$$

where $\theta = k_и [M]/k_X [R]$, $\phi_t = W_t/W_\infty$; W_t is the polymerization rate at time t, W_∞, and $[R]$are the stationary rate and the concentration of polymer radicals.

The graphical dependence of the reduced polymerization rate ϕ_t on time is an S-shaped curve with an inflection point. The parameter θ is determined by the reduced polymerization rate at this point $\phi_{тп}$ by the equation:

$$\theta^3 - \frac{2\theta_{TII}}{1+\phi_{TII}^2}\theta^2 - \frac{3\phi_{TII}^4 - \phi_{TII}^2}{1+\phi_{TII}^2}\theta - \frac{\phi_{TII}^5 - 2\phi_{TII}^3 - \phi_{TII}}{1+\phi_{TII}^2} = 0$$

The inhibition constants k_X/k_p are within $10^3 \div 10^4$ and vary widely depending on the nature of the cation, solvent, monomer, and temperature. In addition, they decrease with increasing concentration of the inhibitor while the additional initiation constants increase.

The interaction of TCNQ anion radicals with macroradicals occurs by several mechanisms. Along with addition, electron transfer from a radical anion to a macroradical with more pronounced acceptor properties is possible. Inhibition by the electron transfer mechanism is most probable in the polymerization of those monomers whose polymer radicals have a higher electron affinity, that is, the reaction rate increases in the monomer order styrene < methylmethacrylate < methylacrylate. As a result of electron transfer, the radical anion is converted into neutral TCNQ, which itself is able to terminate the reaction chains or, reacting with the electron-donor solvent (e.g., dimethylformamide), to convert into a radical anion again.

Thus, for inhibiting the polymerization of vinyl monomers with TCNQ radical anions, inhibitor regeneration reactions are typical, which results in an elongated induction period and a change in the sign of curvature on the plots of the dependence of its value on the anion radical initial concentration.

7.5 USAGE OF STABLE RADICALS FOR STUDYING BIOLOGICAL SYSTEMS

The presence of paramagnetic particles in liquid or solid objects opens up new possibilities to study them by EPR. Ready free radicals and the substances forming paramagnetic solutions due to spontaneous homolyzation of their molecules in liquid and solid media can be sources of paramagnetic particles. The synthesis and use of spin labels and probes based on piperidine and pyrrolidine nitroxyl stable radicals are a major problem of chemistry and molecular biology. Their high resistance to heat, oxygen, and other chemicals allow their use in science and technology as stabilizers for monomers and polymers, antioxidants, voltage sensors, and the working substances of optical quantum generators (lasers). Stable iminoxyl and nitroxide radicals covalently bound to a macromolecule (spin labels) or incorporated into the molecules of biological systems in relatively small amounts (spin probes)

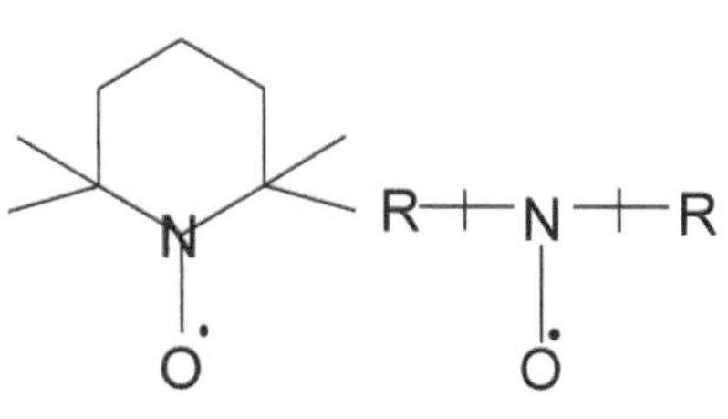

provide unique information of the structure and dynamics of the systems under study.

The compounds of these classes enter into common, long-known free radical reactions: recombination, disproportionation, addition to multiple bonds, isomerization, and β-splitting.[69] All these reactions proceed with the obligatory participation of the radical center and invariably lead to complete paramagnetism loss. These non-functional stable radicals play a very important role, and their use shows that no pronounced delocalization of the unpaired electron in the system of multiple bonds is obligatory for the stability of the paramagnetic material. The discovery of long-living radicals of non-aromatic nature[70] did not change the preexisting ideas about the reactivity of the known organic paramagnetic materials. In the 1960s, E.G. Rozantsev started a new lead in the chemistry of free radicals, namely, the synthesis and reactivity of functional stable radicals with a pronounced localized paramagnetic center.[73] Functional free radicals have quickly found widespread use as paramagnetic probes to study the molecular motion in condensed phases of different nature. With their usage the emergence of the spin label method (a covalently bound paramagnetic probe) is associated, whose idea is based on the dependence of the shape of the EPR spectrum of a free radical on the properties of the nearest atomic environment and the mode of interaction of the paramagnetic fragment with its environment. The reaction of free radicals without affecting their paramagnetic center (Neuman–Rozantsev's reaction) got the chemical basis for obtaining spin-labeled compounds.[74] The concept of using non-radical reactions of radicals in macromolecular studies was formulated by Liechtenstein[75] and the theoretical foundations of the method, calculating methods for the correlation times of the rotational mobility of a paramagnetic particle from the EPR spectrum shape are developed in Refs 76–79. The behavior of iminoxyl radicals in various systems was then studied.[80]

Let us highlight a few important aspects of the use of organic paramagnetic materials in studies of biological systems.

7.5.1 APPLICATION OF IMINOXYL FREE RADICALS TO STUDY IMMUNE GAMMA GLOBULINS

From the physicochemical viewpoint, the mechanism of various immunological reactions is determined by changes in the phase state of the system. To explore the antigen–antibody interaction mechanism, iminoxyl radical-labeled gamma globulins were used.[81] The molecule of gamma globulin consists of four polypeptide chains linked together with disulfide bridges. Upon cleavage of these interchain disulfide bonds, the polypeptide chains continue to be held together. A spin label, the iminoxyl derivative of maleimide

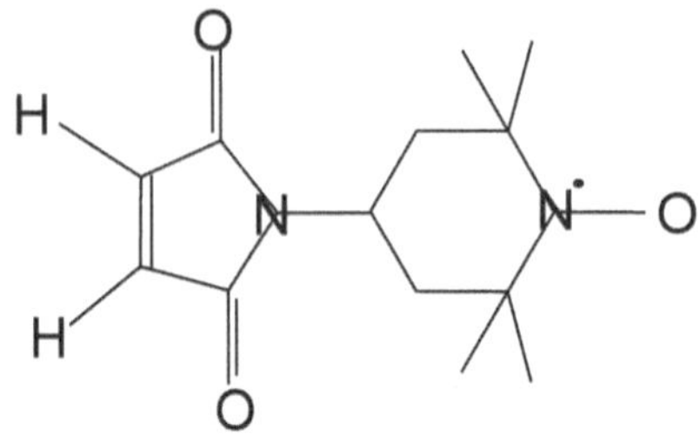

was attached to the sulfhydryl groups produced by reduction of the disulfide bonds with β-mercaptoethanol. These experiments were conducted on the rabbit and human gamma globulins. The EPR spectra in both cases corresponded to the relatively high mobility of free radicals (the correlation time $\tau = 1.1 \times 10^{-9}$ and 7.43×10^{-9} s for the human and rabbit gamma globulin, respectively). Comparing the correlation times in these proteins with the values obtained in experiments with sulfhydryl group-labeled serum albumin treated with urea ($\tau = 1.09 \times 10^{-9}$ s) and dioxane (2.04×10^{-9} s), it can be concluded that the fragments of the polypeptide chains carrying free radicals have no ordered secondary structure. This character of the EPR spectrum of immune gamma globulin, while preserving its specific activity, has allowed exploring the conformational and phase transitions at specific antigen–antibody reactions. Dramatic differences were revealed in the mobility of the spin label when rabbit antibody precipitation by salting out with ammonium sulfate and precipitation with a specific antigen (ovalbumin), resulting in only a slight decrease in the mobility of the paramagnetic labels. But ammonium sulfate-caused precipitation entailed a strong inhibition of the rotational mobility of the free radicals (Fig. 7.12).

These results can be considered as a direct confirmation of the alternative theory,[82] according to which the formation of a precipitate is due to the immunological polyvalency of the antigen and antibody in relation to each other. Indeed, deliberately excluding the location of spin labels at the active site of the antibody, it can be concluded that the relatively mobile state of the spin labels in the antigen–antibody precipitate can be preserved only at no strong antibody dehydration due to intermolecular interactions. Unlike the gamma globulin precipitated with ammonium sulfate, the specific precipitate, in accordance with the lattice theory, has a microcellular structure. At prolonged storage of the antigen–antibody precipitate with no stabilizers added, the mobility degree of iminoxyl radicals decreased sharply.

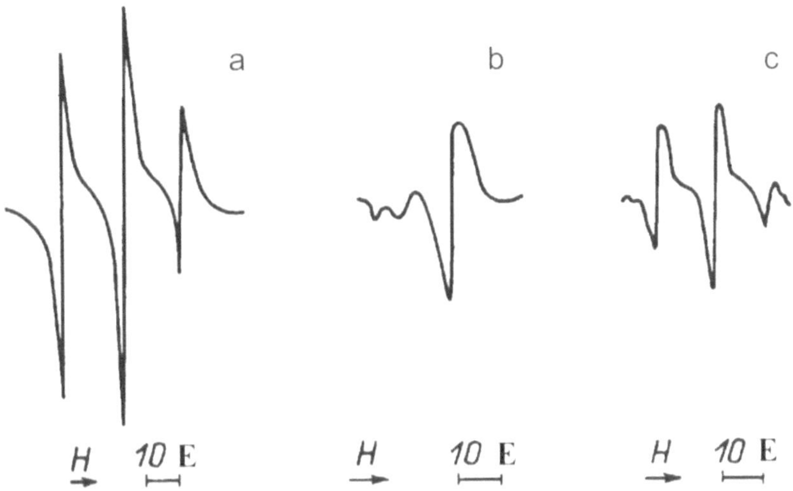

FIGURE 7.12 EPR spectra of gamma globulin labeled with an iminoxyl radical without affecting its free valence: (a) in solution, (b) in the precipitate obtained by salting out with ammonium sulfate, (c) in the specific precipitate. This is apparently the result of the secondary dehydration of the antibodies due to interactions of the protein molecules in the precipitate.

7.5.2 EXPLORATION OF STRUCTURAL TRANSITIONS IN BIOLOGICAL MEMBRANES

Biological membranes, particularly the membranes of mitochondria, play an important role in the redox processes in the cell, as a localization spot of the respiratory chain enzymes. Earlier,[83] significant differences were found in the properties of one of the fragments of the mitochondrial membrane (the complex III) at transition from the oxidized to reduced form: the sulfhydryl groups readily titrated in the oxidized form of this fragment were unavailable in the reduced one. Then, electron microscopy revealed changes in the repeating structural units of the mitochondria in conditions leading to the formation of macroergic intermediates or their synthesis, for example, by active ionic transport.[84] All this has suggested that the redox reaction catalyzed by the enzyme chain of electron transfer is accompanied by a kind of "conformational wave" probably covering not only the protein component of the membrane but also higher levels of organization: fragments of the membrane of a greater or lesser degree of complexity, including its lipid part. To verify this hypothesis, a modification of the spin label method was used, namely, the non-covalently bound paramagnetic probe method. The radical is kept by the matrix (membrane) through weak hydrophobic bonds only. This approach has allowed exploring the weak interactions in the system without significant violations of the biochemical functions of the biomembrane and its structure. The paramagnetic probe was the caprylic ester of 2,2,6,6-tetramethylpiperidin-1-oxyl:

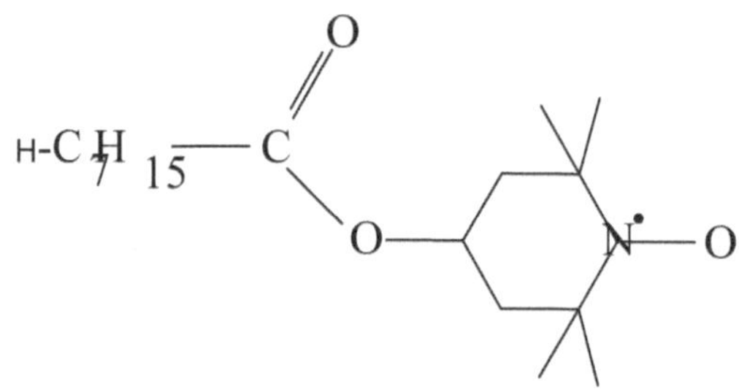

This compound was prepared from the acid chloride of caprylic acid and 2,2,6,6-tetramethyl-4-oxypiperidine-1-oxyl in a triethylamine medium by a radical reaction without affecting the free valency. The paramagnetic probe was introduced into a suspension of electron transport particles (ETP) isolated from bovine heart mitochondria. These fragments of the mitochondrial membrane are characterized by a rather full range of the respiratory chain enzymes with the same molar ratio as in intact mitochondria.[84]

The paramagnetic probe is insoluble in water but solubilized with ETP suspended in a buffer solution. Due to this, the observed EPR spectrum is free from the background created by the radical unbound to the object under study. The presence of a bulky hydrocarbon chain provides "integration" of the probe molecules into the lipid part of ETP. The EPR spectrum therefore reflects the state of this particular fraction of the membrane. To detect conformational transitions, the ESR spectrum was recorded before and after introduction of the oxidation substrates (succinate and nicotinamide adenine dinucleotide's reduced form (NADH)), as well as after oxidation of the previously reduced respiratory chain with potassium ferricyanide. Figure 7.13 shows typical results.[84] The enhancement of the anisotropy of the EPR spectrum of iminoxyl after introduction of the substrate is clearly visible. The spectrum in ferricyanide-oxidized ETP is not visibly different from that in intact ETP.

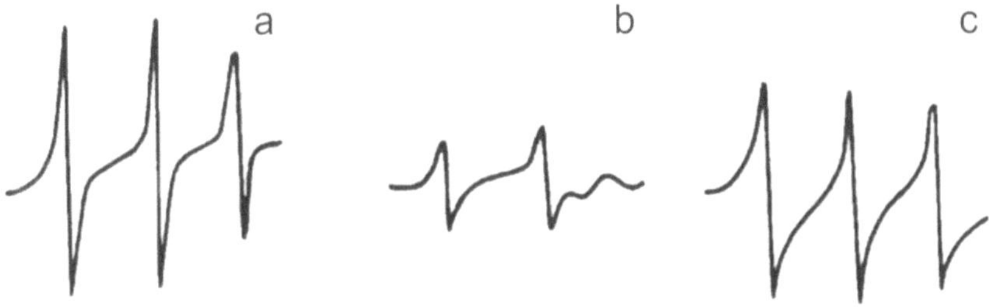

FIGURE 7.13 Changes in the anisotropy of the EPR spectrum of the hydrophobic iminoxyl radical in a suspension of electron transport particles from bovine heart mitochondria on addition of oxidation substrates: (a) before introduction of oxidation substrates, (b) after succinate introduction, (c) after oxidation of the previously reduced respiratory chain with potassium ferricyanide.

ETP inactivation by prolonged storage at room temperature or inhibition with cyanide eliminated this effect. Comparing the shapes of signals and correlation times shows that the ESR spectrum of iminoxyl in intact ETP consists of two signals differing by anisotropy. A weakly anisotropic signal is given by the radical localized in that part of the membrane, where the effective free volume available for radical movement is rather large. A strongly anisotropic (inhibited) spectrum belongs to the radicals localized in other parts of the system with a smaller effective free volume. The reduction of the respiratory chain with substrates leads, due to a conformational transition of cooperative type, to a decrease in the fraction of the sites with a large free volume (i.e., to an increased microviscosity of the inner environment of the radical). The correlation time of the whole spectrum varies from 20×10^{-10} (oxidized ETP) to 4×10^{-10} s (reduced ETP).

Simultaneously with the increasing anisotropy of the signal, its intensity decreases: iminoxyl reduces, apparently, to hydroxylamine derivatives. Potassium ferricyanide reverses this process. It should be noted that the oxidation substrates themselves do not interact appreciably with iminoxyls. Obviously, the conformational transition leads not only to a change in microviscosity but also eliminates the steric hindrances for radical reduction. In principle, this indicates a possible new, proper chemical aspect of application of the paramagnetic probe method.

7.5.3 INVESTIGATION OF THE STRUCTURE OF SOME MODEL SYSTEMS

The application of paramagnetic probe in such systems as biological membranes entails a number of issues relating to the behavior of hydrophobic labels in media with an ordered arrangement of hydrophobic chains. Mixtures of the nonionic detergent Tween 80 and water were studied as a first step. Tween 80 is an olein sorbitan polyethylene glycol and classified as a nonionic surfactant based on polyethylene oxide.[84] The selection of this object is determined by some methodological convenience, as well as by the fact that the literature contains data on the structure of Tween aqueous solutions obtained by classical methods (viscometry, refractometry, etc.). The properties of this detergent are of interest, since it finds fairly widespread use in biological membrane fragmentation. The esters of 2,2,6,6-tetramethyl-4-oxypiperidine-1-oxyl and saturated acids of normal structure with hydrocarbon chain lengths of 4, 7, and 17 carbon atoms, or the corresponding amides were used as the paramagnetic probe. For comparison, the behavior of the hydrophobic labels IV and V was also studied in these systems.

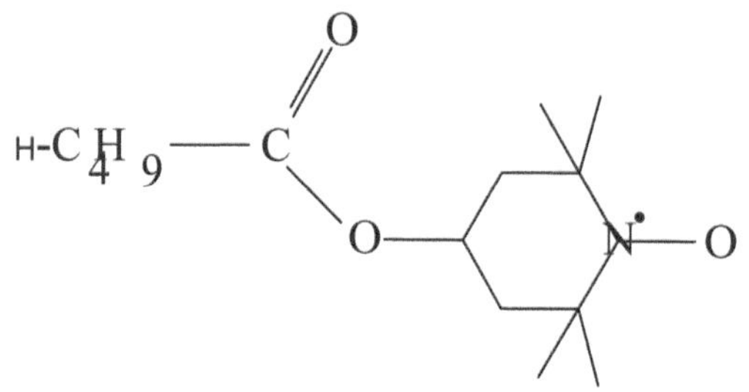

I

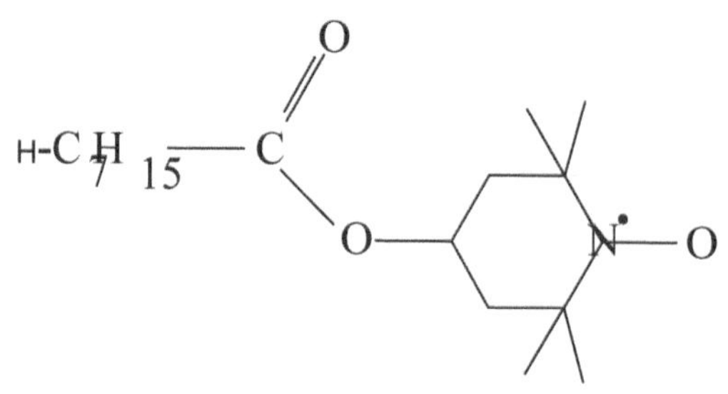

II

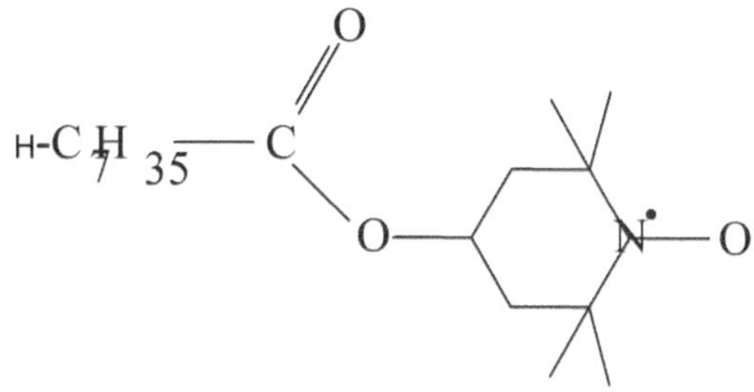

III

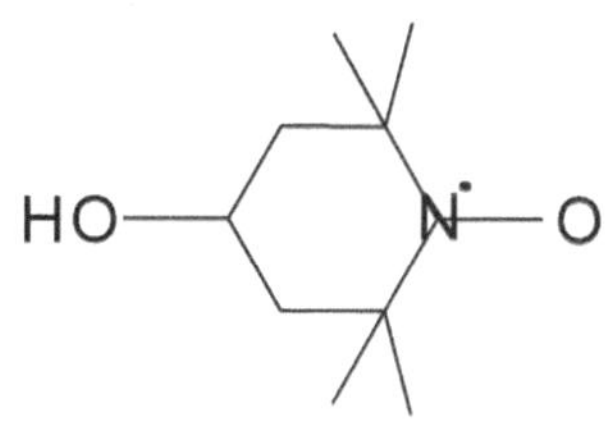

IV

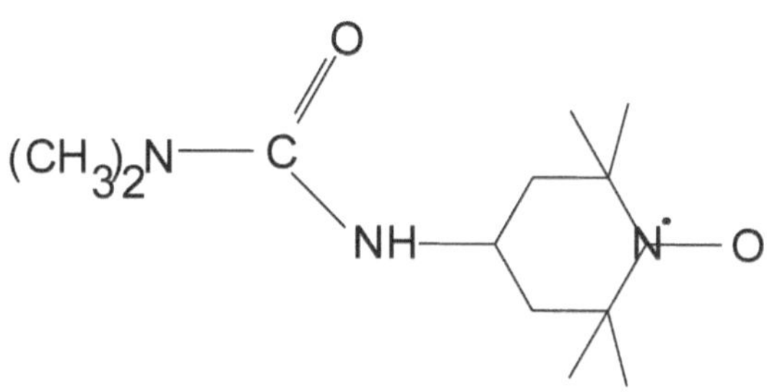

V

The course of the correlation time of radical rotation depending on the Tween concentration is shown in Figure 7.14. Several areas can be distinguished, apparently corresponding to different types of structures. The initial fragment, discernible for the lightest radicals only, corresponds to an unsaturated Tween solution in water. This area is better detected for detergents with a higher critical micelle concentration (CMC), for example, for sodium dodecylsulfate (Fig. 7.15).[85] Then, micelle formation begins. The correlation time of the water-insoluble labels increases and reaches a plateau; with further increase in the surfactant concentration, it passes through a maximum and then monotonically increases up to its value in pure Tween. It is advisable to compare these data with the results of viscometric measurements by common macro methods. Figure 7.16 shows the viscosity having one extremum around 60% of Tween. Therefore, the effective volume available for rotation of the probe molecule and the macroviscosity do not correlate at high Tween concentrations.

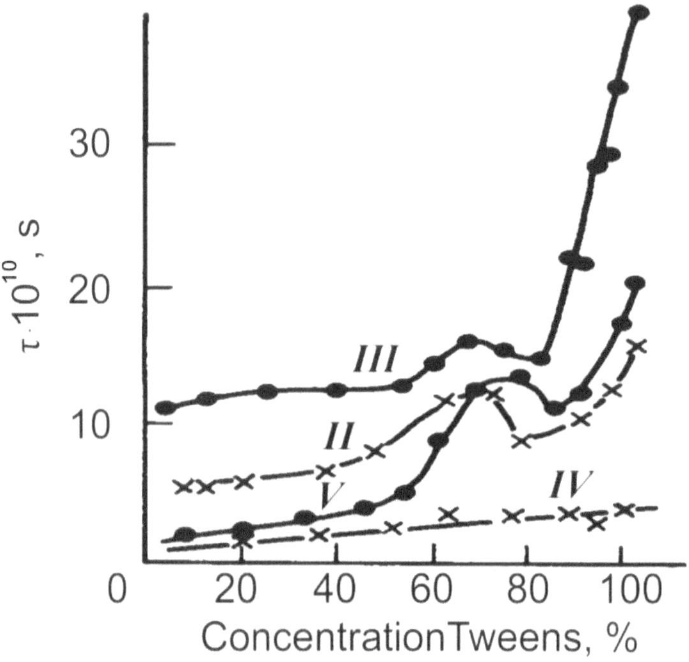

FIGURE 7.14 Changes in the correlation times in the water–detergent system for nitroxyl radicals with a strongly localized paramagnetic center.

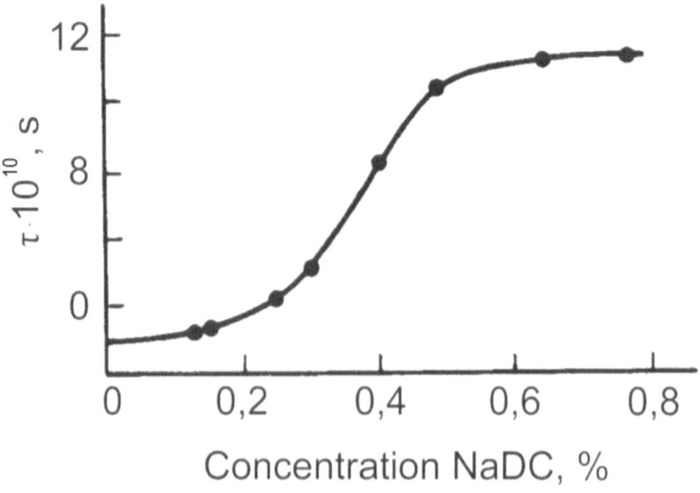

FIGURE 7.15 Determination of the critical micelle concentration of sodium dodecyl using 2,2,6,6-tetramethyl-4-hydroxipiperidyl-1-oxyl valerate.

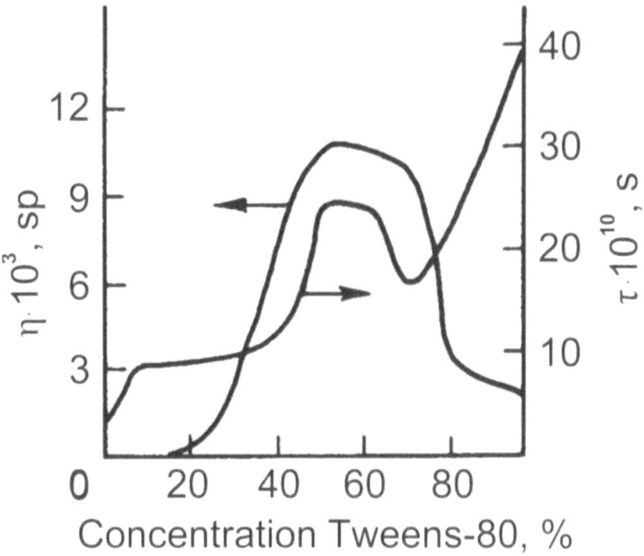

FIGURE 7.16 Changes in the microviscosity and macroviscosity of the water–Tween80 system.

These results can be explained as follows: in pure Tween, the lamellar structure provides easy layer-by-layer slipping, so the macroviscosiry of this system is low. However, in the absence of water, the interaction of the polar groups is strong, the hydrocarbon chains are arranged, and the effective free volume in the radical localization region is small. Small amounts of water lead to the formation of defects in the layered structure. Slipping becomes difficult, so the viscosity increases. However, hydration breaks the close interaction of the Tween polar groups. They are deformed and the hydrocarbon chains are disordered. The microviscosity of the hydrocarbon layer decreases. At the end of the hydration of the polar groups, water-filled cavities are formed. They are the building blocks (micelles), which the system is built from. Structure formation is manifested as an increase in the microviscosity and macroviscosity. In the region of the maximum, phase inversion is possible. A new type of structural units (Tween micelles in water) is formed, passing, on further dilution, to a colloidal solution. The course of microviscosity at high Tween concentrations is strikingly similar to the change in the correlation time at moistening of some lyophilized cell organelles. This similarity confirms that in the region of the maximum of τ, where the restoration of the biochemical activity of chloroplasts starts, a phase transition occurs of the same type as in the liquid crystal "detergent–water" systems.

Some information about the behavior of radical particles in colloid systems is produced by the results of temperature measurements. Figure 7.17 shows the temperature dependence of the correlation time in Arrhenius' coordinates for several iminoxyl radicals of varying hydrophobicity degree.

The strong dependence of the pre-exponential factor on the chain length seems to confirm that the observed correlation time truly reflects rotation of the radical molecule as a whole. However, this result allows another interpretation, namely, depending on the hydrocarbon chain length; the radical is introduced into a detergent micelle at a greater or lesser depth. This may lead to a change in the rotation frequency of the iminoxyl group around single bonds in the molecule depending on the environment of the radical's polar end. If the polar group of the radical is at the micelle–solvent interface, the measured frequency will depend on the surface charge (potential) of the micelle. In our opinion, this circle of colloid chemical problems, closely linked to the issues of transmembrane transport in biological systems, will constitute another field of applications of the paramagnetic probe method.

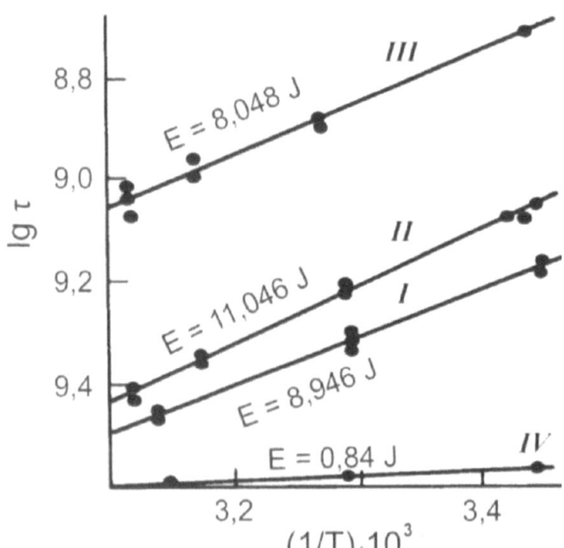

FIGURE 7.17 Activation energy of the rotational diffusion of free iminoxyl (nitroxyl) radicals in the Tween–water system (30% of Tween80).

7.5.4 SOLVING OTHER BIOLOGICAL PROBLEMS USING STABLE RADICALS

Further progress in the use of stable paramagnetic materials to solve a variety of biological problems is reflected in numerous review articles and monographs.[86–98] For dynamic biochemistry, of undoubted interest are local conformational changes of protein molecules in solution. The distances between specific loci in biomacromolecules, in principle, can be quantified using stable paramagnetic materials. After introduction of iminoxyl fragments into certain areas of a native protein, the distance between the neighboring paramagnetic centers can be calculated from the efficiency of their dipole–dipole interaction in vitreous solutions of a spin-labeled preparation (the ESR method). In an earlier work, evaluating the distance between the paramagnetic centers in spin-labeled mesozyme and hemoglobin, the promise of this approach was shown. In this connection, there appeared the need to identify a simple empirical parameter in EPR spectra to quantify the dipole–dipole interaction between paramagnetic centers.

A convenient empirical parameter was found in a study of vitrified solutions of iminoxyl radicals. It is the ratio of the total intensity of the extreme components of the spectrum to the intensity of the central one (Fig. 7.18). To establish the correlation of d_1/d with the average distance between the localized paramagnetic centers, the corresponding calibration curves were plotted (Fig. 7.19). Calculations showed that d_1/d depends on the value of the dipole–dipole broadening and is in satisfactory agreement with the experimental results obtained independently. The methods for quantifying the distances between paramagnetic centers in biradicals and spin-labeled biomolecules subsequently became a reliable tool for structural studies.

In determining the relaxation rate constants of various paramagnetic centers in solution, it has been found that the values of these constants strongly depend on the chemical nature of the functional groups in iminoxyl radicals. The changes of sign of the electrostatic charge of the substituent and its distance from the paramagnetic iminoxyl group have the strongest influence on the constants. The electrostatic effect is a more significant factor due to the magnitude and sign of the charge of the paramagnetic particles interacting in solutions. An increased ionic strength leads to a change in the value of k in qualitative agreement with Debye's theory. The substituent's mass is another factor that significantly affects the value

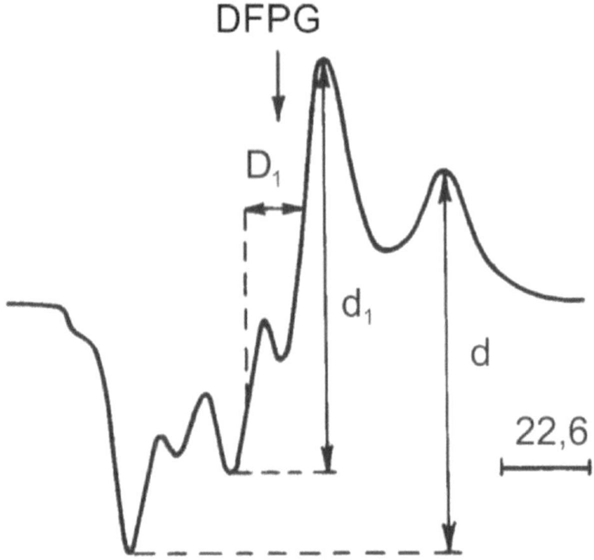

FIGURE 7.18 EPR spectrum of the free iminoxyl biradical (2,2,6,6-tetramethyl-4-hydroxypiperidine-1-oxyl phthalate) vitrified in toluene at 77 K.

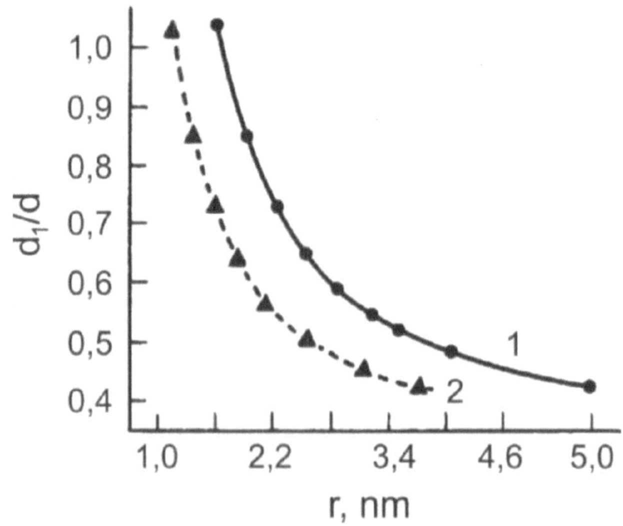

FIGURE 7.19 Dependence of d_1/d of the EPR spectrum on the mean distance *r* between the interacting paramagnetic centers in iminoxyl radicals *1* and biradicals *2* at 77 K.

of the constant. If the substituent is a protein macromolecule, the value of *k* reduces twice in the limit. The possibilities of applying the paramagnetic probe method for detection of anionic and cationic groups and estimation of distances in studies of the protein microstructure were analyzed. The dependences obtained in experiments show that Debye's equation with $D = 80$ can be successfully used to analyze experimental data and, in particular, to estimate the distance from the iminoxyl group of the spin label on a protein to the nearest charged group, provided this distance does not exceed 1.0–1.2 nm.

Considering the role of free radical processes in the radiation therapy of cancer, it was proposed to investigate the effects of iminoxyl radicals on the body of laboratory animals. Pharmacokinetic studies have shown that the simplest functionalized derivative of 2,2,6,6-tetramethylpiperidine-1-oxyl is of relatively low toxicity and exhibits pronounced antileukemic activity. The highest values of the inhibition rates were observed for hemocytoblasts in peripheral blood and bone marrow. This fact stimulated further structural and synthetic studies aimed at obtaining more effective and less toxic cancerolytics and sensitizers for cancer radiotherapy.

KEYWORDS

- **organic paramagnets**
- **magnetic properties**
- **antioxidants**
- **stable radicals**
- **polymerization kinetics**
- **vinyl monomers**

REFERENCES

1. Karlin, R. Magneto-chemistry. *M. World*, **1989**. [Russ].
2. McConnell, H. M. *J. Chem. Phys*. **1963**, *39*, 1910.
3. Buchachenko, A. L. Reports of USSR Science Academy. **1979**, *244*, 1146. [Russ].
4. Izuoka, A.; Murata, S.; Sugavara, T.; Iwamura, H. *J. Amer. Chem. Soc*. **1987**, *109*, 2631.
5. Williams, D. *Vol. Phys*. **1969**, *16*, 1451.
6. Mukai, K.; Mishina, T.; Ishisu, K.; Deguchi, Y. *Bull. Chem. Soc. Jpn*. **1977**, *50*, 641.

7. Agawa, K.; Sumano, T.; Kinoshita, M. *Chem. Phys. Lett.* **1986**, *128*, 587.
8. Agava, K.; Sumano, T.; Kinoshita, M. *Chem. Phys. Lett.* **1987**, *141*, 540.
9. Azuma, N.; Yamauchi, J.; Mukai, K.; Ohya-Nisciguchi, H.; Deguchi, Y. *Bull. Chem. Soc. Jpn.* **1973**, *46*, 2728.
10. Ballester, M.; Riera, J.; Castaner, J.; Vonso, J. *J. Amer. Chem. Soc.* **1971**, *93*, 2215.
11. Rozantsev, E. G.; Goldfein, M. D.; Pulin, V. F. *The Organic Paramagnets*; Saratov State University, 2000. [Russ].
12. Rozantsev E. G., Sholle V. D. The Organic Chemistry of Free Radicals. *M. Chem.* **1979**. [Russ].
13. Karimov, U. S.; Rozantsev, E. G. *Phys. Solid State*. **1966**, *8*, 2787. [Russ].
14. Nakajima, A.; Ohya-Nisciguchi, H.; Deguchi, Y. *Bull. Chem. Soc. Jpn.* **1985**, *107*, 2560.
15. Benelli, C.; Gatteschi, D.; Carnegie, J.; Carlin, R. *J. Amer. Chem. Soc.* **1983**, *105*, 2760.
16. Zvarykina, A. V.; Stryukov, V. B.; Fedutin, D. N.; Shapiro, A. B. Letters in JETPh. **1974**, *19*, 3. [Russ].
17. Zvarykina, A. V.; Stryukov, V. B.; Umansky, S. Ya.; Fedutin, D. N.; Shibaeva, R. P.; Rozantsev, E. G. Reports in USSR Academy. **1974**, *216*, 1091. [Russ].
18. Stryukov, V. B.; Fedutin, D. N. *Phys. Solid State*. **1974**, *16*, 2942. [Russ].
19. Rassat, A. *Pure and Appl. Chem.* **1990**, *62*, 223.
20. Nakajima, A. *Bull. Chem. Soc. Jpn.* **1973**, *46*, 779.
21. Nakajima, A.; Yamauchi, J.; Ohea-Nisciguchi, H.; Deguchi Deguchi, Y. *Bull. Chem. Soc. Jpn.* **1976**, *49*, 886.
22. Caneschi, A.; Gatteschi, D.; Grand, A.; Laugier, J.; Rey, P.; Pardi, L. *Inorg. Chem.* **1088**, *27*, 1031.
23. Caneschi, A.; Laugier, J.; Rey, P.; Zanchini, C. *Inorg. Chem.* **1987**, *26*, 938.
24. Caneschi, A.; Gatteschi, D.; Laugier, J.; Pardi, L.; Rey, P.; Zanchini, C. *Inorg. Chem.* **1988**, *27*, 2027.
25. Laugier, J.; Rey, P.; Bennelli, C.; Gatteschi, D.; Zanchini, C. *J. Amer. Chem. Soc.* **1986**, *108*, 5763.
26. Caneschi, A.; Gatteschi, D.; Laugier, J.; Rey, P. *J. Amer. Chem. Soc.* **1987**, *109*, 2191.
27. Caneschi, A.; Gatteschi, D.; Rey, P.; Sessoli, R.*Inorg. Chem.* **1988**,*27*,1756.
28. Caneschi, A.; Gatteschi, D.; Laugier, J.; Rey, P.; Sessoli, R. *Inorg. Chem.***1987**,*27*, 1553.
29. Caneschi, A.; Gatteschi, D.; Renard, J.; Sessoli, R. *Inorg. Chem.* **1989**, *28*, 2940.
30. Caneschi, A.; Gatteschi, D.; Renard, J.; Rey, P.; Sessoli, R. *Inorg. Chem.* **1989**, *28*, 1976.
31. Caneschi, A.; Gatteschi, D.; Renard, J.; Rey, P.; Sessoli, R. *Inorg. Chem.* **1989**, *28*, 3314.
32. Caneschi, A.; Gatteschi, D.; Renard, J.; Rey, P.; Sessoli, R. *J. Amer. Chem. Soc.* **1989**, *111*, 785.
33. Caneschi, A.; Ferraro, F.; Gatteschi, D.; Rey, P.; Sessoli, R. *Injrg. Chem.* **1990**, *29*, 1756.
34. Miller, J.; Epstein, A.; Reiff, W. *Chem Rev*. **1988**, *88*, 201.
35. Miller, J.; Epstein, A.; Reiff, W. *Acc. Chem. Res.* **1988**, *21*, 114.
36. Miller, J.; Calabrese, J.; Rommelmann, Y.; Chittapeddi, S.; Zhang, J.; Reiff, W.; Epstein, A. *J. Amer. Chem. Soc.* **1987**, *109*, 769.
37. Chittapeddi, S.; Cromack, K.; Miller, J.; Epstein, A. *Phys. Rev. Lett.* **1987**, *22*, 2695.
38. Candela, G.; Swartzendruber, L.; Miller, J.; Rice, M. *J. Amer. Chem. Soc.* **1979**, *101*, 2755.

39. Miller, J.; Epstein, A. *J. Amer. Chem. Soc.* **1987**, *109*, 3850.
40. Miller, J.; Calabrese, J.; Harlow, R.; Dixon, D.; Zhang, J.; Reiff, W.; Chittapeddi, S.; Selover, M.; Epstein, A. *J. Amer. Chem. Soc.* **1990**, *112*, 5496.
41. Miller, J.; Ward, M.; Zhang, J.; Reiff, W. *Inorg. Chem.* **1990**, *29*, 4063.
42. Sugamara, T.; Bandow, S.; Kimura, K.; Itamura, H.; Itoh, K *J. Amer. Chem. Soc.* **1986**, *108*, 368.
43. Sugamara, T.; Bandow, S.; Kivura, K. H.; Itamura, H.; Itoh, K. *J. Amer. Chem. Soc.* **1984**, *106*, 6449.
44. Itamura, H. *Pure and Appl. Chem.* **1986**, *58*, 187.
45. Teki, Y.; Takui, T.; Itoh, K.; Iwamura, H.; Kobayaschi, K. *J. Amer. Chem. Soc.* **1983**, *105*,3 722.
46. Teki, Y.; Takui, T.; Yagi, H.; Itoh, K.; Imamura, H. *J. Chem. Phys.* **1985**, *83*, 539.
47. Sugamara, T.; Murata, S.; Kimura, K.; Itamura, H. *J. Amer. Chem. Soc.* **1985**, *107*, 5293.
48. Sugamara, T.; Tukada, H.; Izuoka, A. *J. Amer. Chem. Soc.* **1986**, *108*, 4272.
49. Magata, N. *Theor. Chim. Acta.* **1968**, *10* 372.
50. Itoh, K. *Bussei.* **1971**, *12*, 635.
51. Torrance, J.; Oostra, S.; Nazzal, A. *Synth. Met.* **1987**, *19*, 709.
52. Torrance, J.; Bugus, P. *J Appl. Phys.* **1988**, *63*, 2962.
53. Breslow, R. *Pure and Appl. Chem.* **1982**, *54*, 927.
54. Breslow, R.; Jaun, B.; Kluttz, R. *Tetrahedron.* **1982**, *38*, 863.
55. Breslow, R. *Pure and Appl. Chem.* **1982**,*54*, 927.
56. Breslow, R.; Maslak, P. *J. Amer. Chem Soc.* **1984**, *106*, 6453.
57. Le Pade, T. Breslow, R. *J. Amer. Chem. Soc.* **1987**, *109*, 6412.
58. Breslow, R. *Mol. Cryst. Liquid cryst.* **1985**, *125*, 261.
59. McConnell, H. *Proc. R.A. Welch Found. Chem. Res.* **1967**, *11*, 144.
60. Miller, J.; Glatzhofer, D.; et al. *Chem.Mater..* **1990**, *2*, 60.
61. Steyn, D. L. In a World of Science. **1989**, *9*, 26. [Russ].
62. Emsley, J. *New Scion.* **1990**, *127* (1727), 29.
63. Dulog, L.; Kim, J.S. *Angev. Chem.* **1990**, *29*, 415.
64. Rozantsev, E. G.; Stepanov, A.P. The Geophysics Apparatus; SPb: Nedra, 1966; Vol. 29, p. 35. [Russ].
65. Rozantsev, E. G. *Free Iminoxyl Radicals*; Chemistry, 1970. [Russ].
66. Rozantsev, E. G.; Neyvan, M.B.; Lihtenshtain, G.I.; Sertificate Authorship of USSR No. 1661331964. [Russ].
67. Bukin, I. I.; Kagarmanov, N. F.; Rozantsev, E. G. *Over Solid Mater.* **1983**, (6), 23–29. [Russ].
68. Rozantsev, E. G.; Goldfein, M. D.; Pulin, V. F. *Organic Paramagnerics*; Saratov State University: Saratov, **2000**. [Russ].
69. Rozantsev, E. G. *Chemical Encyclopedic Dictionary*; Soviet Encyclopedia: Moscow, 1983. [Russ].
70. Rozantsev, E. G.; Lebedev, O. L.; Kazarnovsky, S. N. Diploma for Discovery, No 248 of 10/05/1983. IR. 1982. No 6. P. 6.
71. Rozantsev, E.G. Paramagnetic derivatives of nitric oxide. Doctoral thesis,Institute of Chemical Physics, Academy of Sciences of the USSR, 1965.

72. Zhdanov, R. I. Nitroxyl Radicals and Non-radical Reactions of Free Radicals. In *Bioactive Spin labels*. Zhdanov, R. I., Ed.; Springer: Berlin, Heidelberg; NewYork, 1991, pp. 24.
73. Liechtenstein, G. I. *Spin Labeling In Molecular Biology*; Nauka, 1974. [Russ].
74. McConnell, H. J. *Chem. Phys.* **1956**, *25*, 709.
75. Freed, J.; Fraenkel, G. J. *Chem. Phys*. **1963**, *39*, 326.
76. Kivelson, D. *J. Chem. Phys*. **1960**, *33*, 4094.
77. Stryukov, V. B. *Stable Radicals in Chemical Physics*; Znanie, 1971. [Russ].
78. Goldfeld, M. G.; Grigorian, G. L.; Rozantsev, E. G. *High-molecular-weight compounds (Collected preprints)*; Institute of Chemical Physics, Academy of Sciences: USSR, 1979, p. 269. [Russ].
79. Grigorian, G. L.; Tatarinov, S. G.; Cullberg, L. Ya.; Kalmanson, A. E.; Rozantsev, E. G.; Suskina, V. I. *Dokl. USSR Acad. Sci*., **1968**, *178*, 230. [Russ].
80. Pressman, D. *Molecular Structure and Biological Specificity*; Washington. D.C., 1957.
81. Baum, H.; Riske, J.; Silman, H.; Lipton, S. *Proc. Nature Acad. Sci. US*. **1967**, *57*, 798.
82. Rozantsev, E.G.*Biochemistry of Meat and Meat Products (General Part). A Textbook*; DePi print, 2006. [Russ].
83. Goldfeld, M. G.; Kolkover, V. K.; Rozantsev, E. G.; Suskina, V. I. *Kolloid. Zs*. **1970**, *15* (1), 321–326.
84. Smith, I. C. P.; Schreier-Muccillo, S.; Marsh, D.; Spin Labeling: Free Radicals in Biology. **1976**, *1*, 149–197.
85. Shokova, E. A.; Rozantsev, E. G.; Ozhogina, O. A.; Rozov, A. K. *Dokl. USSR Acad. Sci*., **1990**, *314* (4), 875. [Russ].
86. Henry, Y.; Guissani, A. *Analysis*. **2000**, *28* (6), 445–454.
87. Rozantsev, E. G.; Goldfein, M. D.; Pulin, V. F. *Organic paramagnets*; Saratov Univ. Press: Saratov, 2000.
88. Hanafy, K. A. *Med. Sci. Monit*. **2001**, *7* (4), 801–819.
89. Urushitani, M. Shimohama, S. ALS and Other Neuron Disorders. **2001**, *2* (2), 71–81.
90. Davies, M. J. Electron Paramagnetic Resonance. J. *Royal Soc. Chem*. **2002**, *18*, 47–73.
91. Petukhov M. S.; Rychkov, G. A.; Serrano, L. Bresler's Readings: Molecular Genetics, Biophysics and Medicine Today. **2002**, *1*, 148–168.
92. Bobodzhanov, P.Kh. Study of the structure and molecular dynamics of cotton cellulose and globular proteins by spin labels and probes. Doctoral Thesis, Institute of Chemistry named after V.I. Nikitin, Academy of Sciences of Tajikistan, 2003.
93. Hausmann, O. N. b Spinal Cord. **2003**, *41* (7), 369–378.
94. Koltover, V. K. Carbon. **2004**, *42* (5), 1179–1183.
95. Furukawa, Y.; O'Halloran, T. V. *Antioxid. Redox Signal*. **2006**, *8* (5), 847–867.
96. Vanin, A. F.; Bevers, L. M.; Mikoyan, V. D.; Poltorakov, A. P.; Kubrina, L. N.; van Faassen, E. Nitric Oxide. **2007**, *16* (1) 71–81.
97. Goldfein, M. D.; Ivanov, A. V.; Malikov, A. N. *Concepts of Modern Natural Sciences. A lecture course*; Goldfein M. D. Ed.; RGTEU Press: Moscow, 2009. [Russ].
98. Goldfein, M. D.; Ursul, A. D.; Ivanov, A. V.; Malikov, A. N. *Fundamentals of the Natural-Scientific Picture of the World*; Goldfein M. D. Ed.; SI RGTEU Press: Saratov, 2011. [Russ].

CHAPTER 8

PROBLEMS OF ENVIRONMENTAL POLLUTION WITH POLYMERS

CONTENTS

8.1 Foundations of Modern Strategies in Solving Problems Associated with Earth's Contamination with Macromolecular Compounds 330

8.2 Kinetics and Mechanism of the Processes to Obtain Environmentally Friendly Polymers and to Reduce Plastic Waste 335

8.3 Stabilization of Polymers as a Basis for the Design of New High-Molecular-Weight Compounds and as an Effective Way to Reduce Waste When They Are Used[47–51] 347

8.4 Features of the Synthesis and Use of PVC and Polyethylene[52–54] 354

8.5 Environmental Pollution with Acetate-Containing Wastes of the Production of Polymeric Materials [55,56] 362

8.6 Poly (Alkylene Guanidines) as Environmentally Friendly Biocidic Polymers[57,58] 364

8.7 Biodegradable Polymers 1, [59–61] 368

8.8 Problems of Recovery and Recycling of Polymeric Materials[62–64] 371

Keywords 402

References 402

ABSTRACT

The problems of environmental pollution caused by polymeric materials are raised. The production of environmentally friendly polymers is considered as a basis for addressing macromolecular pollution. The aging, stabilization, and biodegradation of polymers are described. Several sections are devoted to individual members of polymers and their problems.

8.1 FOUNDATIONS OF MODERN STRATEGIES IN SOLVING PROBLEMS ASSOCIATED WITH EARTH'S CONTAMINATION WITH MACROMOLECULAR COMPOUNDS

A characteristic feature of modern ecology (or megaecology) is that it has turned from a strictly biological science to a complex science, a series of knowledge-absorbed parts of various sciences and humanities. At present, the strategy of solving environmental problems caused by the influence of high-molecular-weight compounds on the Earth properties involves the study of treatment methods for plastic industry waste, the study of the kinetics and mechanisms of polymer synthesis with high environmental purity, the development of low-waste technologies to produce and use polymers, and so forth.

The topicality of environmental issues associated with the production and application of polymers is determined, on the one hand, by their indispensability (due to their unique physicochemical and consumer properties) and, on the other hand, by large volumes of their production and their products, which are eventually converted into the category of waste hazardous to the environment. In particular, a large amount of polyethylene (PE) is processed into film materials to be dealt with in everyday life, at every turn. This is packaging bags for products, various containers for storage of various liquids (from water to mineral oils), which, having carried out their functions, are discarded and disturb the natural circulation of substances in nature. The problem is compounded by the fact that many industrial synthetic polymers such as PE, polypropylene, PS, PAs, polyesters, and so forth are quite stable chemical compounds in normal conditions. Many of them are able to withstand exposure to solar radiation and atmospheric oxygen in combination with heat and moisture in the natural environment for tens of years without noticeable chemical degradation.

Polymers undergoing destruction form fragments, which are stored for a fairly long time in the environment and pollute it for many years. What approaches are used to combat the environmental pollution associated with the production of polymers?

First, it is destroying used and discarded polymers. Oxidation of these organic substances at high temperature (burning) could be most natural. However, valuable substances and materials are thus lost. Water and carbon dioxide are combustion products at the best; this means that even the initial monomers (by whose polymerization the destroyed polymers have been obtained) cannot be recovered. In addition, the release of large amounts of carbon dioxide (CO_2) to the atmosphere leads to undesirable global effects, in particular, the greenhouse effect. The combustion forms harmful volatile substances to pollute the air, water, and soil (in some cases they may extend to depths of Earth's crust). For example, in the case of polyvinylchloride, this is a variety of low-molecular-weight chlorinated organic substances characterized by high toxicity, including carcinogenicity; gaseous hydrogen chloride is also formed, which dissolves in water to give hydrochloric acid. Even burning of PE, whose chains consist of carbon and hydrogen atoms and their combustion releases just carbon dioxide and water, is not completely safe. Not to mention numerous additives, including dyes and pigments, various compounds are released to the environment, including heavy metals used as catalysts in the synthesis of PE, they are harmful to human health. Neutralizing the products of combustion is not very efficient from an economic viewpoint. Recycling of worn-out polymers and their products is much more promising and a reasonable way to reduce the environmental pollution.

This problem, however, is not as simple as it may seem at first glance, because we deal, as a rule, with dirty waste including, for example, sand particles. This eliminates any possibility of using high-tech equipment utilized in the primary processing of original polymers. This equipment just would quickly go out of service due to the abrasive action of solid particles of mineral origin. Even during processing, if it is, in principle, possible, "dirty" products are obtained, whose marketability and consumer properties cannot compete with primary products. However, it is possible to use recycled materials for other purposes, with substantially lowered requirements assumed. In particular, contaminated articles of PE can be processed into plates of a few millimeters thick to be used as a roofing material having a number of advantages over conventional ones, such as their low density and, hence, low weight, corrosion resistance and flexibility, low thermal conductivity, and therefore, good thermal insulation properties. The greatest success was achieved by recycling of large-scale rubber products (RP; details are given below). And "secondary" tires are almost as good as original ones. This approach allows us to significantly reduce damage to the environment due to its pollution with useless products and, at the same time, to significantly save the consumption of rubber obtained by polymerization of refined petroleum products or from *Hevea* latex sap.

In recent years, new ideas of the synthesis of "green" plastics and articles thereof have begun to be implemented. First of all, this concerns such polymers and materials which are capable of more or less rapid decomposing under natural conditions. It should be noted that all biological polymers, that is, those synthesized by living organisms (including plants), which include primarily proteins and polysaccharides, are susceptible to destruction in varying degrees, catalyzed by enzymes. In these cases, the principle works: whatever nature creates, it can destroy. Without this principle these polymers in large quantities produced by microorganisms, plants, and animals would remain, after their death, on the ground. But this does not happen, and high-effective biological catalysts (enzymes) do their job and successfully cope with this task. As to synthetic polymers, nature has not yet developed effective means and mechanisms for their destruction. At best, we could expect that some of the enzymes responsible for the rapid and selective destruction of natural organic substances can do the same for synthetic polymers containing appropriate functional groups.

Polymerization processes in the absence of surfactants (e.g., polymeric dispersions containing no emulsifiers) can be an effective method of producing environmentally friendly polymers.

Polyesters are synthetic polymers prone to biodegradation. The number of links in their polymeric chain, called the degree of polymerization, is about 100. If these polymers are buried for 4 weeks and then removed and weighed, it would appear that their weight would have reduced by an average of 20%. This is typical for enzyme-catalyzed hydrolysis whose products are low-molecular-weight substances soluble in water and diffusing into the space surrounding the sample. This process takes place on the surface of a solid sample, since the enzymes, being proteins (natural polymers), are not able to penetrate into the bulk of a polyester sample prepared as a film. Fungi living in soil are the source of these enzymes.

Simultaneously, chemical hydrolysis occurs in the bulk of the sample as well, that is, destruction of the ester bonds in the polymer chains by water molecules able to penetrate into the polymer sample. This process is accompanied by chain cleavage with an equal probability anywhere. As a result of chemical destruction, chain fragments are formed, having their molecular weight (or the polymerization degree) of the same order of magnitude as the source macromolecules, and insoluble in water. As a result, a decreased molecular weight of the polymer and deterioration of its mechanical properties are observed.

If there are no "weak links" in the polymer chain, which can be broken by enzymes or subjected to chemical degradation in the environment, the situation is much more complicated. A typical example of this kind is the PE chain formed by carbon atoms

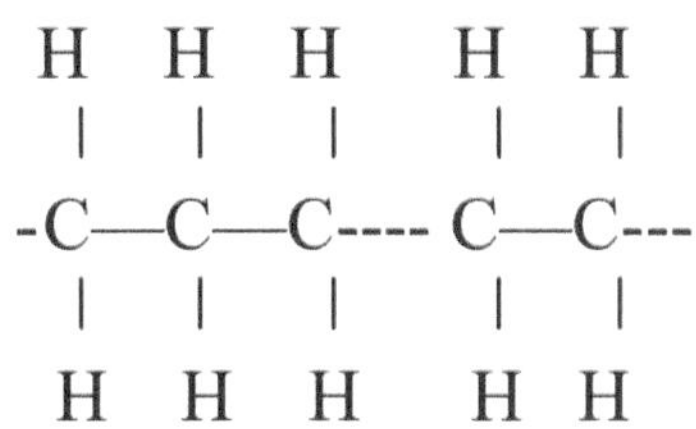

This polymer, if its degree of polymerization exceeds 20–30, is not subject to enzymatic cleavage and is not used by microorganisms as food. At the same time, short PE chains ($n < 20$) in reality are known to be food (substrates) for a specific group of microorganisms comprising over 100 species. Such a process, from the chemical viewpoint, is a sequence of chemical conversions catalyzed by a specific enzyme each. It is assumed that, as a result of these transformations, acetic acid is formed and used by microorganisms as food and source building material. Clearly, if high-molecular-weight PE (used for producing a variety of materials and having the polymerization degree $n > 1000$) is split into shorter fragments with $n < 20$, its further splitting can be done by the microorganisms living in the environment. To enable a long PE chain to be split into relatively short fragments under natural conditions, "weak bonds" are introduced into it during synthesis; under the influence of sunlight, its short-wave component (called ultraviolet), in the presence of atmospheric oxygen and moisture they can be destroyed with a noticeable rate. The resulting fragments, in turn, are capable of biodegradation.

In recent years, other techniques have been developed, which allow obtaining biodegradable polymeric materials (PMs) with no intervention in the synthesis of macromolecules. Other approaches are based on obtaining mixtures of a stable polymer, such as PE, and a biodegradable one, for example, the well-known starch. If enough starch is introduced into such a mixture at making products from PE, this starch will biodegrade by microorganisms in case of contact with the soil. In this case, PE will remain in the product (film) to be dispersed into fine particles and to disappear. Obviously, the best solution to the problem of environmentally friendly PMs would be the use of natural polymers whose destruction proceeds by effective mechanisms developed by nature. Such polymers are found in plants and living organisms in abundance. In the first case, it is the well-known cellulose, a high-molecular-weight polymer, which accounts for about half of wood. In the case of living systems they are mostly proteins, that is, α-amino acid polymers. The remains of α-amino acids (hundreds and thousands) are located in their chains in an order very specific for each type of protein.

Wool, silk, and leather are widely known PMs made of proteins. Both cellulose and wool, entering the environment, are exposed to chemical attacks, decompose and are used as food by a variety of microorganisms, such as bacteria and fungi, which are integral components of the "living" soil. In other words, they are biodegradable substances. Biodegradation is chemical cleavage caused by biochemical reactions, primarily catalyzed by enzymes synthesized by microorganisms. These reactions can proceed in the presence or absence of air oxygen. These natural polymers, however, cannot replace a wide variety of modern synthetic ones.

Therefore, scientists are constantly looking for opportunities to expand the range of polymers synthesized by animate nature. We note in passing that, in addition to the "environmental purity" of such polymers, they have certain advantages over synthetic ones, since they are obtained from so-called renewable raw materials, plants and organisms which are continuously reproduced, while synthetic polymers are produced from oil products, limited on the Earth.

The search for new PMs of natural origin in recent years has led to apparent success. Bacteria living in the soil and capable of synthesizing polymers as their intracellular reserve material have been discovered.

The properties of these polyesters, according to the length of the side group -R, vary from those characteristic of rigid plastics (small x) to typical rubbers (large x). One can change the properties of this polymer within these limits by mixing units having different pendant groups -R in one macromolecule. For the bacteria to synthesize such polymers with various groups -R at a predetermined ratio, one needs to correctly compose the bacteria's diet.

Why do bacteria synthesize such polymers?

It turns out that they prepare them for the future, for possible "severe" conditions, when they would have to fight for survival. If the living conditions of the bacteria are normal, they use a wide range of organic materials by recycling them to produce the necessary energy for life and the creation of materials and substances needed to build cells—and stock nothing for future use. If the wrong food is fed to these bacteria, they will start to synthesize the above polymers and accumulate them inside the cell. Polymer granules inside such cells can be easily detected with an electron microscope.

Following disruption of these cells, the polymer contained therein is separated from the cellular mass and processed. It is important that such bacterial polyesters can be processed into films, fibers, and articles using the same equipment as that for conventional processing of synthetic polymers. It is easy to make the same bacteria produce a wide range of polymer products with desired properties. Currently, industrial fermenters are used in all developed countries, that is, devices in which the bacteria are grown on a commercial scale.

What does a bacterium do with the accumulated polymer and when? The bacterium consumes its polymer as a substrate, when cannot find food in the environment. It subjects the reserve polymer to enzymatic cleavage and uses the hydrolysis products as a source of energy and food in order to survive. In our environment, including soil, many microorganisms live, which produce enzymes capable (as well as intracellular ones) to digest bacterial polyesters. Therefore, worn-out and discarded goods made of these polymers undergo complete biodegradation in nature.

8.2 KINETICS AND MECHANISM OF THE PROCESSES TO OBTAIN ENVIRONMENTALLY FRIENDLY POLYMERS AND TO REDUCE PLASTIC WASTE

The design and widespread use of synthetic PMs is one of the areas of chemistry. This leads to the emergence of new environmental challenges associated with environmental pollution with these materials, as well as the production waste of monomer synthesis. The results of studies of the kinetics and mechanism of vinyl monomer polymerization in various physical and chemical conditions indicate that they can be used for scientific substantiation of the optimization of technological modes of monomer and polymer synthesis, as well as for solving both regional and global environmental problems. To obtain reliable experimental data and their correct interpretation, such physicochemical and analytical methods were used as dilatometry, viscometry, ultraviolet (UV) and infrared (IR) spectroscopy, electron paramagnetic resonance (EPR), electron microscopy, liquid and gas–liquid chromatography, as well as some methods of chemical analysis (bromination and iodometric titration). To analyze the properties of polymeric dispersions, the turbidity spectrum method was used, and the efficiency of flocculants was evaluated by the sedimentation speed of suspended particles in water, with special simulators (copper oxide), and gravimetrically. Polymerization was carried out in air and under oxygen-free conditions.

8.2.1 HOMOPOLYMERIZATION AND COPOLYMERIZATION OF ACRYLONITRILE IN AN AQUEOUS SOLUTION OF SODIUM THIOCYANATE[2–6]

It is known that in the preparation of fiber-forming polymers based on acrylonitrile (AN) by polymerization in a solvent, spinning solution is formed, ready for forming fibers therefrom. Some organic solvents or aqueous solutions of inorganic salts are used as solvents in these cases. A comparative study was made of the kinetics and mechanism of AN polymerization in dimethylformamide (DMF) and in an aqueous solution of sodium thiocyanate (ASST) initiated by azo-isobutyric acid dinitrile (AAD) and some newly synthesized azonitriles (azo-*bis*-cyanopentanol, azo-*bis*-cyanovaleryanic acid, and azo*bis* dimethylethyl amidoxime). It has been found that the polymerization rate in ASST is significantly higher as compared to the reaction in DMF, despite the fact that the initiation rate in the presence of all these azonitriles is significantly lower than in DMF. The relatively low initiation rate in ASST is due to amplification of the "cellular effect" (due to the higher viscosity of the water–salt solvent), to the ability of water to form hydrogen bonds (which reduces the efficiency of initiation), and to a reduced decay constant of the initiator due to its donor–acceptor interaction with the solvent. The ratio $k_p / k_o^{0,5}$ of the rate constants of chain growth and termination in ASST exceeds this value in DMF by about 10 times. This corresponds to a significantly higher molecular weight of the resulting polymer. These differences are due to the influence of the medium's viscosity on k_o, as well as the formation of hydrogen bonds with the nitrile groups of the end link of the macroradical and the joining monomer. However, the main contribution to the increase in k_{δ} is made by sodium thiocyanate, which forms charge-transfer complexes with an AN molecule or a radical, leading to their activation.

The kinetics of AN copolymerization with methyl acrylate (MA) or vinyl acetate (VA) in ASST is qualitatively similar to AN homopolymerization in the identical conditions. However, the initial reaction rate, the molecular weight of the copolymer, the effective activation energy (E_{ef}), and the orders by initiator and by the total monomer concentration differ. For example, the reduced E_{ef} is caused by the presence of a more reactive monomer (MA) and that in the AN–VA system is due to the influence of the nonterminal monomeric units in macroradicals on the rate constant of cross-chain termination. For the binary systems discussed, the copolymerization constants were evaluated, whose values indicate higher MA contents in the copolymer relative to the initial monomer mixture. The same is observed for the AN copolymerization with VA.

The production of synthetic fibers such as polyacrylonitrile (PAN) fiber (nitron) is preceded by the formation of a spinning solution by AN copolymerization with MA (or VA) and itaconic acid. Acrylic or methacrylic acid and methallyl sulfonate are used as a third monomer. When this third monomer is introduced into the reaction system, the rate of the process and the molecular weight of the copolymer reduce, which makes these kind of comonomers ineffective inhibitors. When copolymerizing ternary monomer systems based on AN in ASST (51.5%), the rate of chain origin increases with the total concentration of monomers. The order of the initiation reaction relative to the total monomer concentration varies from 0.5 to 0.8, depending on the degree of decelerating influence of the third monomer. Furthermore, it is shown that AN in the ternary system is less involved in the reaction of chain initiation than in its homopolymerization and copolymerization with MA or butyl acrylate (BA).

Thus, these results allow adjusting the kinetics of copolymerization and the structure of the copolymer formed, which is a way of chemical modification of synthetic fibers.

8.2.2 EFFECT OF SULFURIC ACID ON THE POLYMERIZATION OF (METH)ACRYLIC ESTERS[6–7]

Sulfuric acid is one of the components in the synthesis of acrylic and methacrylic monomers and causes their premature polymerization in synthesis conditions. To elucidate the mechanism of the relevant processes, the polymerization of (meth)acrylic esters was investigated in the presence of sulfuric acid at elevated temperatures without releasing the monomer from the atmospheric oxygen dissolved therein, which generally corresponds to the industrial conditions.

Acrylic and methacrylic monomer polymerization in the air is preceded by induction periods due to oxygen inhibition. The chemistry of such inhibition is the copolymerization of the monomer with oxygen to form inactive peroxide radicals. Polymeric peroxides (an alternating monomer–oxygen copolymer) are the copolymerization products. Introduction of sulfuric acid into the reaction medium dramatically reduces the induction period and increases the polymerization rate. The formation of monomer–acid complexes has been established, whose composition depends on the concentration ratio of the components. Free sulfuric acid acts as an inhibitor, leading to the emergence of extreme dependencies for the rate and the induction period.

Kinetic studies of the polymerization show a significant increase, in the presence of acid, of the decomposition rate of polymeric peroxides, which are major contributors to initiation of the polymerization reaction in the air. Furthermore, the ratio $k_p / k_o^{0,5}$ increases, due to a decrease in k_o (owing to an increased viscosity) and an increase in k_p due to monomer activation as a result of its protonation with acid. The effective activation energy of the thermal polymerization in the presence of sulfuric acid is 50–55 kJ mole^{-1}, which allows the reaction to be carried out at relatively low temperatures. As the acid concentration increases, not only the rate increases drastically but also the molecular weight of the resulting polymer. Thus, the polymerization of acrylic and methacrylic monomers in the presence of sulfuric acid can be regarded as an effective method for polymer synthesis, allowing high-molecular-weight products to be obtained with high rates, and, in the case of the reaction in the air, with short induction periods.

8.2.3 INFLUENCE OF METAL SALTS OF VARIABLE VALENCY[6,7]

In a certain concentration range, additives of iron (Fe-St), copper (Cu-St), cobalt (Co-St), zinc (Zn-St), and lead (Pb-St) stearates increase the polymerization rate of styrene and methyl methacrylate (MMA) compared with thermal polymerization. By initiating activity they can be arranged in the following order: Pb-St < Co-St < Zn-St < Fe-St < Cu-St. The observed decrease in the values of effective activation energy, activation energy of initiation and kinetic reaction order by monomer indicates the active participation of the monomer in the chain initiation reaction. IR spectroscopy data allowed speaking of the formation of an intermediate monomer–stearate complex capable of disintegrating into active radicals, leading to polymerization initiation. The benzoyl peroxide (BP) + Fe-St (or Cu-St) systems can be used for efficient polymerization initiation. Concentration inversion of the catalytic properties of stearates has been detected, which depends on the salt concentration and the degree of conversion. The efficiency of the accelerating influence decreases with increasing temperature, the BP + Fe-Ss system possessing a higher initiating activity. Initiation occurs due to stearate radicals formed during decomposition of the complex consisting of one BP molecule and two stearate molecules.

The phenomenon of concentration and temperature inversion of the catalytic properties of gold, platinum, palladium, and osmium chlorides at the thermal end initiated polymerization of styrene and MMA was discovered. The mechanism of the dual action of noble metal salts is due to competition of the initiation influence of the complexes of the monomer with colloidal metal particles and the inhibition reaction proceeding through ligand transfer.

8.2.4 STABILIZATION OF MONOMERS

The inhibition (stabilization) of vinyl monomer polymerization was mentioned in Section 7.4, as well as in refs 6, 8, 9, and 20. The practical significance of inhibitors is often associated with their use to stabilize monomers and to prevent various spontaneous and unwanted polymerization processes. In industrial conditions, polymerization can occur in the presence of air oxygen, and thus with peroxide radicals being active sites of the chain reaction. In these cases, compounds with labile hydrogen atoms, such as phenols and aromatic amines, are used to stabilize monomers. They inhibit polymerization in the presence of oxygen only, that is, are antioxidants. Aromatic amines known as polymer stabilizers were tested as polymerization inhibitors of (meth) acrylates in a flowing air atmosphere, namely: dimethyldi-(*n*-phenylaminophenoxy) silane, dimethyldi-(*n*-β-naphthylaminophenoxy) silane, 2-oxy-1,3-di-(*n*-phenylamino-phenoxy)-propane, and 2-oxy-1,3-di-(*n*-β-naphthyl-aminophenoxy)-propane. These compounds are more effective stabilizers in comparison with the commonly used hydroquinone (HQ), as evidenced by the high values of stoichiometric coefficients of inhibition (by 3–5 times higher than for HQ).[9–28] Inhibition of the thermal polymerization of acrylic and methacrylic acid esters at relatively high temperatures (100°C or higher) is characterized by a sharp increase in the induction periods when a certain critical inhibitor concentration $[X]_{cr}$ is exceeded. This is due to the fact that the formation of polymeric peroxides by copolymerizing monomers with oxygen must be considered during polymerization in the air. The decay of these polyperoxides occurs during the induction period and can be considered as degenerative branching. The presence of critical phenomena is characteristic of branched chain reactions. However, in early papers, degenerative chain branching on polymeric peroxides was not considered while describing inhibition of thermooxidative polymerization. The results indicate that the value of the critical inhibitor concentration $[X]_{cr}$ may be one of the main characteristics of its effectiveness.

Inhibition of the spontaneous polymerization of (meth) acrylates is necessary not only for monomers during storage but also in their synthesis, which is conducted in the presence of sulfuric acid. In this case, stabilization is even more relevant, since sulfuric acid inactivates many inhibitors and, moreover, can intensify the process of polymer formation. The concentration dependence of induction periods in these conditions has a pronounced nonlinear character. In contrast to bulk polymerization, in the presence of sulfuric acid the decomposition of polymeric peroxides is observed at relatively low temperatures as well, and the values of $[X]_{cr}$ for the studied amines is about 10 times lower than that of hydroquinone. EPR in solutions of the studied amines containing sulfuric acid has found radical products. Analysis of the EPR signal structure indicates a partial conversion of the aromatic amine into a stable radical cation, which gives it the ability to inhibit polymerization in the absence of oxygen as well.

The preparation of MMA from acetone cyanohydrin is a common method of its industrial synthesis. The process is carried out in the presence of sulfuric acid in several stages, during which the formation and interconversion of various monomers occur. Separate steps of the synthesis were simulated by reaction systems comprising not only MMA but also methacrylamide and methacrylic acid as well as sulfuric acid and water in various ratios. As both homogeneous and heterogeneous systems appeared, inhibition was studied in static and dynamic conditions. It was found that the above aromatic amines effectively inhibit polymerization at various stages of the synthesis and purification of MMA. Their advantages over hydroquinone are more pronounced in the presence of sulfuric acid in homogeneous conditions, and in biphasic reaction systems while stirring. Moreover, polymerization inhibitors are most necessary in dynamic conditions at the esterification step.

The use of monomer stabilizers to prevent various spontaneous polymerization processes involves the further release of the monomer from the inhibitor prior to processing into polymer. This is usually achieved by monomer distillation, often with preliminary extraction or chemical deactivation of the inhibitor, which requires high energy expenditure, results in large losses of the monomer, and additional pollution of the environment. An optimal solution would be to develop such a method of stabilization, whereby the inhibitor effectively inhibits polymerization of the monomer during storage, but virtually has no adverse effect on it at the stage of polymer synthesis. One possibility is the use of low-soluble inhibitors in the monomer. When monomer is stored and the polymerization initiation rate is low, the amount of the dissolved inhibitor may be sufficient for stabilization. In addition, as the inhibitor is consumed, its constant replenishment can occur due to additional dissolution of the previously non-dissolved substance. The poorly soluble ammonium salt of *N*-nitroso-*N*-phenylhydroxylamine (cupferron) and some cupferronates were studied as such inhibitors. The solubility of these compounds in acrylates, its dependence on the degree of water content of the monomers, and the effect of the amount of inhibitor and the time of its dissolution on subsequent polymerization were studied. The differences in the action of cupferronates are due to their solubility in monomers, different stability in solution, and the ability to deactivation; all this leads to the reduced influence of the inhibitor on the monomer polymerization during its processing into polymer.

Another option for solving this problem is to use, as a polymerization inhibitor, such substances which are inactivated when a polymerization initiator is introduced into the reaction medium or even enhance its effect. Anion–radical complexes of tetracyanoquinodimethane can be used as such an inhibitor. It is shown above that they act as an efficient inhibitor of the polymerization of vinyl monomers, but, when certain peroxide initiators (e.g., BP) are added, the reaction proceeds without an induction period and at a much higher rate, much greater than the increase resulting from the initiating additional action of this initiator.

8.2.5 SOME FEATURES OF THE KINETICS AND MECHANISM OF EMULSION POLYMERIZATION[29–41]

Emulsion polymerization is a method for synthesizing polymers, which allows to implement the process at a high rate and to form a polymer with a high molecular weight, to prepare high-concentrated latexes with a relatively low viscosity, to use polymeric dispersions at their processing without separation of the polymer from the reaction mixture, and to greatly increase the fireproof of the product. At the same time, the mechanism and kinetics of emulsion polymerization differ due to several specific factors such as the multiphase nature of the reaction system and the variety of kinetic parameters whose values depend not only on the reactivity of the reagents but also on their distribution over the phases, the topochemistry of the reactions, and the method and mechanism of nucleation and particles stabilization. The obtained results inconsistent with classical concepts can be characterized by the following effects: (1) recombination of radicals in the aqueous phase, leading to a decreased number of particles and to the formation of surfactant oligomers capable of acting as an emulsifier; (2) the presence of several growing radicals in the polymer–monomer particles, giving rise to the gel effect and an increased polymerization rate at high conversion degrees; (3) a decreased number of latex particles with increasing conversion, which is associated with their flocculation at different stages of polymerization; and (4) an increased number of particles in the reaction course when monomer-soluble emulsifiers are used, and also due to the formation of an "own" emulsifier (the oligomers).

Surfactants (emulsifiers of various chemical natures) are generally used as stabilizers for dispersed particles, which are rather stable, poorly decompose under the influence of environmental factors, and pollute the environment. The fundamental possibility to synthesize emulsifier-free latexes was demonstrated. In the absence of an emulsifier (but under the conditions of emulsion polymerization), with the usage of persulfate-type initiators (e.g., ammonium persulfate), acrylate latex particles can be stabilized by the ionized end groups of the macromolecules. The ion radicals $^{\bullet}M_nSO_4$appearing in the aqueous phase of the reaction medium, precipitate out of solution when a critical chain length is achieved, to form primary particles which are flocculated to form aggregates with a charge density that ensures their stability. Furthermore, due to recombination of radicals, oligomeric molecules are formed in the aqueous phase; they possess properties of surfactants and are capable of forming micelle-like structures. The monomer and oligomer radicals are then absorbed by these "micelles" where chain growth occurs. In the absence of a specially introduced emulsifier, all major kinetic regularities of emulsion polymerization are observed, and any differences only relate to the stage of particle generation and the mechanism of their stabilization, which can be enhanced by the copolymerization of hydrophobic monomers with highly hydrophilic comonomers. An increased temperature raises the polymerization rate and the number of latex particles in the dispersion formed, a reduction in their size and number of the formed coagulum, and to improved dispersion stability.

Accounting for these factors, influencing the mechanism and kinetics of emulsion polymerization (in both the presence and absence of an emulsifier) has allowed explaining the influence of comonomers on the processes of formation of polymer dispersions based on (meth)acrylates. Changes in some reaction conditions affect the nature of the influence of other ones. For example, an increased concentration of methacrylic acid (MAA) at its copolymerization with MA at a relatively low initiation rate reduces the rate and the particle number but increases the amount of coagulum. At high initiation rates, the number of particles in the dispersion in the presence of MAA increases and their stability is improved. The same effects were observed in the case of the emulsifier-free polymerization of BA as well, when at high temperatures its partial replacement by MAA leads to better stabilization of the dispersion, an increased reaction rate and an increased number of particles (while in the presence of an emulsifier their decrease was observed). Similar effects were also observed in the case of AN, which impairs the dispersion stability at relatively low temperatures but increases it at high temperatures. An increased AN concentration in the ternary monomer system with a high MAA content leads to an increased number of particles and an increased dispersion stability even at relatively low temperatures.

To explore the possibility of synthesis of such dispersions whose particles contain reactive polymer molecules with free multiple C=C bonds, the emulsion copolymerization of acrylic monomers and nonconjugated dienes was studied. The use of such latexes for top-finishing textiles and other materials creates some conditions of chemical bonds between the deposited substance and the surface to be treated, which allows obtaining strong indelible coatings.

The kinetics and mechanism of the emulsion copolymerization of ethyl acrylate (EA) and BA with allyl acrylate (AlA) (initiators being ammonium persulfate (APS) or the APS–sodium thiosulfate system) were studied. The copolymerization constants (ratios) were found for AlA with EA (r_{AlA} = 1.05, r_{EA} = 0.8) and for AlA with BA (r_{AlA} = 1.1, r_{BA} = 0.4). The degree of unsaturation of the copolymer and Ala units in the copolymer were estimated. The first value has turned out to linearly increase with increasing content of AlA in the reaction system, and the second one (equal to 70–75%) is almost independent of the synthesis conditions. This indicates an approximate equality of the rates of the interaction of Ala radicals with the acrylic bond of the "own" and "foreign" monomer, which agrees well with the found values of reactivity ratios. The temperature independence of the degree of unsaturation of allyl acrylate units indicates the equality of the activation energies of the reactions of interaction of Ala radicals with its own allyl double bond (cyclization) and with the acrylic double bond of the comonomer.

The emulsion copolymerization of multicomponent monomer systems on the basis of BA was also studied, including water-soluble monomers (AN, MAA) and the nonconjugated diene β-acryloxyethyl maleate (AOEM). Copolymerization with AOEM depends on the reaction conditions. For example, AOEM reduces the polymerization rate in MAA-free systems. In the presence of MAA, the rate of the process at high conversions increases with increasing AOEM concentration, which enhances the gel effect due to partial cross-linking of the polymer chains in the polymer–monomer particles by the side groups with C=C bonds. Double bonds are primarily preserved in the maleic group of AOEM molecules. The degree of unsaturation of the diene units in the copolymer depends on the monomer composition, AOEM concentration, and temperature. In the BA–AOEM system, an increased diene concentration leads to an increased degree of unsaturation. This means that the radical of the diene is attached to its "own" monomer with a higher rate than to BA (r_{AOEM} = 7.7). The increased unsaturation degree of diene units in the MAA-containing systems indicates that the AOEM$^\bullet$ radical interacts with MAA with a higher rate than with BA; therefore the probability of cyclization decreases, the unsaturation degree of the polymer increases, and the inhibitory effect of diene on polymerization decreases. To increase the content of reactive groups of polymer molecules on the surface of the latex particles, we propose to carry out seed polymerization with AOEM introduced into the reaction zone after the formation of polymer–monomer particles.

8.2.6 COMPOSITIONS FOR PRODUCING RIGID POLYURETHANE FOAM[39,42–44]

Studies related to the implementation of the decisions of the Montreal Protocol (1987) (according to which it is necessary to significantly reduce the production and consumption of chlorofluorocarbons (CFC), and even to replace ozone-depleting substances by non-ODS) involved the replacement of trichlorofluoromethane (CFC-11). This compound has been long time used as a foaming agent in the synthesis of rigid polyurethane foam (PUR/PIR), used as thermal insulation in refrigerators and building structures. On the basis of experimentally obtained dependence of the kinetic parameters of the foaming process (starting time, structuring time, and end foam rise), the values of the density and thermal conductivity of foam prototypes on the concentration of the components of the reaction mixture and the physicochemical conditions of polyurethane foam synthesis, optimum compositions (formulations) were designed, containing an ozone-friendly azeotrope mixture of dichlorotrifluoroethane and dichlorofluoroethane instead of CFC-11, its ozone depletion potential being more than 10 times lower compared with CFC-11. An important advantage of these developments is that their practical implementation requires no fundamental changes in the known technological methods and no use of new chemicals.

8.2.7 SCIENTIFIC BASIS OF THE SYNTHESIS OF A HIGH-MOLECULAR-WEIGHT FLOCCULANT[28,29,45,46]

Increasing volumes of water consumption and reduction of its quality (due to anthropogenic influences) entail the problem of purification of natural and industrial waste waters from various contaminants, including suspended and colloidal-dispersed particles. Flocculants are used for these purposes, which are high-molecular-weight compounds with the ability to be adsorbed onto dispersed particles to form rapidly precipitating aggregates. Polyacrylamide (PAA) is most active of them. Due to the fact that in many countries (including Russian Federation) acrylamide (AA) is scarce, we have developed a modification of the synthesis of a PAA flocculant by using AN and sulfuric acid through hydrolysis and polymerization reactions. In the presence of a radical polymerization initiator and sulfuric acid, AN participates in these two processes simultaneously, and as AA is being formed from AN, their joint polymerization begins. The required properties of the polymer were obtained by polymerizing AN in aqueous sulfuric acid solution to a certain conversion degree followed by hydrolysis of the polymerate (a two-stage synthesis scheme), or when an optimum ratio of the rates of these reactions (proceeding in one step) is achieved.

The influence of the nature and concentration of the initiator, the content of sulfuric acid, temperature and the duration of the reaction on the amount and molecular weight of the polymer contained in the final product, its water solubility, and flocculating properties was studied. The desired monomer conversion at the first stage of our two-stage synthesis scheme is determined by the concentrations of AN and the aqueous solution of sulfuric acid. At the one-step synthesis, changes in temperature and the amount of the monomer approximately equally affect the rates of hydrolysis and polymerization and do not have much influence on the composition of the copolymer. But changes in the acid or initiator concentration affect mainly one of the reactions by altering not only the molecular weight but also the composition of macromolecules, which leads to extremal dependences of the flocculation activity on these factors and facilitates selection of optimum conditions for obtaining the flocculant.

8.3 STABILIZATION OF POLYMERS AS A BASIS FOR THE DESIGN OF NEW HIGH-MOLECULAR-WEIGHT COMPOUNDS AND AS AN EFFECTIVE WAY TO REDUCE WASTE WHEN THEY ARE USED[47–51]

8.3.1 *INTRODUCTION*

The need in polymer stabilization is due to several causes. The main one is preserving the physical and chemical properties that can be varied during their usage, under the influence of temperature, light, mechanical stress, chemicals, and even microorganisms. Under the influence of these factors such physical (including structural) and chemical processes frequently occur in a PM, which lead to its aging and, to some extent, to destruction (degradation). Aging processes can have physical and chemical character. Physical aging is usually reversible, since it leads to no breaking or cross-linking of polymer chains. For example, the processes of crystallization, recrystallization, or unwanted solvent penetration may result in deterioration of the mechanical properties. Chemical aging processes are irreversible, since accompanied by breakage of chemical bonds, changes in the chemical structure, and quantitative changes in the molecular weight.

Various methods of polymer stabilization can significantly increase the duration of their "life," and thus reduce the amount of plastic wastes. The need for stabilizers arises already in the processing of polymers, wherein the polymer material should be transferred into a fluid state, within a temperature range between the melting point and the thermal decomposition point of the polymer. The wider this range and the higher the processing temperature (above the melting point), the easier the processing is. It seems more advantageous to extend this temperature range by raising the temperature of polymer decomposition via stabilization rather than by lowering the melting point, since the latter option would lead to a lowered heat capacity of the material. That is why most polymers are processed with a stabilizer added. Of key importance is the polymer stabilization directed against the combined action of light and oxygen. And, despite the diversity of phenomena occurring during polymer destruction, the major role is played by the processes of oxidation and destruction of chain macromolecules. The action of inhibitors on the free radicals formed at the ends of destructible chains stops further development of chain processes, but cannot prevent decreasing the molecular weight. This undesirable effect can be reduced by using, for example, nonvolatile monomers as inhibiting and plasticizing substances, which, terminating chain decomposition processes, increase the length of the broken parts of macromolecules. The use of such inhibitors where functional groups are inside polymer molecules is more efficient. In this case, the act of inhibition is recombination of a decaying molecule and a polymeric inhibitor one. Inhibition by interaction with colloid-sized particles has several advantages in choosing substances and in improving the bonding between the polymer and the surfaces of these particles (if they are pigments or fillers). It should be noted that the structure of PMs changes and phenomena analogous to metal aging are observed under mechanical impacts and during storage. In this regard, there appears a problem of not only the creation but also the stabilization of the most favorable structures in PMs, which should be solved by completely different methods than the problem of chemical stabilization.

8.3.2 AGING AND STABILIZATION OF POLYOLEFINS

Polyolefins (PE, polypropylene, and copolymers thereof) are currently among the PMs of particular importance. They have a high mechanical strength, low density, flexibility at low temperatures, high impact strength, moisture resistance, electrical insulating properties, and so forth. However, like most other high-molecular-weight compounds, polyolefins are susceptible to oxidation-destructive processes caused by weathering, heat, light, and aggressive media. In these cases, polyolefins lose their elasticity, become brittle, crack, lose mechanical strength and dielectric properties, and sometimes they change color, that is, aging. Studies of polyolefin oxidation are usually carried out under conditions similar to the conditions of their processing. As will be shown below, conventional processing techniques are pressing at temperatures 10–15°C higher than the fluidization temperature, injection molding at 200–250°C, and extrusion. Any processing in the air dramatically deteriorates the properties of the material. In most cases, methods based on oxygen absorption in a closed system are used. This allows one not only to trace changes in the chemical composition of the polymer but also to isolate and identify the volatile reaction products. Oxygen absorption by the polymer occurs with an appreciable induction period, when the oxygen absorption rate is very low. Therefore, the value of the induction period can be used as oxidation rate characteristics. The induction period sharply reduces with increasing temperature and with a change in the oxygen pressure. Numerous studies have established that peroxides (primarily hydroperoxides) are the primary products of oxidation. It has been shown that at the initial stages of the process, the amount of peroxides formed almost exactly corresponds to the amount of absorbed oxygen. The kinetic rate curve coincides with the curve of hydroperoxide accumulation. This proves that the basic oxygenated volatile products are generated from hydroperoxides.

As the reaction proceeds with free radicals involved, it is natural to use inhibitors (antioxidants) to suppress the development of a chain process. Therefore, to stabilize polyolefins, compounds such as aromatic hydrocarbons are used, having a movable hydrogen atom: aromatic amines, substituted phenols, mercaptans, and so forth. These inhibitors react with radicals according to the scheme:

$$ROO + AH \rightarrow ROOH + A,$$

where the formed radical of inhibitor is unable to continue the oxidation chain.

Differentiation of oxidation inhibitors on chain terminators and destroying peroxides allows one to recommend the use of mixtures of these two substances. It was found that, when mixtures of antioxidants are used, various manifestations of their inhibitory activity may occur: mutual enhancement of their actions (e.g., synergies), the addition of efficiencies, and weakening of the effect of the strong antioxidant. In the case of synergy, the dependence of the induction period on the molar composition of the antioxidant mixture is expressed by a curve with a maximum. The value of the induction period at the maximum for effective compositions exceeds that of the individual compounds taken in the concentration equal to the total mixture concentration.

Properties of polyolefins greatly vary under the influence of UV radiation (particularly in an oxygen atmosphere): the relative elongation decreases, the tensile strength varies, the brittleness increases, and the color changes. The wavelength range of 300–400 nm is most dangerous for polyolefins. Therefore, the problem of photostabilization mainly reduces to the introduction of such compounds into the polymer which would absorb this wavelength range. Benzophenone-based substances are considered the most effective light stabilizers.

8.3.3 STABILIZATION OF PVC AND VINYL CHLORIDE COPOLYMERS

A lot of works are devoted to the investigation of the decomposition and stabilization of the homopolymers and copolymers of vinyl chloride. The main conclusions about destruction are as follows:

- thermal decomposition begins with the unsaturated end groups and is accompanied by the formation of conjugated systems; radical chain reactions are possible, that lead to polymer chain destruction and macroradical recombination;
- thermooxidative decomposition is a radical chain process with degenerate branching in which initiation due to the unsaturated end groups is considerably less important than when decay in the absence of oxygen;
- some features of the thermal and thermooxidative decay of the (co) polymer depend on the conditions of its synthesis;
- when studying destruction processes occurring under the influence of radiation, photodecomposition and radiative decay should be distinguished;
- dehydrochlorination, oxidation, destruction, and structuring are the main variants of polymer degradation;
- the rate of photodecomposition in the presence of oxygen is higher than that in an inert atmosphere or in vacuum;
- in the presence of oxygen, destruction reactions predominate, resulting in reduced molecular weights;
- in an inert atmosphere or in vacuum, structuring processes predominate, resulting in a three-dimensional insoluble polymer;
- the stability of a polymer is in direct proportion to the degree of its crystallinity.

Stabilization features of polyvinylchloride and its (co)polymers are directly related to the above conclusions of the specifics of the macromolecular decay.

The activity of heat stabilizers is evaluated by qualitative or quantitative comparison of the rates of dehydrochlorination, oxidation, destruction, and structuring of both unstabilized and stabilized polymer at temperatures from 150 to 190°C. Any quantity characterizing the influence of a heat stabilizer on the rate of dehydrochlorination is most commonly taken as the criterion of its activity. The influence of a stabilizer on the decomposition initiation temperature, the induction period duration prior to hydrogen chloride release (which is called thermal stability), and on the HCl evolution rate at the end of the induction period is typically estimated. The evaluation of the electrical insulation properties of finished products and the degree of discoloration at heating sample are also used to evaluate the activity of thermostabilizers by their effect on the dehydrochlorination rate. Optical methods of analysis, such as infrared spectroscopy, are widespread. In assessing the effectiveness of stabilizers, their preliminary testing is conducted under conditions different from their operating conditions.

Metal salts of organic fatty and aromatic acids belong to a most common type to stabilizers. The lead, tin, barium, calcium, cadmium, strontium, lithium, and sodium salts of such acids as formic acid, oxalic acid, maleic acid, caprylic, lauric, stearic, and so forth are used as stabilizers. Mixtures rather than individual salts are generally applied, which allow strengthening the stabilizing effect. For example, the activity of organotin stabilizers can be significantly improved when used in a mixture with additives giving a synergistic effect. Esters of orthosilicic or orthoformic acid, phosphoric acid salts, epoxy compounds, barium, and cadmium salts are recommended as the second components. Glycidyl ethers (epichlorohydrin condensates with aliphatic alcohols and phenols), epoxidized esters of higher fatty acids, and so forth are the major types of epoxy stabilizers. Soap, metal salts of inorganic acids, of phosphoric and phosphonic acid esters are used for enhancing the activity of epoxy stabilizers. Of the organic nitrogen-containing compounds for stabilizing chlorine-containing homopolymers and copolymers, mainly used are aliphatic and aromatic amines, of organic and inorganic acid amides. The stabilizing effect of sulfur-containing compounds is owing to their ability to destruct peroxide and hydroperoxide groups with no formation of free radicals and their properties to break reaction chains while polymer decays. Sulfur-containing substances give a synergistic effect with amines and phenols.

8.3.4 STABILIZATION OF POLYAMIDES

Materials made of PAs age very rapidly during operation since they are vulnerable to heat, sunlight, the air oxygen, moisture, and other abiotic environmental factors.

Heating of PAs in the presence of oxygen or air leads to significant changes in their chemical composition and physicochemical properties. Large amounts of organic antioxidants like amines and phenols were studied as aging stabilizers. Aromatic amines and diamines, for example, *N*-phenyl-2-naphtylamine (neozone D) and *N,N'*-di-β-naphthyl-*n*-phenylenediamine and aromatic oxycompounds (containing one or more OH groups and cyclohexyl groups), amines having condensed nuclei, and so forth were most effective. Among the inorganic inhibitors of PA oxidation, most effective are metallic copper, bromide and iodide salts of alkali metals; two-component systems (copper salt + a salt of an organic base), a phosphorus compound with an inorganic halogen; and ternary mixtures, for example, a copper compound + a halide compound + a phosphorus compound.

Products made of PAs (fibers, films, etc.) change their properties when exposed to UV rays (they become brittle, non-meltable, and yellow). Changes in the mechanical properties strongly depend on the irradiation conditions. The destruction mechanism was ascertained by studying the composition of the gases evolved during irradiation. To improve the light fastness of PAs, a fairly large number of inorganic and organic stabilizers are recommended. Salts of two-, three-, and hexavalent chromium should be primarily mentioned as inorganic additives to PAs. Divalent chromium is oxidized and hexavalent one is reduced to trivalent when PA is treated by the stabilizer. Lightfast PAs are obtained by adding aqueous suspension of carbon black containing water-soluble chromium salts prior to polycondensation. The action of certain salts of other metals (Al, Mn, Cu, Co, and Ni) reduces to strengthening the amide bond (depending on conditions). Furthermore, the protection against light aging can be carried out by introducing organic compounds capable of absorbing UV light stronger than the polymer. Compounds of the oxybenzophenone series, benzotriazole, and some salicylic acid derivatives are used as UV absorbers.

8.3.5 STABILIZATION OF CAOUTCHOUC AND RUBBER

The presence of impurities and special ingredients (sulfur, carbon black, and plasticizers) in technical caoutchoucs has an impact on their aging and the behavior of inhibitors. In most cases, they weaken and suppress the action of antioxidants. The impurities contained in the monomers enter the structure of polymer chains at polymerization, break their regularity and are frequently the weak points to provoke their thermal or thermal-oxidative process. The development of a three-dimensional spatial network in rubbers during their vulcanization creates an additional source of rubber destruction as the transverse (especially polysulfide) bonds can be broken under the influence of heat and oxygen.

The oxidation process is dramatically slowed under the influence of small amounts of amines and phenols. Slowing oxidation, the inhibitor reduces the rate and changes the character of structural changes in polymers. It should be noted that the regularities found often contradict the known concepts of the mechanism of oxidation, whereby a peroxide radical is the chain-leading radical, and inhibition occurs as a result of a reaction of this radical with an inhibitor molecule. For example, it has been shown that the inhibitor is consumed at a constant rate. This can be explained by assuming that the inhibitor is only consumed by its reaction with the oxidizable hydrocarbon, leading to chain termination. In this case, the inhibitor consumption rate is equal to the initiation rate (a stationary process). However, such a pattern occurs only at low inhibitor concentrations (below 1%); at concentrations above 1%, the consumption or addition of the inhibitor to the polymer proceeds nonlinearly, and the consumption rate increases with increasing concentration of the inhibitor.

Experimental evidence indicates that the application of various classes of compounds, the mechanism of inhibition is different and is dependent on the chemical nature (structure) of the comonomers, the cross-linking nature, and the optical properties of the ingredients contained in the copolymer and impurities. Analysis of these data shows that the most effective stabilizers of caoutchouc and rubbers are: secondary amines, diamines, and derivatives thereof; quinoline compounds; salts of dithiocarbamic acid derivatives; monohydric and polyhydric alkyl phenols; and phosphorous acid esters.

8.3.6 *CONCLUSION*

The chemical physics of molecular degradation and stabilization of polymers is a scientific basis of such an applied field as PMs science, the main achievements of which are as follows.

The chemical nature and the electronic structure of the active sites responsible for molecular polymers destruction and the mechanisms of migration of these active sites, their decontamination and neutralization were established. Answers to few fundamental questions were obtained, namely, whether the destruction occurs within one macromolecule or it spreads throughout the PM; in which conditions the destruction process develops in the amorphous phase only and when the reaction begins to penetrate into polymer crystals. Principles of polymer stabilization were formulated, the mechanisms of stabilizer action to protect polymers against heat, oxygen, light, radiation, and combustion were revealed, and their classification is given. The main practical result of this trend is the chemical physics is the design of polymer materials with long-term antioxidant, antiozonic, and photochemical stability, thereby reducing the amount of wastes in the plastics industry.

8.4 FEATURES OF THE SYNTHESIS AND USE OF PVC AND POLYETHYLENE[52–54]

Nearly 800,000 tons of polymer wastes are annually produced in Russian Federation. About 10% of them are recycled. Production wastes are mainly recycled, while only some consumer wastes are recycled. What are the causes of this and what are ways to solve this problem? Recycling plastic waste is no less complex and costly than the production of articles from polymers. Emissions of pollutants into the atmosphere from plastic molding pollute the environment due to the outdated pollution-control equipment or owing to its failure. PMs are usually multicomponent systems as various components (ingredients) except the polymer itself are used to produce them. The preparation of PMs that meet certain performance requirements in any industry is a task of the PM manufacture technique. The multicomponent nature of PM often leads to the fact that the production of PM and their practical use in some cases are complicated by the undesirable process of emission of harmful low-molecular-weight substances from the material. Depending on operating conditions, their amount may be up to several percent by weight. Tens of compounds of different chemical nature can be detected in the media in contact with PM. The designing and applying PM is directly or indirectly related to the effect on the human body, on the environmental production environment, and the human environment, as well as on the environment as a whole. The latter is especially important after using PM and products made of them, when waste materials are disposed in the soil, and the pollutants released during decomposition of the PM contaminate soil, waste water, thereby worsening the environment. It is therefore necessary to ensure control of the environmental safety of the process of designing polymers and PMs, their exploitation, and disposal of PM waste after their use by people.

8.4.1 *PRODUCTS MADE OF POLYVINYL CHLORIDE (PVC)*

There are four international PVC brands: PVC-S-7059M, PVC-S-6358M, PVC-S-6768M, and PVC-S -5868PG, each of which is used for the production of specific things:

PVC-S-7059M: critical plasticized products (light-heat-stable cable plastic, high-strength pipes, special linoleum, plasticized films, and artificial leather).

PVC-S-6358M: plasticized and semirigid products for general use (linoleum, artificial leather, and plasticized films) and sheets of special purpose.

PVC-S-6768M: pipes, profile moldings, and other plasticized materials (the basic PVC for the production of window designs).

PVC-S-5868PG: films and bulk polymeric packaging for food and consumer goods.

8.4.2 *BRIEF DESCRIPTION OF THE PROCESS OF PRODUCING PVC*

PVC resin is a universal thermoplastic polymer produced from a petrochemical product (ethylene) and sodium chloride (table salt) by vinyl chloride polymerization. The production of PVC is most complex and highly technological, which requires the latest technology and high-quality equipment. PVC resin is obtained by polymerization, a process in which the monomer molecules (low-molecular-weight compound: vinyl chloride) are combined, thus forming a high-molecular-weight compound.

The chemical formula of PVC is:

$$\begin{array}{c}(-H_2C-\underset{\displaystyle |\atop \displaystyle Cl}{CH}-)_n\end{array}$$

Bulk polymerization is a way that takes place by a periodic pattern in two steps. This technology is developed by the only company—Peshine Sant Gobain (France). Recently, the interest to this method has fallen, since the PVC resin thus obtained has a sufficiently narrow application and is difficultly freed from residual vinyl chloride. This method has the following disadvantages: (1) in the course of the reaction it is necessary to maintain a certain temperature, but removal of the reaction heat is difficult due to some design features; (2) the formation of a crust on the walls of the equipment affects the uniformity of the resulting PVC; and (3) the resulting resin has a relatively low heat resistance and homogeneity.

Emulsion polymerization is a process occurring according to a continuous or intermittent pattern. The continuous scheme is highest performing, but the resulting particles of the emulsion resin are too inhomogeneous, so the periodic scheme is more often used. According to the batch PVC technology, the emulsion resin is obtained with a required particle size distribution, which is important for its processing. The production of emulsion resin has a significant drawback, namely, additives (various adjuvants), which, on the one hand, accelerate and increase the formation of PVC emulsion resin, and, on the other hand, pollute it. Because of this, the emulsion PVC resin is ultimately used for the production of pastes and plastisols only. PVC emulsion resin has a broad molecular weight distribution, a high content of impurities, and a relatively high water absorption, worse dielectric characteristics, and heat resistance and light resistance. PVC emulsion resin is processed into articles by extrusion, injection molding, pressing, and through paste—into soft products.

Suspension polymerization is a way occurring by the periodic scheme and is the most common method of PVC resin production. This method combines some advantages not enjoyed by the previous two ones, for example, easy heat removal from the reaction, the high performance; PVC suspension resin is much cleaner relative to emulsion PVC, PVC resin is pretty well combined with other components in processing, and its properties can be easily modified. In addition, PVC suspension resin has a relatively narrow molecular weight distribution, a low branching degree, high purity, low water absorption, good dielectric properties, better heat and light resistance as compared with the emulsion resin. PVC suspension resin is processed into articles by extrusion, rolling, pressing, and injection molding; it is used for the production of rigid, semi-soft and soft (plasticized) plastic mass. The proportion of PVC resin emulsion is gradually reduced, although it still finds applications. The proportion of PVC suspension resin constantly grows and already accounts for 80% of the total world output. PVC of various brands of the general formula $[-CH_2-CHCl-]_n$, $n = 100–2500$ is prepared by suspension polymerization of vinyl chloride monomer during 8–14 h at a temperature within 30–70°C and a pressure of 0.4–1.2 MPa by the periodic scheme in a reactor made of stainless steel HE1X18H10T or a glass-lined one with a stirrer of a 10–25 m^3 volume. Four basic components are used: vinyl chloride, water, a suspending agent, and a polymerization initiator.

The monomer (vinyl chloride $CH_2{=}CHCl$) is obtained by four ways:

1. from acetylene:acetylene (100%) + HCl (102–110%) are mixed, diluted with nitrogen, and transferred into a contact device with a HqCl2 catalyst at 120–220°C;
2. from ethylene:ethylene + Cl_2 in 1:1.2 to 1:3 ratios within 200–700°C (with a catalyst—within 45–60°C) in the presence of a diluent (dichloroethane);
3. from dichloroethane: (by pyrolysis or dehydrochlorination);
4. a new cost-effective way—from light gasoline, Cl_2 and O_2 .

The flow of light gasoline after pyrolysis with a 1:1 ethylene:acetylene ratio is first hydrochlorinated by HCl and then chlorinated to dichloroethane, from which vinyl chloride is obtained by pyrolysis. Then the following operations are performed: separation from impurities (partial condensation); drying; liquefaction; distillation (under pressure); storage (in stainless steel cylinders at (−30)–(−50)°C under nitrogen, without inhibitors; at higher storage temperatures inhibitors should be added—hydroquinone, etc.). Water is demineralized, without impurities and oxygen. The suspending agent is a protective colloid made of metal hydroxides, phosphates or carbonates, kaolin, and so forth or a water-soluble polymer (resins, cellulose derivatives). The polymerization initiator is monomer-soluble benzoyl peroxide, lauroyl peroxide, chloroacetyl peroxide, and so forth. A typical formulation of loading components into a 10 m^3 reactor is: the monomer (vinyl chloride) 3000 kg; water 6000 kg; a suspending agent 4 kg; a polymerization initiator 4.2 kg; the monomer consumption is 1050–1070 kg ton^{-1} of dry product, and auxiliary products 1–2 kg. Vinyl chloride is suspended in water under vigorous stirring in the presence of a monomer-soluble polymerization initiator and a suspending agent. After completion of the polymerization cycle (10–20 h, including 2–3 h for loading, unloading, and ancillary work), a PVC suspension with 75–150 μm-sized particles (sometimes up to 600 μm) is obtained. After separation of the unreacted monomer in a separator, the suspension is fed through a mixer into a centrifuge where it is squeezed to 25–30% moisture. The separated PVC is dried by speed drying or in a drying drum heated by hot (65–150°C) air. Then PVC is sorted on shakers by particle sizes and packaged in paper bags. PVC can be supplied to largest consumers in special tanks.

8.4.3 *SAFE AND CLEAN WORKING WITH SUSPENSION PVC*

Suspension PVC is a white powder, tasteless and odorless. Because of its effects on the human body, it belongs to the 3rd class moderately hazardous substances. When heated above 150°C, suspension PVC partially decomposes to release hydrogen chloride and carbon monoxide. The maximum permissible concentrations in the working area are: vinyl chloride and hydrogen chloride—5; carbon monoxide—20 mg m^{-3}. The maximum allowable concentration of PVC dust in the working area of industrial premises is 6 mg m^{-3}. PVC is combustible. Its ignition temperature is 310–330°C and the self-ignition temperature is 470–490°C. Dust-air PVC mixtures are explosion-proof. Flame propagation over a dust-air mixture is not observed up to a concentration of 300 g m^{-3} at any dispersion degree. In contact with water, acids, alkalis and atmospheric oxygen, PVC does not burn and explosion-proof. Indicators of fire and explosion hazards are identified by a standard procedure. When this standard is revised, the indicators of fire and explosion hazards are checked. Production facilities associated with drying, sieving, crushing, and packaging PVC shall conform to a certain category of fire safety. Fire-fighting tools are sprayed with water, foam, felt mat, and sand. Dust in industrial premises is removed by vacuum dust cleaning. Production facilities must be equipped with a ventilation system and comply with sanitary standards. Sampling places must be equipped with additional local ventilation. Apparatus and reactors must be grounded to protect against static electricity. Personal protective equipment is: overalls, respirators, gas masks, and biological gloves (silicone cream). PVC forms no toxic compounds in the air and sewage.

8.4.4 *POLYETHYLENE*

PE is the cheapest nonpolar synthetic polymer from the class of polyolefins, representing a white solid with a grayish tinge. Owing to a broad range of its properties, PE is used in many industries and the national economy: cable, electronic, chemical, light industry, medicine, and so forth. Various products are manufactured from PE for engineering purposes, pipes, cable insulation, packaging materials, household items, and greenhouse coverings. For contact with food, only low-density PE is allowed because high-density PE may contain residues of the catalysts, heavy metals harmful to human health. About 70% of PE is produced for this purpose as film and sheeting. The world production of PE is over 30 million tons. Practically, all major petrochemical companies produce PE. The main raw material for them is ethylene. PE is synthesized under low, medium, and high pressure. PE is mainly produced in pellets of 2–5 mm diameters, very seldom as powder. There are four basic methods of PE production to prepare low-density polyethylene (LDPE), high-density polyethylene (HDPE), medium-density polyethylene (MDPE), and linear low-density polyethylene (LLDPE). Each of them differs in structure and properties and they are used for different tasks.

8.4.5 *PRODUCTION OF LOW-DENSITY POLYETHYLENE (LDPE)*

In industry, LDPE is prepared under high pressure by ethylene polymerization in an autoclave or a tubular reactor. The process in the reactor occurs by a radical mechanism under the influence of oxygen, organic peroxides (lauroyl, benzoyl), or mixtures thereof. Mixed with an initiator, heated up to 700°C and compressed by a compressor to 25 MPa, ethylene initially enters the first part of the reactor where it is heated up to 1800°C and then the second one, for polymerization at temperatures of 190–300°C and pressures from 130 to 250 MPa. On average, ethylene stays in the reactor from 70 to 100 s. The conversion degree is below 20%, depending on the type and amount of the initiator. Unconverted ethylene was removed from the resulting PE, and then it is cooled and granulated. The granules are dried and packaged. Commercial LDPE is produced as uncolored and colored pellets.

8.4.6 *PRODUCTION OF HIGH-DENSITY POLYETHYLENE (HDPE)*

HDPE is commercially produced at low pressure. For this purpose, three main technologies are used: suspension polymerization, solution polymerization (in hexane), and gas phase polymerization. Solution polymerization is the most common method carried out at a temperature between 160 and 250°C and a pressure from 3.4 to 5.3 MPa, contact with the catalyst takes place within 10–15 min. PE is isolated from the solution by solvent removal: firstly in an evaporator, then in a separator, and then in the vacuum chamber of a granulator. Granulated PE is steamed (at a temperature above the melting point of PE). Commercial HDPE is produced as granules, unstained and stained, and sometimes as powder.

8.4.7 *PRODUCTION OF MEDIUM-PRESSURE POLYETHYLENE (MDPE)*

MDPE is commercially produced under a medium pressure by ethylene polymerization in solution. It is formed at a temperature of 150°C and a pressure of 4 MPa; in the presence of Ziegler–Natta's catalyst. MDPE falls out of solution in the form of flakes. The PE thus obtained has an average molecular weight up to 400,000; the degree of crystallinity of up to 90%.

8.4.8 *PRODUCTION OF LINEAR LOW-DENSITY POLYETHYLENE (LLDPE)*

LLDPE is prepared by chemical modification of LDPE (at a temperature of 150°C and pressures of 30–40 atm). LLDPE is similar to HDPE by structure, but has longer and numerous lateral branches. Linear PE production is implemented in two ways: first, it is gas phase polymerization. The second method takes place in a reactor with a liquefied layer. Ethylene is supplied to the bottom of the reactor, the polymer is discharged continuously while constantly keeping the level of liquid in the reactor bed. The conditions are: the temperature of about a 100°C, the pressure from 689 to 2068 kPa. The efficiency of this polymerization process in a liquid phase is lower (a 2% conversion per cycle) than for the gas phase (up to a 30% conversion per cycle). However, this method has its advantages: the plant is significantly smaller than the gas-phase polymerization equipment, and the capital investment is significantly lower. There is another method, nearly identical to the above, in a reactor with a stirring device using the Ziegler–Natta catalysts; the highest yield is thus achieved.

8.4.8 *SAFETY AND ENVIRONMENTAL FRIENDLINESS WHEN DEALING WITH POLYETHYLENE OF LOW AND HIGH PRESSURE*

No harmful substances above their maximum allowable concentrations shall be released from PE into the ambient air, water, and food: formaldehyde—0.52; acetaldehyde—5.03; carbon monoxide—20.04; organic acids (in terms of acetic acid)—5.03; and PE aerosol—10.03 mg m^{-3}. The PE of basic grades and compositions releases no toxic substances into the environment at room temperature and has no impact on the human body at direct contact. Working with them requires no special precautions. When working with powdered PE, personal respiratory protection (a universal respirator, etc.) must be used. Upon heating over 140°C during processing, the volatile products of thermooxidative destruction, containing organic acids, carbonyl compounds including formaldehyde, acetaldehyde, and carbon monoxide may be released into the air. The threshold limit value (TLV) of the products of thermal-oxidative destruction in the working area of production facilities must comply with the requirements of GOST 12.1.005-88. The products of thermooxidative destruction of PE and its compositions in the working area in concentrations exceeding the TLV may cause acute or chronic poisoning. High-density PE is a combustible material. The aerosol flashpoint lies not lower than 280°C. Aerosol is explosive: the lower flammable limit of aerosol is 36–42 g m^{-3}; the maximum explosion pressure is 0.83–0.86 MPa; the average rate of pressure rise explosion is 9.5–10.5 MPa s^{-1}, and the maximum one is 22.5–28.0 MPa s^{-1}. The autoignition temperature of aerosol is 340–352°C, the minimum ignition energy is not less than 5.6 mJ, and the minimum explosive oxygen content when diluting a dust-air mixture with nitrogen is not less than 9 vol. %. The air exchange multiplicity in the room should be at least 8. The general exchange exhaust ventilation is assumed as 0.5 times of the local one at the air velocity in ventilation of 2 m s^{-1}. The ignition temperature of PE is about 300°C and the autoignition temperature is about 400°C. In the event of a fire, all known fire extinguishing agents must be utilized. The maximum explosion pressure of PE dust of 0.071 mm dispersion is 50 kPa and the maximum rate of pressure rise at explosion is 13,100 kPa s^{-1}. In accordance with the rules of static electricity protection, all equipment must be grounded, the relative humidity in the working premises must not be below 50%. Workplaces must be equipped with rubber mats.

8.5 ENVIRONMENTAL POLLUTION WITH ACETATE-CONTAINING WASTES OF THE PRODUCTION OF POLYMERIC MATERIALS[55,56]

Of the wastes produced in large quantities (hundreds of thousands of tons) is the waste of the polymer production using acetic anhydride and acetic acid as acylating agents, for example, acetyl cellulose production (films, etc.), the production of copolymers based on VA and vinyl chloride (followed by their hydrolysis into a hydroxyl-containing polymer), and the production of PMs such as spandex. In all these processes widely implemented in our country and abroad, acetic acid salts (acetates) are formed as wastes. The known methods of processing acetate-containing wastes into useful products require significant material and energy costs, almost new productions. Therefore, these wastes go to the dump. Local accumulations of this waste adversely affect the environment. Localization of even 4th hazard class substances, which include acetates, is a threat to the environment, since they are highly soluble and, therefore, exerts a depressing effect on the bacterial population of the soil, on the surrounding vegetation, and water reservoirs in high concentrations. This removes large areas of land from the economic use. In this regard, one of the fields of environmental protection is preventing pollution directly within the technological cycle and the use of the by-products in various fields of economy.

The significant amount of this waste makes important the methodology of application of this waste, in particular, in building to replace the currently used antifreeze additives in concrete preparation. The volume of used concrete with additives is about 100 million m^3 in the whole country, almost 40% is monolithic and precast concrete, and reinforced concrete. By the use of antifreeze additives, Russian Federation is one of the leading positions in the world. Many of these additives are highly toxic substances, which makes the process of preparation of concrete mixtures hazardous to the environment and requires protective measures. For example, sodium nitrate commonly used as an antifreeze additive to concrete is classified as the 3rd hazard class. In particular, getting into the food chain, nitrates are converted into nitrites and other more hazardous nitro compounds, having a mutagenic effect. Acetates are involved in the biological activity of microorganisms; they are added to pet food and used as a diuretic (mainly when edema associated with blood circulation). The daily dose for adults is 5–10 g. The replacement of nitrates as antifreeze additives by acetate-containing waste would allow neutralizing a large part of them (without additional energy and material resources).

Another object where the large-scale use of acetate-containing waste is possible is the replacement of chloride-containing substances used as anti-icing agents to fight winter icing on roads. Chlorides, accumulating in the soil, exert a depressing effect on the roots of plants. Furthermore, they have the ability to be accumulated over time, to increase the electrical conductivity of the soil, thereby promoting corrosion processes in communication networks (waterline, gas pipe, etc). The use of any of the proposed methods of environmental protection can simultaneously solve the problem of acetate-containing waste utilization even in the event of a significant increase in production volumes related to their formation.

The use of acetate-containing waste as an additive in concrete in either liquid or solid form, instead of highly toxic nitrates, leads to a significant improvement in the environmental performance of the technological process, to a reduced water consumption and improved physicomechanical properties of concrete: the increase of the compressive strength when the solid or liquid form of the waste is introduced is 21 or 65%, respectively; the increase of the flexural strength when the solid or liquid form of the waste is introduced is 11 or 17%, respectively; the rheological properties of cement paste are improved; the cement consumption decreases by 15–20%; the water consumption reduces by 10–15%; all this allows us to consider this technology as energy and resource saving. The additive increases the mobility of the concrete mix, shortens the duration of laying, acts as an antifreeze, protecting the concrete mixture from freezing within −15–0°C. It was established experimentally that acetate-containing waste and the concrete produced with its use have a permissible effective specific activity of natural radionuclides.

8.6 POLY (ALKYLENE GUANIDINES) AS ENVIRONMENTALLY FRIENDLY BIOCIDIC POLYMERS[57,58]

The increasing environmental pollution, the increasing number of man-made disasters, the risk of terrorist attacks of various kinds increased sharply in recent years, including bioterrorism, pose the task of ensuring the environmental well-being of the population as one of the most pressing and urgent problems facing the society. This problem is particularly important in view of geographical features of our country, the recent social and political problems, and global trends related to the emergence of new infectious diseases. Only intensive efforts on prevention and prophylactics would help us to prevent or minimize possible human and material losses from such threats. We need to have a supply of efficient and environmentally friendly antiseptics for effective and safe (for humans and the environment) localization of epidemic-prone areas and the normalization of the situation, to reduce the risk of spreading centers of infectious diseases. The design of such an antiseptic, as well as technologies of their application for both disinfection in the broadest sense of this word and combating the biodegradation of various materials is one of the pressing challenges facing chemists, microbiologists, and engineers.

According to the current requirements to biocides, the chemical compounds used as biocidal active ingredient in the composition of disinfectants and composite materials must possess a broad spectrum of biocidal action and, at the same time, be low hazardous for man and the human environment. Furthermore, they must be well combined with various materials and, protecting against biodeterioration, not to cause corrosion damage. High-molecular-weight biocidal products named *polyalkylene guanidines* (PAGs) are most promising biocides of the last time, meeting these requirements. Conventional biocidal agents have been hitherto prevailed: chloro-active, oxygen-containing, quaternary ammonium compounds, and compounds containing heavy metal (copper, tin, etc.) salts. Chloro-active compounds, though suppressing the vast majority of microorganisms, are, however, insufficiently effective or completely ineffective against spore forms (bacilli), viruses, *Pseudomonas aeruginosa*, protozoa cysts; while oxygen-containing compounds are significantly less active. The majority of these classes of chemical compounds are very aggressive and toxic; their use poses a serious threat to human health, they are unsafe for the environment, cause corrosion of equipment, damage, and discolor materials. Quaternary ammonium compounds actively suppress various kinds of bacteria, but are not effective against viruses and not always harmless to humans. Compounds containing heavy metals which are commonly used in agriculture, wood preservative, antifouling paints, are highly toxic and environmentally unsafe; they easily enter the body of humans and animals via the food chain, causing serious consequences.

The main representatives of polyguanidines are the high-molecular-weight salts of polyhexamethylene guanidine (PHMG) of the general formula:

$$[\text{—}(CH_2)_6\text{—}NH\text{—}\underset{\underset{\displaystyle NH_2^+Cl^-}{|}}{C}\text{—}NH\text{—}]_n$$

where $n = 5–90$, A^- is an acid residue of a mineral or organic acid.

A comparative evaluation of a series of chemical compounds (derivatives of chlorinated bisphenyl, triazine, silatranes, stanylthiosilanes, chlorothiobenzoic and dithiocarbamic acids, acryl-substituted pyran, compounds from the furan series, etc.) has shown that PAGs are the most promising antiseptics by complex criteria of efficiency, toxicity and hazards, the availability of raw materials, the technological convenience and environmental safety of production, physicochemical properties. The range of the biocidal action of polyguanidines is very wide: even in small concentrations, they are effective against gram-positive and gram-negative bacteria (including *Mycobacterium tuberculosis*), various kinds of fungi (mold, yeast, dermatophytes, etc.). Moreover, in contrast to bigluconate chlorhexidine (a low-molecular-weight analog) widely used abroad, PAGs simultaneously affect not only the aerobic and anaerobic flora but also suppress viruses. These drugs not only have a detrimental effect on a wide range of the pathogens of many diseases, but also destroy, for example, keratophagous insects (larvae of moths, carpet beetle, and woodworm beetle). The antimicrobial activity of polyguanidines against the pathogens of some especially dangerous infections (glanders, plague, and legionellosis) was established.

In an aqueous medium, PAGs effectively suppress unwanted flora and algae. Institute of Applied Microbiology has found that, for protection against biological corrosion of cooling systems, of over 20 various classes of biocidal compounds (Progress, Tween-40, Triton X-100, Brij 35, sodium dodecylsulphate, glutaraldehyde, dichloroglyoxime, *N*-cetylpyridinium chloride, catamine AB, etc.), PHMG chloride is one of the most promising agents, which is effective against a community of 45 microorganisms biodestructors isolated from these systems. Special studies have shown that PAGs effectively suppress up to 30 species of mold fungi, staining and destroying wood, so that their solutions can be used to handle both raw wood and wood products.

When a number of chemical compounds were tested as fungicides to protect film materials, among which were quaternary ammonium compounds, commonly known fungidines (salicylanilide, etc.), polyethylenimine and its derivatives, antibiotics, propylene oxide, parapharm, trichloro phenolates, pentachloro phenolates, it was found that only few chemical compounds, including PAGs, provided the necessary biostability, maintaining the optical characteristics and the technical state of the surface of photographic films.

The wide range of biocidal effect of polyguanidines is due to the presence of guanidine groups in the repeating units of the polymeric macromolecules, which are the active ingredient of some natural and synthetic medicines and antibiotics (sulgin, Faringosept, streptomycin, etc.). The biocidal effect of polyguanidines is caused by the fact that the phospholipid cellular membranes of microorganisms having a total negative electric charge effectively sorb the biocidic polycation which destroys the cellular membrane, inhibits the exchange function of enzymes, violates the reproducing ability of nucleic acids and proteins, and inhibits the respiratory system; such effects along with the destruction of cellular walls leads to death of the microorganism.

Leading Russian poison control centers have carried out extensive studies of the toxicity and danger of polyguanidines, which underlies their hygienic regulation. The median lethal dose when various polyguanidines are administered into the body through the skin and stomach is 10,000–15,000 and 815–3200 mg kg^{-1}, respectively. No mutagenic and carcinogenic effects of such drugs have been found. By the results of these studies, the tested preparations are classified as to the class IV of low-hazard compounds for administration through the skin and to the class III of moderately dangerous compounds when administered through the stomach. As a result of dynamic observations (25 tests), it was found that the threshold dose of the substance in the body is 1 mg kg^{-1} and the non-acting dose is 0.15 mg kg^{-1}. In water, a dose of 3 mg L^{-1} (by the sanitary-toxicological hazard basis) is accepted for PHMG chloride as occupational exposure limit (OEL). The low toxicity of polyguanidines is also due to the fact that the body of warm blooded has enzymatic systems capable of destructing guanidine-containing polymers. The first stage of PHMG chloride or phosphate metabolism in vivo is the replacement of the chloride or phosphate anion by a gluconate anion; further, hydrolysis of the guanidine groups proceeds to convert them into urea ones, as well as destruction of the polymer chains into separate fragments.

The solid forms of polyguanidines are extremely stable against oxidative and thermal degradation, aging (their shelf life is 15 years). Aqueous solutions of polyguanidines are also stable and long retain their physicochemical properties and biocidal activity. PAGs are colorless and odorless, nonirritating to the skin and mucous membranes, do not discolor fabrics, cause no corrosion of equipment, have surfactant properties. After drying of a solution, a thin intangible polymer film is formed on the surface, which provides long-lasting protection of the surface against the attack of microorganisms. Thus, the combination of biocide, toxicological, and physicochemical properties makes PAGs promising for the use as a standalone disinfectant as well as biocidal additives and auxiliaries.

Furthermore, by their chemical properties, PAGs largely resemble the properties of polyamines and quaternary ammonium compounds, are high-molecular-weight cationic polyelectrolytes, organic bases stronger than polyethyleneimine. The properties of PAG preparations can be varied by changing the chemical nature of the A^- anion, and the length or composition of the hydrocarbon chain in the biocidal macromolecule. PAGs are recommended as cationic polyelectrolytes for use in the paper, rubber technical industry, in optics, and electroplating. Large opportunities of modifying polyguanidines are associated with the relatively high reactivity of guanidine groups. While low-molecular-weight compounds substantially lose their biocidal properties on any chemical transformation, the biocidal properties of polyguanidines are preserved in many chemical reactions because the guanidine groups are combined into a common polymeric chain and any chemical reaction always involves only a part of them; wherein the groups remaining unchanged attach biocidal properties to the reaction product. PAGs readily undergo various chemical reactions with low-molecular-weight and high-molecular-weight compounds to form both soluble and cross-linked interpolymeric complexes and covalently linked interpolymers. Such a way of chemical modification has led to the design of organomineral sorbents, as well as organo-soluble drugs, which form water-resistant polymeric films on the surface with high strength characteristics and a prolonged biocidal effect after drying.

8.7 BIODEGRADABLE POLYMERS[1,59–61]

The most important environmental issue is the annual increase in the wastes of the plastic industry, formed mainly when using synthetic plastics. The popularity of them is due to their physicochemical and consumer properties. While glass containers are usually locked in the consumer cycle and the paper packaging is degradable in natural conditions, the package made of synthetic polymer (40% of household waste) is practically not subjected to degradation. One solution to the problem of plastic waste disposal can be the design of biodegradable polymers which retain their performance during the period of consumption only, and then undergo physicochemical and biological transformations under the influence of environmental factors and are easily included in the circulation of substances.

Biodegradable polymers can be prepared in two ways: either on the basis of natural organic substances (oligosaccharides, cellulose, corn, etc.) or by biotechnology. Biopolymer decomposition occurs due to natural factors: light, temperature, humidity, and with the participation of living microorganisms (bacteria, yeasts, fungi, etc.). High-molecular-weight substances are decomposed into low-molecular-weight ones (water and carbon dioxide), humic substances, and biomass. Natural circulation of substances is thus made capable of maintaining the ecological balance in nature.

Currently, the most widely used method for manufacturing bioplastics is based on the introduction of substances of vegetable origin into a synthetic polymer, which are a breeding ground for microorganisms initiating polymer degradation under certain environmental conditions. Potatoes, beets, cassava, legumes and grains, cellulose (wood, cotton, lignin), and so forth are the raw material for producing biopolymers. A significant part in the production of packaging materials is given to a copolymer of ethylene and VA, with starch added as a biodegradable component. It is well decomposed by the action of water and microorganisms without polluting the soil. Effective microorganisms (biodestructors) are proposed to destruct this material. The ability of polymers to degrade and to be assimilated by microorganisms depends on their structural characteristics. The most important of them are the chemical nature of the polymer, its molecular weight, the branching of macrochain (the presence and nature of side groups), and supramolecular structures. An important factor that determines the resistance of a polymer to biodegradation is the size of its molecules. While monomers or oligomers can be used by microorganisms as carbon sources, polymers of high-molecular-weight are resistant to the action of microorganisms. The biodegradation of most technical polymers usually occurs by thermal and photooxidation, thermolysis, mechanical degradation, etc.). For example, PE whose chains are formed by carbon atoms with a polymerization degree above 20–30 is not subjected to enzymatic cleavage and not used by microorganisms in food. At the same time, short PE chains (less than 20 monomer units) may serve as a food (substrate) for a specific group of microorganisms. As a result of biochemical reactions, the formation of acetic acid is possible, which is used by microorganisms as a source of energy and the source building material.

If high-molecular-weight PE used for producing a variety of materials and having a degree of polymerization $n > 1000$ is destructed onto shorter fragments with $n = 20$, then its further splitting can be carried out by the microorganisms living in the environment. To enable long PE chains to be split into relatively short fragments under natural conditions, "weak links" are introduced therein in the synthesis of the polymer, which are destroyed under the influence of sunlight, specifically UV light, in the presence of atmospheric oxygen and moisture, with the formation of fragments subject to further biodegradation.

Other techniques to obtain biodegradable PMs have been developed as well, which require no intervention into the process of macromolecular synthesis. It is exemplified by the preparation of stable mixtures of polymers (PE and starch). If a sufficient amount of starch is added into such a mixture when PE processing in articles, it will be biodegraded in contact with the soil. In this case, PE will remain in the product (film), which is dispersed into fine particles and disappear.

The main advantages of producing and using such quasibiopolymers are:

1. the ability of processing on standard equipment like conventional polymers;
2. a low barrier for oxygen and water vapor penetration (optimized for use in food packaging);
3. resistance to degradation under normal conditions;
4. rapid and complete degradability under specially created or natural conditions;
5. independence from petrochemical resources.

However, some problems arise in the production and consumption of biodegradable materials:

1. the high cost (currently 2–5 Euros per kg on average); however, it should be noted that the economic cost, in addition to the price of the product, includes the costs of disposal and use; it is also important to note that the high price of the material is a temporary phenomenon, when the production of biopolymers is not become widespread, and the process of their manufacturing is not fully adjusted;
2. limited opportunities for large-scale production;
3. difficult regulation of the rate of decomposition in landfills under the influence of environmental factors;
4. technological problems, and so forth.

For these reasons, the design and use of rapidly degradable materials should have a limited and controlled application.

The main goal of research in the field of biodegradable polymers is to identify general regularities of the process of obtaining biopolymer materials, to select components and to determine the technological parameters of manufacturing materials with specific physical and chemical properties (strength, low gas permeability, density required, etc.) and biodegradability.

8.8 PROBLEMS OF RECOVERY AND RECYCLING OF POLYMERIC MATERIALS[62–64]

8.8.1 INTRODUCTION

Polymeric wastes are classified into production (technological) waste and consumer waste. Consumer waste is produced in the residential sector, in factories, shopping malls, various commodities and goods are packed and unpacked. Let us consider several types of waste classification.

According to the complexity and costs of utilization. With good properties: pure production waste (gates, trimmings, burrs, and spoilage), relatively pure consumer waste produced in places where the collection and sorting are either tuned up or not required (medical disposable products and systems, film, plastic boxes, and PET bottles). Their recycling provides a relatively high profitability of their processing. The percentage of the total number of polymeric waste is within 5–12%. The usage is 70–90%.

With medium properties: the same types of production and consumption wastes containing allowable amounts of pollutants as well as waste from the production of edible products. The collection and processing of such waste are associated with some costs for sorting, cleaning and using more sophisticated equipment for processing and manufacturing products. However, their use can be cost-effective. The percentage of the total number of polymeric waste is within 10–25%. The usage is 20–30%.

Hardly recyclable waste: highly contaminated and mixed waste production and consumer waste, waste from composite materials, household items and automotive equipment. The collection and processing of hardly recyclable waste is usually unprofitable. That is why this waste is practically not collected and recycled in Russian Federation. To cover the costs for disposal of such wastes, external financial resources (tax incentives, targeted investments, and subsidies) and noneconomic measures are needed. The percentage of the total plastic waste is within 60–85%. Recycling (except burial) is up to 3%.

By the kinds and types of polymers, wastes can be divided into two groups: the wastes of large and expensive engineering plastics. These wastes have a developed market for recycled materials in Russian Federation, are partly provided (technological waste) with equipment for their processing into secondary materials: low- and high-density PE (film and lumpy waste); polypropylene (PP: lumpy waste and film, production waste, disposable tableware, and fiber); polystyrene plastics (PS, Acrylonitrile butadiene styrene (ABS): lumpy waste, sheets, disposable tableware, and production waste); PAs, polycarbonate (PC); polyethylene terephthalate (PET); plasticized PVC (primary production waste); hard PVC (primary production waste), and so forth.

Reuse of the source polymers involves the introduction of fillers, reinforcements into them, manufacturing various products (the so-called material method). However, only about 10% of the total weight of polymer waste can be reused. Even if polymer waste is carefully separated from other ones, they are almost impossible to be processed into polymeric recyclate with satisfactory properties due to the inherent feature of polymers—their inability to mix with each other (thermodynamic incompatibility). Therefore, practically only one type of polymers can be sent to recycling, which require sorting and correspondingly high costs. The quality and properties of secondary polymers are different from those of the primary materials. This is due to reduction of their strength, thermal stability, plasticity, and other parameters. A scheme is used for recycling, which includes the following steps: collection and transport of plastic waste; manual sorting and initial separation of pollutants; metal detection and separation; grinding; metal separation; washing in baths and centrifuges; floatation sorting; drying in drum, tubular or contact dryers; air purification in a cyclone; stamping on a press; cleaning of polymers on continuous or discontinuous filters; granulating with or without filtration by means of water or air granulators; and the production of finished products.

Recycling of wastes of most polymers into monomers and artificial fuel by depolymerization followed by the synthesis of various PMs does not find industrial applications. However, wastes of some PMs can be subjected to thermal recycling to produce useful products of non-polymeric nature. For example, PET can be depolymerized to the original components. Polyurethane waste can be subjected to similar processing because obtaining synthetic liquid fuel is a very promising direction of their recycling. Recently developed technologies allow producing high-quality brands of gasoline, kerosene, diesel, and fuel oil. However, the main drawback of these techniques is the high cost of the equipment used and, accordingly, the high cost of the synthetic liquid fuel produced. There are ways of separating polymers from a mixture of collected waste: froth flotation, separation dissolving, detector dissolving with monitoring, and finally, a noteworthy processing method for mixtures of incompatible polymers by sonication, allowing obtaining block copolymers. Work is underway to cover PET bottles with a thinnest barrier layer of silicon oxide, nitride, carbide, and so forth, a more barrier PET is introduced for beer. Obsolete polyolefins can be added to coal, carbonated, and entered into coke for smelting iron.

Burning of plastic waste to produce heat and power energy (the energy method) attracts increasing attention due to the continuous price rise for fossil fuels. There is no need to sort; only crushing waste to sufficiently large pieces is required to ensure their effective mixing with the addition of carbon-based fuels, most often, coal, and free access to oxygen necessary combustion. The danger of environmental pollution with supertoxicants during plastic waste combustion is largely exaggerated and applies rather to old or obsolete incinerators. At temperatures of 1200–1400°C, typical for modern systems, these substances are irreversibly decayed and the undecomposed portion is absorbed in absorbent filters. Dioxin emissions reach only 0.6 μg per ton. When a ton of coal or gasoline is burned, 1–10 or 10–2000 μg of dioxin is released, respectively.

Plastics are high-molecular-weight compounds, whose modifications have various physicochemical and consumer properties. Their production at the present stage of development is annually increasing by 5–6% on average. Their per capita consumption in industrialized countries has doubled over the past 20 years, reaching almost 100 kg. Packaging is one of the fastest growing trends of plastics use—about 40%, of which 50% is spent on food packaging. Convenience and safety, low price and high aesthetics are critical conditions of the accelerated growth in the use of plastics when making packaging. Plastics are serious competitors to metals, glass, and ceramics. For example, the manufacture of glass bottles requires approximately 20% more energy than in the case of plastics. But, along with this, there appears a problem of waste disposal, there being over 400 different kinds of waste. However, solving the issues related to the protection of the environment requires significant capital investment. The cost of treatment and disposal of waste plastics is about eight or almost three times higher than that of processing the majority of industrial and household waste, respectively. This is due to specific characteristics of plastics, which significantly hinder or render useless the known methods of solid waste disposal. The use of waste polymers allows saving primary raw materials (especially oil) and electricity. There are quite a lot problems associated with plastic waste disposal.

8.8.2 CLASSIFICATION OF POLYMER WASTE AND BASIC METHODS OF THEIR DESTRUCTION

Waste plastics can be divided into three groups:

(a) *technological waste products* which arise in the synthesis and processing of thermoplastics. They are divided into *unavoidable* and *avoidable* technological waste. *Unavoidable* wastes are edges, cuttings, trimmings, gates, burrs, fins, and so forth. In the industries engaged in the production and processing of plastics, 5–35% of such waste is generated. *Unrecoverable waste,* substantially representing high-quality feedstock, does not differ from the original primary polymer by properties. Their processing into articles requires no special equipment and is done at the same plant. *Removable* technological waste is formed when non-compliance with technological regimes during the synthesis and processing, that is, this is a technological scrap, which can be minimized or even eliminated. Process waste is processed into various articles, used as an additive to the feed, and so forth; (b) *industrial consumption waste* is accumulated as a result of failure of products made of PMs, used in various sectors of the economy (depreciated tires, packaging, machine parts, agricultural film waste, bags from fertilizers, etc.). These wastes are most homogeneous, low contaminated and, therefore, of the greatest interest from the viewpoint of their recycling; (c) *public consumption waste* which accumulate in our homes, in catering, and so forth, and then get on municipal landfills; eventually they move into a new category of waste, namely, *mixed waste*. The greatest difficulties are associated to the processing and use of mixed waste. The cause is in the incompatibility of thermoplastics contained in household waste, which requires their stepwise selection. In addition, the collection of used polymeric products from the population is a very difficult business from an organizational viewpoint. The basic amount of waste is destroyed by burial in the ground or incineration. However, the disposal of waste is uneconomical and technically difficult. Moreover, burial, flooding and burning of plastic waste lead to environmental pollution, reduce land (numerous landfills), and so forth. However, both burial and incineration are still fairly common ways of waste plastic disposal. The heat released during combustion is most often used to produce steam and electricity. But the calorific power of combusted materials is low, so incinerators are usually economically inefficient. Furthermore, incineration is accompanied by the formation of carbon black from incomplete combustion of polymeric products, the release of toxic gases and hence recontamination of air and water, rapid furnace failure due to severe corrosion.

At the early 1970s, work on designing bio-, photo- and water-destroyed polymers began to be intensively developed. The preparation of biodegradable polymers caused a true sensation, and this method of disposal of defective or worn-out plastic products was considered as an ideal. However, subsequent work in this area has shown that it is difficult to combine high physical and mechanical characteristics, a beautiful appearance, the ability to rapid destruction, and a low cost in one product. In recent years, research in the field of auto destructive polymers has significantly decreased, mainly because the production costs while producing such polymers are generally much higher than in the synthesis of conventional plastics so this method of disposal is uneconomical. The main way of using plastic waste is their utilization (reuse). It is shown that the capital and operating costs of the main methods of waste disposal do not exceed (and in some cases even lower than) the cost of their destruction. A positive side of recycling is also that an additional amount of useful products is gotten for various industries and no recontamination of the environment occurs. For these reasons, recycling is not only an economically desirable but environmentally preferable solution to the problem of using plastic wastes.

The *main methods of recycling plastics* are: (1) thermal decomposition by pyrolysis; (2) decomposition to recover the initial low-molecular-weight products (monomers, oligomers); and (3) recycling.

1a. Pyrolysis is the thermal decomposition of organic products in the presence of oxygen or without it. Pyrolysis of plastic waste provides a high-energy fuel, raw materials and semi-finished products used in various industrial processes, as well as monomers to be used for the synthesis of polymers.

1b. The gaseous products of thermal decomposition of plastics can be used as fuel to produce a working steam. The liquid products are used for heat transfer. The range of application of the solid (waxy) products of plastic waste pyrolysis is wide enough (components of various kinds of protective compounds, greases, emulsions, impregnating materials, etc.).

1c. Catalytic hydrocracking processes for the conversion of polymer waste to gasoline and fuel oil have also been developed.

Many polymers, as a result of reversibility of the formation reaction can again decompose into initial substances. Methods to decompose PET, PAs, and foamed polyurethanes are important for practical use. The cleavage products are used again as a raw material for polycondensation or as a supplement to the primary material. However, the contaminants contained in these products often do not allow producing high-quality plastic products, such as fiber, but their purity is sufficient for the manufacture of molded masses, fusible, and soluble adhesives.

2a. Hydrolysis is the reaction reverse to polycondensation. With its help, polycondensates are destroyed down to the starting compounds by the directed action of water at the joints of the components. Hydrolysis occurs under the action of extreme temperatures and pressures. The depth of the reaction depends on pH of the medium and the catalysts used. This method of using waste is energetically more favorable than pyrolysis as high-quality chemical products return to their turnover.

2b. As compared with hydrolysis, another way (glycolysis) for the cleavage of PET is more economical. Degradation occurs at high temperatures and pressures in the presence of ethylene glycol and with catalysts to produce pure diglycol terephthalate. According to this principle, the carbamate groups in polyurethane can also be re-esterified.

2c. Yet, the most common method of thermal processing of PET waste is its cleavage with methanol (methanolysis). The process takes place at temperatures above 150°C and a pressure of 1.5 MPa, is accelerated by the transesterification catalysts. This method is very economical. In practice, a combination of glycolysis and methanolysis is used.

3. Currently, *mechanical recycling* of waste plastics is most acceptable in Russian Federation, because this method requires no special expensive processing equipment and can be implemented anywhere the waste is accumulated.

8.8.3 BASIC METHODS OF RECYCLING MOST COMMON POLYMERIC MATERIALS

8.8.3.1 DISPOSAL OF WASTE POLYOLEFINS

Polyolefins are the most large tonnage kind of thermoplastics. They are widely used in various branches of industry, transport, and agriculture. Polyolefins include HDPE, LDPE, and PP. The most effective method of disposing of waste polyolefins is their reuse. Polyolefin waste treatment methods depend on the grade of a polymer and its origin. Technological waste, that is, such waste products that have not been exposed to intense light during usage, is most easily recycled. Consumer HDPE and PP wastes require no elaborate processing techniques as well, because, on the one hand, the products made of these polymers undergo no significant effects due to their construction and purpose (thick-walled parts, containers, furniture, etc.), and, on the other hand, the basic polymers are more resistant to weathering than LDPE. Such wastes prior to their reuse need only grinding and granulation.

8.8.4 STRUCTURAL AND CHEMICAL PROPERTIES OF RECYCLED POLYETHYLENE

The choice of technological parameters of waste processing and the usages of products obtained is determined by their physicochemical, mechanical, and technological properties, which differ significantly from those of the primary polymer. The main features of secondary LDPE (SLDPE), which determine the specificity of its processing, include low bulk density, peculiarities of the rheological behavior of its melt due to the high gel content, and increased reactivity due to structural changes occurring during the processing of the initial polymer and the exploitation of the obtained products. During processing and usage, the material is subjected to mechanochemical stress, thermal destruction, photooxidative destruction, and thermooxidative destruction, which leads to the appearance of active groups, which could initiate oxidation reactions in subsequent processing. Changes in the chemical structure begin just in the primary processing of polyolefins, in particular, by extrusion, when the polymer undergoes considerable mechanochemical and thermooxidative influences. Photochemical processes mainly contribute to the changes occurring during the usage of polymeric things. These changes are irreversible, while the physicomechanical properties, for example, PE film served one or two seasons to cover greenhouses, are almost completely restored after re-pressing and extrusion. The formation of a significant amount of carbonyl groups in a PE film during its usage leads to an increased SLDPE ability to absorb oxygen, resulting in the formation of vinyl and vinylidene groups in secondary raw materials, which significantly reduces the thermal-oxidative stability of the polymer during subsequent processing, initiates the process of photoaging of such materials and products from them, and reduces their service life. The presence of carbonyl groups determines neither mechanical properties (their introduction to the original macromolecule in 9% does not significantly affect the mechanical properties of the material) nor film solar light transmittance (light absorption by carbonyl groups is within the range of wavelengths less than 280 nm, and the light of this composition is substantially not contained in the solar spectrum). However, it is the presence of carbonyl groups in PE which determines its very important property, the resistance to light. Hydroperoxides formed even in the processing of raw material by mechanochemical degradation are the initiator of PE photoaging. Their triggering action is particularly effective at early aging stages, whereas carbonyl groups have a significant effect at later stages. It is known that competing degradation and structuring reactions proceed during aging. The former leads to the formation of low-molecular-weight products, while the latter gives an insoluble gel fraction. The rate of formation of low-molecular-weight products is maximum at the beginning of aging. This period is characterized by low gel content and decreased physicomechanical properties. Further, the rate of formation of low-molecular-weight products reduces, a sharp increase in the gel content and a decrease in elongation are observed, which indicates the occurrence of a structuring process. Then (after peaking), the gel content in SPE when it is photoaging reduces, which coincides with the complete consumption of the vinylidene groups in the polymer and the achievement of the maximum permissible values of elongation. This effect is attributed to the involvement of spatial structures formed in the process of degradation and cracking at the border of the morphological formations, which reduces the physicomechanical characteristics and degradation of the optical properties. The rate of change of the physicomechanical characteristics of secondary polyethylene (SPE) hardly depends on the content of gel fraction. However, the gel content should always be considered as a structural factor in deciding how to recycle, modify, and identify areas of use of the polymer. Table 8.1 shows the characteristic properties of LDPE before and after aging during 3 months and those of SLDPE prepared by extrusion of an aged film.

The nature of changes in the physicomechanical characteristics of LDPE and SLDPE varies: a monotonic decrease of both the strength and elongation is observed for the primary polymer, which is 30 and 70%, respectively, after 5-month aging. For secondary LDPE, the nature of change of these indicators is somewhat different: the breaking stress is almost not changed and the elongation reduces by 90%. The cause for this may be the presence of a gel fraction in SLDPE, which functions as an active filler in the polymer matrix. The presence of such a "filler" is the cause of considerable stresses, resulting in embrittlement of the material, sharp decreases in elongation (up to 10% of the initial values for PE), the resistance to cracking, tensile strength (10–15 MPa), elasticity, and stiffening. In PE during aging, not only oxygenated groups (including ketonic) and low-molecular-weight products are accumulated, but the physicomechanical characteristics significantly reduce and cannot be restored after recycling of an aged polyolefin film.

TABLE 8.1 Characteristics of LDPE Properties Before and After Aging

Characteristics	LDPE		SLDPE
	Source	**After usage**	**Extrusional**
Content of C—O groups, mol	0.1	1.6	1.6
Content of low-molecular-weight products, %	0.1	6.2	6.2
Gel content, %	0	20	20
Tensile strength, MPa	15.5	11.4	10
Elongation at break, %	490	17	125
Resistance to cracking, h	8	–	1
Light resistance, days	90	–	50

Structural and chemical transformations mainly occur in the amorphous phase of SLDPE. This leads to a weakened interphase in the polymer, whereby the material loses its strength, becomes brittle and prone to further aging during both recycling into articles and the usage of such ones characterized by their low physicomechanical properties and durability.

To find the optimal regimes of processing of secondary PE raw materials its rheological characteristics are of great importance. SLDPE is characterized by low flow at low shear stresses, which increases with increasing stress, and the flow increase for SPE is greater than for the primary one. The cause is the presence of gel in SLDPE, which greatly enhances the activation energy of viscous flow. Fluidity can be adjusted by varying temperature in the processing: as temperature increases, the fluidity of the melt increases.

So, under recycling is a material whose prehistory has a very significant impact on its physicomechanical and technological properties. In the process of recycling, the polymer undergoes additional mechanochemical and thermal-oxidative impacts, and changes in its properties depend on the multiplicity of processing. Three- to fivefold processing has little effect on the properties of the resulting product (much less than for the original one). A dramatic decline in strength begins at 5–10-fold processing. In the process of repeated SLDPE processing, it is recommended to increase the casting temperature by 3–5%, or to increase the number of extrusion screw revolutions by 4–6% to destruct the gel formed. It should be noted that in the process of repeated processing, especially when exposed to oxygen, the molecular weight of polyolefins reduces, which results in sharp material embrittlement. Repeated processing of another polymer from the class of polyolefin, PP, usually leads to an increase in the melt flow index (MFI), although the strength characteristics of the material do not undergo significant changes. Therefore, the waste from the manufacture of PP articles and the articles themselves can be reused at the end of exploitation in a mixture with the starting material for new ones. From the above, it follows that secondary raw material should be subjected to modification in order to improve the quality and prolong the service life of its products.

8.8.5 TECHNOLOGY FOR PROCESSING OF SECONDARY POLYOLEFIN RAW MATERIALS INTO GRANULATE

To convert thermoplastic waste into raw material suitable for further processing into articles, its pretreatment is necessary. The choice of the way of this pretreatment depends mainly on the source of waste generation and the degree of their contamination. For example, homogeneous waste of LDPE production and processing is usually processed on-site, which requires a small pretreatment, mainly crushing and granulation. Waste as worn-out products requires more thorough preparation. The pretreatment of waste of agricultural PE film, bags from fertilizers, wastes from other compact sources, and mixed wastes includes the following steps: sorting (coarse) and identification (for mixed waste), grinding, separation of mixed waste, washing, and drying. The material is then subjected to granulation. Presorting involves a gross waste separation on different grounds: color, size, shape, and, if necessary and possible, by types of plastics. The presorting is usually done by hand on tables or conveyor belts; various foreign objects and inclusions are removed from the waste at sorting. Separation of mixed (household) thermoplastic waste is carried out by the following main methods: flotation, separation in heavy media, aeroseparation, chemical methods, and deep cooling techniques. Flotation is most widely used, which allows to separate mixtures of industrial thermoplastics such as PE, PP, PS, and PVC. Plastics separation is done by adding surfactants to water, which selectively modifies their hydrophilic properties. In some instances, an effective way of polymer separation can be dissolution into a common solvent or a solvent mixture. Processing solution with vapor, PVC, PS, and a mixture of polyolefins is recovered; the product purity being no less than 96%. Flotation and separation in heavy media are most effective and cost-effective of all of the above. Obsolete polyolefin waste containing impurities (not more than 5%) is supplied from a warehouse of raw materials to a waste sorting unit, where random foreign inclusions are removed and heavily soiled pieces are discarded. Sorted waste is minced in knife mills of wet or dry grinding to obtain friable masses with 2–9 mm particles. The performance of a comminuting device is determined not only by its design, the number and length of blades, and rotor speed but also by the type of waste. For example, the recycling of plastic foams which occupy a very large volume and are difficult to be compactly loaded has the lowest performance. Better performance is achieved by recycling films, fibers, and blown products. All knife grinders feature an increased noise associated with the specifics of the grinding process of secondary materials. To reduce noise, a chopper with its motor and fan is enclosed in a noise protection cover which can be designed as split and has special windows with shutters to load crushed material. Grinding is a very important stage of waste preparation for recycling, as its fineness determines the bulk density, flowability, and particle size of the product. Adjustment of the grinding degree allows one to mechanize the process of processing, improve the quality of the material by averaging its technological characteristics, shorten the duration of other technological operations, and simplify the design of the processing equipment. Cryogenic grinding is very promising, which produces powders from waste with a dispersion degree of 0.5–2 mm. The use of powder technology has several advantages: a reduced duration of mixing; a reduced energy consumption and shorter working time for routine maintenance of the mixers; better distribution of the components in the mixture; lower macromolecule degradation, and so forth. Mechanical grinding is most suitable of the known methods of obtaining powdery polymer materials used in chemical technology for grinding thermoplastic wastes. Mechanical grinding can be carried out in two ways: by a cryogenic process (grinding in liquid nitrogen or another cryoagent) and at ordinary temperatures in a medium of deagglomerating ingredients, which are less energy-consuming. The shredded waste is further fed into a washing machine for washing. The washing is carried out in several stages with special detergent mixtures. The mass wrung out in a centrifuge with a humidity of 10–15% is fed for final dewatering into a drying plant, until a residual moisture content of 0.2%, and then into a granulator. Combined plants with devices for washing and drying are commercially available with a capacity of up to 350–500 kg h^{-1}. In such a plant, crushed waste is charged into a bath, which is filled with a detergent solution. The film is stirred with a paddle stirrer, and mud settles to the bottom while the washed film buoys to the surface. Dehydration and drying of the film is carried out on a vibrating screen and in a vortex separator. The residual humidity is less than 0.1%. Granulating is the final stage of preparation of secondary raw materials for further processing into articles. This step is especially important for SLDPE due to its low bulk density and transportation difficulties. During granulation, the material is compacted, its further processing is facilitated, and the characteristics of recovered materials are averaged, resulting in a material which can be processed on standard equipment. For plasticization of chopped and purified waste polyolefins, a single-screw extruder equipped with a continuous filter and having a degassing zone is most widely used. In such extruders, almost all types of secondary thermoplastics with a bulk density of the particulate material within the range 50–300 kg m^{-3} are processed quite efficiently. However, for the processing of contaminated and mixed waste, a screw press of a special design is needed, with short multiple screws (of length 3.5–5 *D*), having a cylindrical nozzle in the area of extrusion. An extruder with a drive power of 90 kW, a screw diameter of 253 mm, and an *L*/*D* ratio of 3.75 is the basic unit of this system. A goffered nozzle of 420 mm diameter is constructed at the outlet of the extruder. Due to heat generation by friction and shearing effects on the polymer material, it melts in a short time, and fast melt homogenization is provided. By varying the gap between the conical nozzle and the jacket, the shear stress and the frictional force can be controlled, thus changing the mode of processing. Since melting occurs very quickly, no thermal degradation of the polymer is observed. The system is equipped with a degassing unit, which is a prerequisite for the processing of second polymeric raw material. Secondary granulates are prepared according to the sequence of processes of cutting and cooling in two ways: granulation on the head and underwater pelletizing. The selection of the granulation method depends on the properties of the processed thermoplastic, especially its melt viscosity and adhesion to metal. During granulation on the head, polymer melt is extruded through a hole as cylindrical ropes which are cut by moving knives sliding along the spinneret plate. The resulting pellets are cast away from the head with the knife and cooled. Cutting and cooling can be performed in air, in water, or cutting in air and cooling in water. For polyolefins with a high adhesion to metal and an increased tendency to stick, water is used as a cooling medium. When equipment with a large unit capacity is used, the so-called underwater pelletizing is applied. In this method, the polymer melt is extruded as strands through the spinneret plate holes at the head immediately into water and cut into pellets by rotating knives. The temperature of cooling water is kept within 570°C, which promotes more intensive evaporation of residual moisture from the surface of the granules; the content of water is 20–40 m^3 per 1 ton of the granulate. Strands or ribbons are most often molded in the pelletizer's head, which are granulated after cooling in a water bath. The diameter of the obtained granules is 2–5 mm. Cooling should be carried out under optimal conditions, for the granules not to be distorted, not stick together, and removal of residual moisture should be ensured. The head's temperature has a significant impact on the size distribution of granules. To ensure a uniform melt temperature, grids are located between the extruder and the outlet holes of the head. The number of outlets in the head is 20–300. The capacity of the granulation process depends on the type of secondary thermoplastic and its rheology. Studies of SPE granulate show that its viscous properties practically do not differ from those of primary PE, that is, it can be processed under the same conditions of extrusion and injection molding as primary PE. However, the articles obtained are characterized by low quality and durability. Packaging for household chemicals, hangers, details of construction purposes, agricultural tools, pallets for the transportation of goods, chimneys, lining for drainage channels, and non-pressure pipes for reclamation and other products are produced from the granulate. These products are prepared from "pure" secondary raw materials. However, adding 20–30% of secondary materials to the primary one is more promising. Introduction of plasticizers, stabilizers, and fillers into a polymer composition can increase this figure up to 40–50%. This improves the physicomechanical characteristics of the articles, but their durability (in harsh climatic conditions) is 0.6–0.75 of the service life of the primary polymer products. A more efficient way is modification of secondary polymers, as well as the design of highly filled secondary PMs.

8.8.6 METHODS FOR MODIFYING SECONDARY POLYOLEFINS

Results of the studies of the mechanism of the processes occurring in the usage and processing of polyolefins and their quantitative description allow one to conclude that the semifinished products obtainable from recycled raw should not contain more than 0.1–0.5 mol of oxidized active groups and have an optimum molecular weight and MWD, and have reproducible physicomechanical and technological parameters. Only in this case, this intermediate can be used to produce articles with a guaranteed service life instead of the primary (deficit) polyolefin materials. However, the currently produced granulate does not satisfy these requirements. Modification of the granulate is a reliable way to solve the problem of making high-quality polymer materials and products from recycled polyolefins, the purpose of which is screening of functional groups of the active centers by chemical or physicochemical methods and the creation of a material with a uniform structure with reproducible properties. The methods of modification of the secondary raw polyolefin material can be divided into *chemical* (cross-linking, introduction of various additives, mainly of organic origin, processing with silicone-organic fluids, etc.) and *physicomechanical* (filling with mineral and organic fillers). For example, the maximum gel fraction (80%) and the highest physicomechanical characteristics of cross-linked SLDPE are achieved by introducing 2–2.5% of dicumyl peroxide on mills at 130°C for 10 min. The elongation at break of this material is 210%, its melt index is 0.1–0.3 g/10 min. The degree of cross-linking decreases with increasing temperature and duration of milling as a result of a competing destruction process. This allows adjusting the degree of cross-linking, the physicomechanical and technological characteristics of the modified material. A method of forming articles from SLDPE by dicumyl peroxide introduction directly during the processing has been designed, and test samples of pipes and molded articles containing 70–80% of the gel fraction have been obtained. Introduction of wax and elastoplastic (below 5 wt. %) significantly improves the processability of SPE, improves its physicomechanical properties (particularly the elongation at break and the resistance to cracking by 10% and from 1 to 320 h, respectively), and reduces their dispersion, demonstrating the improved homogeneity of the material. SLDPE modification with maleic anhydride in a disk extruder also increases its strength, heat resistance, adhesiveness, and the resistance to photoaging. The modifying effect is achieved with a lower modifier concentration and a shorter process duration than when elastoplastic is introduced. Thermomechanical processing with silicon-organic compounds is a promising way to improve the quality of PMs from recycled polyolefins. This method allows one to make products from recycled materials with high strength, elasticity, and resistance to aging. The mechanism of modification is the formation of chemical bonds between the siloxane groups of the silicone-organic fluid and the unsaturated bonds and the oxygen-containing groups of secondary polyolefins. The technological process for producing a modified material comprises the following steps: sorting, crushing, and washing the waste; processing of the waste with a silicon-organic liquid at 90±10°C for 4–6 h; drying of the modified waste by centrifugation; and regranulation of the modified waste. Besides the solid-phase technique, a method is offered of modifying SPE in solution to obtain SLDPE powder with a particle size less than 20 μm. This powder can be used for processing into articles by rotational molding and for coating by electrostatic spraying. The design of filled PMs based on recycled plastic raw material is of great scientific and practical interest. The use of PMs from recycled waste containing up to 30% of a filler will free up to 40% of the primary raw materials and send it for the production of goods which cannot be obtained from the secondary one (pressure pipes, packaging films, reusable transport containers, etc.). This will largely reduce the deficit of primary PMs. Dispersed and reinforcing fillers of mineral and organic origin and fillers which can be produced from plastic waste (shredded waste of thermosets and rubber crumb) can be used for getting filled PMs from recycled materials. Substantially, all thermoplastic and mixed waste can be subjected to filling, whose usage is preferred for this purpose from an economic standpoint. For example, the usefulness of applying lignin is due to the presence of phenolic compounds therein to stabilize SLDPE during operation; mica allows to form articles having low creep, increased heat and weather resistance, and characterized by a small wear of the processing equipment and low cost. Kaolin, coquina, shale ash, coal spheres, and iron are used as cheap inert fillers. When fine-dispersed phosphogypsum granulated in PE wax is introduced into SPE, compositions having increased elongation at break are obtained. This effect can be explained by the plasticizing action of PE wax. For example, the tensile strength of SPE filled with phosphogypsum is 25% higher than that of SPE alone, and the tensile modulus is greater by 250%. The enhancing effect when mica is introduced into SPE is associated with the peculiarities of the crystal structure of the filler and the high aspect ratio (ratio of the diameter of a flake to its thickness); the use of crushed, powdered SPE helps to preserve the structure of the scales with minimal destruction. Compositions containing lignin, shale, kaolin, spheres, and sapropel waste have relatively poor physicomechanical properties, but they are most inexpensive and can find application in the manufacture of building materials.

8.8.7 RECYCLING OF PVC

During processing, polymers are exposed to high temperatures, shear stresses, and oxidation, which alter their structure, processing, and performance properties. The structure of a material is greatly influenced by thermal and thermooxidative processes. PVC is a least stable carbon-chain industrial polymer. PVC degradation (dehydrochlorination) begins just at temperatures above 100°C and proceeds very quickly at 160°C. As a result of the thermal oxidation of PVC, aggregative and deaggregative processes (cross-linking and destruction) occur. PVC degradation is accompanied by changes in the initial color of the polymer due to the formation of chromophore groups and a significant deterioration in its physicomechanical, dielectric, and other operational characteristics. As a result of cross-linking, linear macromolecules are converted into branched ones and, ultimately, into a three-dimensional cross-linked structure, thus significantly deteriorating the solubility of the polymer and its ability to be recycled. In the case of plasticized PVC, its cross-linking reduces the compatibility of the plasticizer with the polymer, increases plasticizer migration, and irreversibly degrades the performance properties of the materials. Along with the influence of the operating conditions and the multiplicity of secondary processing of PMs, it is necessary to evaluate the optimal waste–fresh raw ratio in the composition intended for processing. When products are extruded from mixed raw, there is a risk of spoilage because of the different viscosities of the melts. It is therefore proposed to extrude primary and secondary PVC on different machines. But powdered PVC is almost always likely to mix with the secondary polymer. An important characteristic determining the principled recyclability of PVC waste (allowable processing time, the service life of secondary material or product) and the need to further enhance the stabilizing group is the time of thermostability.

8.8.8 METHODS OF RECYCLING OF PVC WASTE

Homogeneous industrial wastes are usually subjected to a secondary processing, only when thin layers of the material are prone to deep aging. In some cases, it is recommended to use an abrasive tool to remove the degraded layer with the subsequent processing of the material into articles which are not inferior to the properties of those obtained from the raw materials. For separating a polymer from a metal (wires, cables), a pneumatic method is used. Typically, isolated plasticized PVC can be used as insulation for wires with a low voltage or to manufacture articles by injection molding. To remove metal and mineral inclusions, the milling industry experience can be used, based on the application of an induction method, the method of separation by magnetic properties. To separate aluminum foil from a thermoplastic, heating in water at 95–100°C is used. The method of dry preparation of plastic waste using a compactor is energetically economical. It is recommended for recycling artificial leather

(AL) and PVC linoleum wastes and includes a number of technological operations: grinding, separation of textile fibers, plasticization, homogenization, compaction, and granulation; additives can also be introduced. The lining fibers are separated three times: after the first blade crushing, after compaction, and after the secondary blade crushing. A molding composition is prepared which can be processed by injection molding but still comprises fibrous components which do not prevent processing but serve as a filler to reinforce the material.

8.8.9 PVC PLASTICS RECYCLING METHODS

8.8.9.1 INJECTION MOLDING

The main types of waste from unfilled PVC are ungelatinized plastisol, technological waste, and defective articles. The light industry of Russian Federation has the following technology of recycling plastisol by injection molding. Products from secondary PVC materials of satisfactory quality can be obtained by the plastisol technology. The process involves grinding film or sheet waste, making a PVC paste in a plasticization unit, and forming a new product by molding. Ungelatinized plastisol was collected in vessels when cleaning the dispenser and the mixer, subjected to gelation, and then mixed with technological waste and defective products on mills, the resulting sheets were processed on rotary grinders. Thus, obtained plastisol chips were processed by injection molding. Crumb plastisol in amounts of 10–50 wt. % can be used as a composition with rubber to obtain rubber mixtures, and this eliminates softeners from the formulations. For recycling by injection molding, machines are typically used operating like intrusion, with an ever-rotating screw, whose design allows spontaneous seizure and homogenizing wastes. Multicomponent molding is a promising method of using PVC waste. With this method of processing, the product has its outer and inner layers of high quality, stable, colored, and with good appearance. The inner layer is made up of secondary polyvinylchloride raw materials. Processing of thermoplastics by this method saves a lot of scarce primary raw material, reducing its consumption more than twice.

8.8.9.2 EXTRUSION

Currently, one of the most efficient ways of processing PVC-based waste plastics for the purpose of their recycling is the method of elastic-deformation dispersing, based on the phenomenon of multiple fracture under the combined influence of high pressure and shear deformation on the material at elevated temperatures. Elastic-deformation dispersing of pre-crushed materials with particle sizes of 103 μm is held in a single-screw rotary disperser. Used wastes of plasticized duplicate film materials on various bases (linoleum on polyester fabric support, penoplen on a paper base, AL on a cotton base) are processed in a uniform secondary disperse material—a mixture of PVC plastics with a milled base with the most probable particle size of 320–615 μm, preferably asymmetrical in shape, with a high specific surface area (2.8–4.1 $m^2 g^{-1}$). The optimal dispersion conditions for obtaining the most highly dispersed product are: the temperatures in the disperser's zones are 130–150–70°C; the degree of loading is no more than 60%; the minimum screw speed is 35 RPM. An increased processing temperature for PVC materials leads to unwanted intensification of degradation processes in the polymer, expressed in the darkening of the product. Increasing the degree of load and the rotational screw speed degrades the degree of dispersion of the material. Wastes of base-free plasticized PVC materials (agricultural film, insulating film, and PVC hoses) can be processed by elastic-strain dispersion to obtain fine-quality recycled material without technological difficulties by a wider variation of the dispersion mode. A finer product is formed with particle sizes within 240–335 μm, preferably of spherical shape. The elastic-deformation impact when hard PVC materials (impact-resistant material for mineral water bottles, sanitary PVC pipes, etc.) are dispersed must be carried out at higher temperatures (170–270°C), load degrees of no more than 40%, and the minimum speed of screw rotation (35 RPM). In the event of deviations from the specified dispersion mode, technological difficulties are observed and the quality of the secondary product is lowered by dispersion. During the recycling of PVC waste, some modification of the polymer material can be carried out simultaneously with dispersion by introducing 1–3 wt. % of metal stabilizers and 10–30 wt. % of plasticizers into the source raw material. This leads to an increase in thermostability by 15–50 min when using metal stearates and an improved flow index of the melt processed together with ester plasticizers by 20–35%; the processability of the dispersion process is also improved. The resulting secondary PVC materials, due to their high dispersion and developed surface of the particles, have a surface activity. This property of the resulting powders predetermines their very good compatibility with other materials, so they can be used for replacement (up to 45 wt. %) of initial raw materials in the preparation of the same or new PMs.

For recycling of PVC, screw extruders can also be used. They provide perfect homogenization of the mixture, and the plasticizing process is performed under milder conditions. As a twin-screw extruder operates by displacement, the residence time of polymer therein at a plasticization temperature is clearly determined and its delay in the high temperature zone is excluded. This prevents overheating and provides good conditions for degassing in the low-pressure zone, which allows removing moisture, oxidation and degradation products, and other volatiles usually contained in the waste. Methods based on the combination of extrusion preparation and molding by pressing can be used for the processing of combined PMs, including synthetic rubber, cable insulation waste, paper-based thermoplastic coatings, and so forth. To implement this method, a unit consisting of two machines of a 10 kg injection each is proposed. The fraction of the specifically introduced non-PMs present in waste may be up to 25%, and the copper content may even reach 10%. The method of coextrusion of fresh thermoplastic forming wall layers and of polymer waste comprising an inner layer is also applied. As a result, a three-layer article (e.g., film) can be obtained.

8.8.9.3 CALENDERING

The so-called "Regal" process comprising calendering the material and making plates and sheets used to produce containers and furniture is an example of recycling by calendering. The convenience of this process for processing waste of various compositions is the easiness of control by changing the gap between the calender's rolls to obtain a good shear and dispersing effect on the material. A good plasticization and homogenization of the material during processing provides making articles with relatively high strength characteristics. The method is economical for thermoplastics plasticized at relatively low temperatures, mainly soft PVC.

To prepare the AL waste and linoleum, a unit is designed consisting of a knife grinder, a mixing drum, and three-roll refiner rolls. The components of mixture between the rotating surfaces are even more crushed, kneaded, and homogenized. The material acquires a good quality already after one pass through the machine.

8.8.9.4 PRESSING

Pressing is a conventional method of processing waste plastics; in particular, most common is what can be termed "Regal-converter." Milled waste of uniform thickness is fed into a furnace on a conveyor belt and melted. The kneaded mass is then pressed. The proposed method recycles mixed plastics containing foreign matter above 50%. There is a continuous process for recycling synthetic carpets and AL. Its essence is as follows: shredded waste is fed into a mixer, added are 10% of the material, pigments, and fillers (for reinforcement). Plates are pressed from this mixture in a two-tape press. Such plates have a thickness of 8–50 mm at a density of about 650 kg m^{-3}. Due to their porosity, the plates have thermal and sound-insulating properties. They find application in mechanical engineering and in the automotive industry as construction elements. With single- or double-sided lamination, these plates can be used in the furniture industry. In the United States, the pressing process is used to manufacture heavy plates. Another manufacturing method is also applied, based on foaming in a mold. Developed options differ by the methods of introduction of blowing agents into secondary raw materials and heat supply. Blowing agents can be introduced into an internal mixer or an extruder. However, the method of mold foaming is better when the process of pore formation is carried out in a press. A significant disadvantage of pressing sintering of plastic waste is weak stirring of the mixture of components, which leads to decreased mechanical properties of the materials obtained.

8.8.10 UTILIZATION OF WASTE POLYSTYRENE PLASTICS

PS waste is accumulated as obsolete PS products and its copolymers (bread bins, vases, cheese tables, various utensils, grills, pots, hangers, cladding sheets, details of commercial and laboratory equipment, etc.), as well as industrial (technological) waste of general-purpose PS, high-impact PS

(HIPS), and its copolymers. Process PS wastes (as polyolefin ones), by their physicomechanical and technological properties, do not differ from the primary raw material. These wastes are recurrent and mainly used in those enterprises where they are formed. They can be added to the primary PS or used as a raw material in the independent production of various products. A considerable amount of process waste (50%) is formed in the processing of PS plastics by injection molding, extrusion, and vacuum molding, whose return to the technological process can significantly improve the efficiency of polymer materials and design waste-free production in the plastics industry. ABS plastics are widely used in the automotive industry to manufacture large parts of vehicles, in the production of sanitary equipment, pipes, and so forth. During the repeated processing of ABS polymer, two competing processes occur in it: on the one hand, partial degradation of macromolecules, and on the other hand, partial intermolecular cross-linking, both increase with the number of processing cycles. When choosing a method of processing of extrusion ABS, the principal possibility to form products by direct pressing, extrusion, and injection molding was proven. Drying of the polymer is an effective technological stage of recycling ABS waste, which allows to bring the moisture content to a level not exceeding 0.1%. This eliminates the formation of such defects in the material as scaly surface, silver appearance, exfoliation products across the thickness, which arise from excessive moisture; the pre-drying improves the material properties by 20–40%. However, direct pressing is unproductive and polymer extrusion is hindered due to high viscosity.

The processing of technological ABS waste by injection molding seems promising. Introducing processing additives is necessary to improve the fluidity of the polymer. Additives to the polymer facilitate the processing of ABS polymers as they lead to an increased mobility of the macromolecules, better polymer flexibility, and a lower viscosity. Articles thus obtained are not inferior to those made of the primary polymer by performance, and sometimes even surpass them. Defective or worn items can be recycled by grinding and further using as a separate raw. Depreciated cultural and household products, as well as wastes of the industrial, construction, and other heat-insulating PMs can be recycled into products. This applies mainly to products made of high-impact PS. Before reprocessing, block PS is necessary to be combined with high-impact PS (70/30), to be modified in other ways, or its copolymer should be recycled with AN, methyl methacrylate (MS), or terpolymers with MS and AN (MSN). MS and MSN copolymers feature more irregular compositions by structure, which is of great importance in their subsequent processing. Secondary PS can be added to PE. To convert wasted PS films into secondary polymeric raw materials, they are subjected to agglomeration in rotary agglomerators. The low toughness value of PS causes rapid grinding (in comparison with other thermoplastics). However, the high adhesive strength of PS leads, first, to coalescence of the particles of the material to form large aggregates before (80°C) the material becomes ductile (130°C), and, second, to the material sticking to the processing equipment. This greatly hinders PS agglomeration compared to PE, PP, and PVC. Waste PS foam can be dissolved in styrene and then polymerized in a mixture containing comminuted rubber and other additives. The so-obtained copolymers have rather high impact strength. At present, there is a problem faced by the processing industry of recycling of mixed plastic waste. The technology of processing mixed waste includes sorting, grinding, washing, drying, and homogenization. The secondary PS obtained from mixed waste has high physicomechanical characteristics, it can be added (in a molten state) into asphalt and bitumen. This reduces their cost and the strength characteristics increase by about 20%. To improve the quality of secondary raw PS material, its modification is conducted. This requires a study of its properties during heat aging and service. Aging of PS plastics has its own specificity, which is particularly evident for high-impact materials containing rubber in addition to PS. When PS materials are heat treated (at 100–200°C), their oxidation proceeds via the formation of hydroperoxide groups whose concentration rapidly increases at the initial stage of oxidation, followed by the formation of carbonyl and hydroxyl groups. Hydroperoxide groups initiate the process of photooxidation occurring during the use of PS products under solar radiation. The lower resistance to photooxidative degradation of PS products compared with polyolefins is a consequence of the combined effect of hydroperoxide and unsaturated groups at the early stages of oxidation and that of carbonyl groups at the later stages. The presence of unsaturated bonds in the rubber component of the HIPS at its heating leads to autoacceleration of the degradation process. When rubber-modified PS photoages, chain breakage prevails over cross-linking, especially at a large content of double bonds, which has a significant effect on the polymer morphology, its physicomechanical and rheological properties.

8.8.10.1 *RECYCLING POLYAMIDES*

PA wastes formed mainly in the production and processing of fibers (nylon and amide), and obsolete products occupy a significant place among solid plastic waste. The amount of such waste in the production and processing of fiber reaches 15% (including 11–13% in production). As PA is an expensive material with a number of valuable chemical and physicomechanical properties, the rational use of its waste is of particular importance. The diversity of secondary PA requires special methods of processing and, at the same time, opens up opportunities for their choice. PA 66 waste possesses the most stable properties, which is a prerequisite for the design of universal methods of its processing. Some wastes (rubberized cord, trimmings, and worn hosiery) contain non-PA components and require special handling during processing. Worn articles are contaminated, and the amount and composition of pollution are determined by the conditions of their use, the organization of their collection, storage, and transportation. The main areas of processing and use of PA waste are grinding, thermoforming from melt, depolymerization, reprecipitation from solution, and various methods of modification and textile processing to obtain materials with a fiber structure. Of great importance is the molecular weight of the waste, which affects the strength of recycled materials and products, as well as the technological properties of secondary PA. A significant impact on the strength, thermal stability, and processing conditions is rendered by the content of low-molecular-weight compounds in PA 6. To select methods and modes of processing, as well as uses of the waste, it is important to study the thermal behavior of secondary PA. At the same time, structural and chemical properties of the material and its prehistory may play a significant role.

8.8.11 *METHODS OF PA WASTE TREATMENT*

The existing methods of recycling PA waste can be classified into two main groups, namely, mechanical (involving no chemical transformations) and physicochemical. The mechanical methods include grinding and various methods and techniques used in the textile industry to produce materials with a fibrous structure. Ingots, off-grade tape, casting waste, and partially stretched and unstretched fibers can be subjected to mechanical processing. Shredding is not only an operation accompanying the majority of technological processes but also an independent method of recycling. It allows obtaining powder materials and chips for injection molding from ingots, strips, and bristles. Characteristically, shredding virtually unchanges the physicochemical properties of the source raw. To get powder products, in particular, cryogenic grinding processes are used. Fiber and bristle waste is used to manufacture fishing line, sponges, handbags, and so forth, but it requires substantial manual labor. Of mechanical methods of recycling, making nonwoven materials, flooring, and staple fabrics should be considered as most promising and widespread. Of particular value for this purpose is nylon fiber waste, which is easily processed and stained. The physicochemical methods of PA waste utilization can be classified as follows: (1) the depolymerization of waste to obtain monomers suitable for producing fibers and oligomers and their subsequent use in the manufacture of adhesives, paints, and other products; (2) remelting of the waste to produce granules, agglomerates, and products by pressure extrusion; (3) reprecipitation from solution to give powder coating; (4) preparation of composite materials; and (5) chemical modification to produce materials with new properties (varnishes, adhesives, etc.). Depolymerization is widely used in industry to produce high-quality monomers from uncontaminated technological waste. Depolymerization is conducted in the presence of catalysts which can be neutral, basic, or acidic compounds. The method of remelting PA waste is widespread, which is carried out mainly in vertical machines for 2–3 h and in extrusion plants. Under prolonged thermal exposure, the specific viscosity of a PA-6 solution in sulfuric acid reduces by 0.4–0.7%, and the content of low-molecular-weight compounds increases from 1.5 up to 5.6%. Melting in superheated steam, moistening and melting in vacuum improve the properties of the recovered polymer; however, it does not solve the problem of obtaining sufficiently high-molecular-weight products. During extrusion, PA is significantly less oxidized than at prolonged melting, which contributes to preservation of high physicomechanical properties of the material. Increasing the moisture content of source raw (to reduce oxidation) leads to destruction of PA. Making powder from PA waste by reprecipitation from solution is a way of cleaning polymers, obtaining them in a form suitable for further processing. Powders can be used, for example, for cleaning dishes, as a component of cosmetics, and so forth. Filling with fibrous materials (glass fibers, asbestos fibers, etc.) is a common method of regulating the mechanical properties of PA. The highly efficient waste usage is exemplified by the production of the material ATM-2, which has high strength, wear resistance, and dimensional stability. Physical modification of molded articles by their bulk-surface treatment is a promising direction of improving the physical and performance properties of recycled PA. The bulk-surface treatment of samples of secondary PA filled with kaolin and plasticized with a shale softener in heated glycerin leads to an increase in toughness by 18%, in the failure stress at bending by 42.5%, which can be explained by forming a more perfect structure of the material and removal of residual stresses.

8.8.12 TECHNOLOGICAL PROCESSES OF RECYCLING PA WASTE

The main processes used for the regeneration of secondary polymer raw materials from PA waste are: (1) PA regeneration by extrusion of worn nylon network materials and technological waste to produce granular products suitable for processing into articles by injection molding; (2) PA regeneration from worn articles and nylon process waste containing fibrous impurities (not PA) by dissolution, filtering the solution, and subsequent precipitation of PA as a powdery product. The technological processes of recycling of waste products differ from the technological waste processing by the presence of a stage of preconditioning, including sorting waste, its washing, pressing, and drying of secondary raw materials. Preconditioned worn articles and process waste are fed to grinding, and then sent to an extruder for granulation. Secondary PA fibrous raw containing non-PA materials is treated in a reactor at room temperature with an aqueous solution of hydrochloric acid, filtered to remove non-PA impurities. Powdered PA is precipitated with aqueous methanol. The precipitated product is grinded and the resultant powder is sieved. Currently in the Russian Federation, the technological waste generated in the production of nylon fibers is effectively used to produce nonwovens, floor coverings, and granules for injection and extrusion. The main cause of the poor use of defective PA products is the lack of high-performance equipment for their primary processing and recycling. Developing and implementing processes for recycling of waste products made of nylon fibers (hosiery, network materials, etc.) into secondary materials will achieve saving large quantities of raw materials and send it to the most effective application.

8.8.13 RECYCLING OF POLYETHYLENE TEREPHTHALATE WASTE

Recycling of Dacron™ fibers and worn PET products is similar to recycling nylon waste, so we consider the recycling of PET bottles in this section. For more than 10 years of mass consumption of PET-bottled beverages in the Russian federation, by some estimates, above 2 million tons of used plastic packaging has been accumulated in the landfills, which is a valuable chemical raw. The explosive growth of the production of bottle preforms, the rise in world oil prices and, consequently, in the prices of primary PET have influenced the active formation of the market of PET bottles recycling in Russia since 2000. There are several methods for recycling used bottles. One interesting technique is deep chemical recycling of secondary PET to obtain dimethyl terephthalate during methanolysis or terephthalic acid and ethylene glycol in a number of hydrolytic processes. However, such methods of processing have a significant drawback, the high cost of depolymerization. Therefore, commonly known methods of mechanochemical processing are currently used quite often, during which the end products are formed from polymer melt. A significant range of products made from recycled bottle PET have been developed. The main large-scale production is obtaining Dacron™ fibers (mainly staple), the production of synthetic padding and nonwovens. A large segment of the market is occupied by sheet extrusion for thermoforming on extruders with sheet dies, and finally, the most promising way of processing is, as is widely acknowledged, producing granulate suitable for food contact, that is, a material for repeated casting of preforms. Bottle intermediate can be used for technical purposes: in the processing into articles, recycled PET can be added to the primary material; compounding when recycled PET can be fused with other plastics (e.g., with PC, SPE) and filled with fibers to produce parts for industrial purposes; obtaining dyes (superconcentrated) for the production of colored plastic articles. Also, purified PET flakes can be used directly for manufacturing a wide range of products: textile fibers; stuffing and staple fibers—sintepon or cintepon (thermal insulation for winter jackets, sleeping bags, etc.); roofing materials; films and sheets (colored, metallized); packaging (boxes for eggs and fruit, packaging for toys, sporting goods, etc.); molded articles of construction purposes for the automotive industry; details of lighting devices and appliances, and so forth. In any case, the source raw for depolymerization or processing into articles is pure PET flakes rather than bottle wastes that could have been in a landfill (formless heavily contaminated objects).

Consider the process of recycling bottles into clean plastic flakes. If possible, bottles should be collected in sorted order, without mixing with other plastics and polluting objects. The optimal target for processing is a compressed pile of colorless PET bottles (painted bottles should be sorted and recycled separately). Bottles should be stored in a dry place. Plastic bags with PET bottles are emptied into a loading hopper in bulk. Further, the bottles enter a feed hopper. The pile feeder is used as both a storage bin with a system of uniform supply and as a pile splitter. A transporter located at the bottom of the hopper promotes the pile to three rotating screws, decomposing agglomerates into individual bottles and feeding them to a discharge conveyor. Here it is necessary to separate the bottles of colored and uncolored PET and remove foreign objects (rubber, glass, paper, metal, and other types of plastics). In a single-rotor crusher equipped with a hydraulic pusher, the PET bottles are crushed to form large fractions of up to 40 mm. The shredded material passes through a vertical air classifier. Heavy particles (PET) fall against the air flow on the screen of a vibroseparator. Light particles (labels, film, dust, etc.) are carried upward by the airstream and collected in a dust collector below the cyclone. On the vibroscreen of the separator, the particles are divided into two fractions: the larger PET particles "flow" through the screen, while the small particles (mainly heavier fractions of impurities) extend inwardly of the screen and are collected in a vessel below the separator. A flotation tank is used for the separation of materials with different relative densities. PET particles sink to an inclined bottom, and a screw continuously discharges PET onto a water-separation screen. The screen serves as a separator for the water injected together with PET from the floatation bath and as a separator for fine particles of impurities. The pre-granulated material is effectively washed in a two-stage inclined rotating drum with perforated walls. The flakes are dried in a rotating drum made of perforated sheet. The material is rotated in a hot airstream. The air is heated by electric heaters. Further, the flakes come into a second crusher. At this stage, the large PET particles are crushed into flakes, whose size is about 10 mm. It should be noted that the idea of recycling is that the material is not crushed into flakes of commercial quality at the first stage of grinding. This implementation of the process avoids loss of the material in the system, to achieve optimal separation of labels, to improve the cleaning effect, and to reduce knife wear in the second grinder because glass, sand, and other abrasive materials are removed prior to the secondary grinding stage. The final process is similar to the primary air classification. Label residues and PET dust are removed from the airstream. The final product (clean PET flakes) is put into barrels. Therefore, it is possible to solve the serious problem of secondary recycling of plastic containers to give a product.

Making bottles from bottles is a promising way of PET recycling. The main stages of the classic recycling process for implementation of the "bottle-to-bottle" scheme are: collection and sorting of secondary raw materials; packaging; crushing and washing; separation of the crushed fraction; extrusion to form pellets; and processing the granules in a screw apparatus to increase the viscosity of the product and to ensure the sterility of the product for direct contact with food. But significant capital investment is required for the implementation of this process, since it is impossible to conduct it on standard equipment.

8.8.13.1 BURNING

It is expedient to burn only certain types of plastics, which have lost their useful properties, to generate heat energy. For example, a power station (The West Midlands Whole-Tyres-To-Electricity Facility) that converts used tires (rather than natural gas and fuel oil) into electricity was opened at Wolverhampton (UK) in November 1993. It produces enough power to supply 25,000 homes. Burning of certain polymer types is accompanied by the release of toxic gases: hydrogen chloride, nitrogen oxides, ammonia, cyanides, and so forth, necessitating measures to protect the air. Moreover, the economic efficiency of this process is the lowest in comparison with other processes of plastic waste disposal. Nevertheless, the relative simplicity of combustion determines its rather widespread practice.

8.8.14 RECYCLING OF RUBBER PRODUCTS

According to the latest statistics, about 2 million tons of waste tires are annually produced in Western Europe, about 1 million tons of tires are produced in the Russian Federation, and the same amount of old rubber is given by RP. Factories producing tires and rubber goods produce a lot of waste, whose large proportion is not reused (e.g., waste rubber diaphragms on tire plants, ethylene propylene waste, etc.). Because of the large amount of old tires, incineration still predominates in utilization, while material recycling is still a small proportion, despite the importance of this particular waste disposal to improve the ecology and to conserve natural resources. Material recycling is not widely used due to high energy costs and the high cost of producing fine rubber powder and regenerates. Without economic regulation by the state, tire disposal remains unprofitable. The Russian Federation lacks a system for collecting, depositing, and processing scrap tires and technical RP. No methods of legal and economic regulation and stimulation of solving this problem are developed. For the most part, worn tires are collected in the territories of car depots or removed into woods and open pits. Currently, large amounts of annually generated scrap tires are an environmental challenge for all regions of the Russian Federation. As practice shows, to solve this problem is very difficult at the regional level. A federal program for recycling tires and rubber articles should be developed and implemented in Russian Federation. In this program, it is necessary to lay legal and economic mechanisms to ensure the movement of used tires along the proposed scheme.

Two basic approaches are discussed as the economic mechanism of a waste tire utilization system in the Russian Federation:

1. the disposal of tires is paid directly by their owner (the polluter pays);
2. the disposal of tires is paid by the manufacturer or importer of tires (the producer pays).

The "polluter pays" principle is partially implemented in regions such as Tatarstan, Moscow, St. Petersburg, and others. Realistically assessing the level of environmental and economic nihilism of common consumers, no successful use of this principle can be expected.

The introduction of the "producer pays" principle would be better for the Russian Federation. This principle has been successfully operating in the Nordic countries. For example, its use enables utilization of more than 90% of tires in Finland.

8.8.14.1 CRUSHING WORN TIRES AND TUBES

The initial stage of obtaining regenerate by the existing industrial methods from worn RP (tires, inner tubes, etc.) is crushing them. Shredding tire rubber is accompanied by some destruction of the rubber vulcanization grid, estimated by the change in the degree of equilibrium swelling: with other conditions being equal, the smaller the particle sizes of the obtained crumb rubber, the greater the degree of equilibrium swelling. The chloroform rubber extract changed very little, at the same time. Simultaneously, destruction of the carbon structures occurs. Crushing rubbers containing active carbon black is accompanied by some destruction of chain structures by their carbon–carbon bonds; in the case of low-activity carbon black (thermal), the number of contacts between the carbon particles increases. In general, changes in the vulcanization grid and carbon structures in rubber under crushing must, as in the case of any mechanochemical process, depend on the type of the polymer, the nature and amount of the filler contained therein, the nature and density of cross-links in the vulcanization grid, the process temperature, the degree of rubber grinding, and the type of the equipment employed. The particle size of the crumb rubber is determined by the method of rubber devulcanization, the type of rubber to be chopped, and quality requirements of the final (regenerated) product. The smaller the particle size of crumb, the more quickly and uniformly the destructed material is distributed, which helps to reduce the content of fine, insufficiently devulcanized rubber particles ("grains") in the devulcanizate. It allows regenerating tire crumb, which reduces refinement wastes and improves the performance of refining equipment. However, as the particle size of crumb rubber decreases, its production costs increase.

In this regard, the use of the currently available methods of producing tire rubber crumb having a particle size of 0.5 mm or less seems economically impractical. Since foreign materials (textiles and metal) are contained in worn tires along with rubber, rubber is cleaned from these materials when crushing tires. While the presence of metal in the rubber crumb is inadmissible, the possible content of textile remnants in it depends on the subsequent method of rubber crumb devulcanization and the type of textiles. Rollers (in Russian Federation, Poland, England, and United States) and disc mills (in Germany, Hungary, and the Czech Republic) are most widely used for crushing worn RP. Percussion (hammer) crushers and rotor crushers (e.g., "Novorotor" units) are used for this purpose as well. Rubber is also comminuted by extrusion based on the destruction of rubber under uniform compression and shear. An apparatus is offered in which the material to be comminuted passes between a rotor and a casing wall. The grinding effect is enhanced by changing the size and shape of the gap between the rotor and the casing wall when the rotor rotates. A comparison of a number of existing schemes of crushing waste tires shows that the scheme based on the use of rollers has a better performance than the use of disk mills or a rotary machine with regard to the performance of equipment, and the energy and time consumed.

KEYWORDS

- **environment**
- **pollution**
- **environmentally friendly**
- **polymers aging**
- **stabilization of polymers**
- **biodegradation of polymers**
- **polyolefins**
- **PVC**
- **polyamides**
- **caoutchouc**
- **rubber**
- **polyethylene**
- **polyalkylene guanidines**

REFERENCES

1. Zezin, A. B. Polymers and Environment. *Soros Educ. J.* **1996,** *2,* 57–64. [Russ].
2. Goldfein, M. D.; Kozhevnikov, N. V.; Rafikov, E. A.; Stepukhovich, A. D.; Kosyreva, R. V. Polymerization of Acrylonitrile in an Aqueous Solution of Sodium Thiocyanate. *Polym. Sci. U.S.S.R.* **1975,** *17*(10), 2630–2637. [Russ].
3. Gol'dfein, M. D.; Rafikov, E. A.; Kozhevnikov, N. V.; Stepukhovich, A. D.; Rabinovich, I. S. Copolymerization of Acrylonitrile and Methacrylate in an Aqueous Solution of Sodium Thiocyanate. *Polym. Sci. U.S.S.R.* **1977,** *19*(2), 318–324. [Russ].
4. Gol'dfein, M. D.; Rafikov, E. A.; Kozhevnikov, N. V.; Stepukhovich, A. D.; Rabinovich, I. S.; Kosyreva, R. V. The Influence of Methallyl Sulphonate and Itaconic Acid on the Copolymerization of Acrylonitrile and Methyl Acrylate in an Aqueous Solution of Sodium Thiocyanate. *Polym. Sci. U.S.S.R.* **1977,** *19*(11), 2945–2952. [Russ].
5. Gol'dfein, M. D.; Zyubin, B. A. Kinetics and Mechanism of the Processes of Preparing Fibre-Forming Polymers Based on Acrylonitrile. Review. *Polym. Sci. U.S.S.R.* **1990,** *32*(11), 2145–2166. [Russ].
6. Goldfein, M. D.; Kozhevnikov, N. V.; Trubnikov, A. V. *Kinetics and Mechanism of Regulation of Polymer Formation Processes*; Saratov Univ. Press: Saratov, 1989; 178 p. [Russ].
7. Kozhevnikov, N. V.; Trubnikov, A. V.; Stepukhovich, A. D.; Larina, N. M. Polymerization of Acrylic Monomers in Presence of Sulphuric acid. *Polym. Sci. U.S.S.R.* **1984,** *26*(4), 761–769. [Russ].

8. Goldfein, M. D. *Kinetics and Mechanism of the Radical Polymerization of Vinyl Monomers: A Textbook*; Saratov Univ. Press: Saratov, 1986; 139 p. [Russ].
9. Stepukhovich, A. D.; Kozhevnikov, N. V.; Leont'eva, L. V. The Effect of the Complex of 7,7,8,8-tetracyanquinodimethane and Dimethyl Viologen on the AIBN-Initiated Solution Polymerization of Styrene in DMF. *Vysokomol. Comm. A.* **1974,** *16*(7), 1763–1771. [Russ].
10. Kozhevnikov, N. V. Study of the Influence of the Lithium Complex of 7,7,8,8-tetracyanquinodimethane on the Radical Polymerization of Styrene. *Polym. Sci. U.S.S.R.* **1986,** *28*(4), 749–758. [Russ].
11. Kozhevnikov, N. V. Study of the Effect of the Lithium 7,7,8,8-tetracyanoquinodimethane Complex on the Radical Polymerization of Acrylic Monomers. *Izv. Vuzov. Chem. Chem. Eng.* **1987,** *30*(4), 103–106. [Russ].
12. Kozhevnikov, N. V.; Stepukhovich, A. D. On the Effect of the 7,7,8,8-tetracyanquinodimethane Anionic Radical on the Benzoyl Peroxide Initiated Polymerization of Methyl Methacrylate in Dimethylformamide. *Polym. Sci. U.S.S.R.* **1976,** *18*(4), 996–1003. [Russ].
13. Kozhevnikov, N. V.; Stepukhovich, A. D. The Combined Effect of Some Peroxide Initiators and Tetracyanoquinodimethane Anion-Radicals on Polymerization of Methyl Methacrylate in Solution. *Polym. Sci. U.S.S.R.* **1980,** *22*(5), 1061–1071. [Russ].
14. Gol'dfein, M. D.; Rafikov, E. A.; Kozhevnikov, V. N.; Stepukhovich, A. D.; Trubnikov, A. V. The Effect of Stable Free Radicals of the Kinetics and Mechanism of Polymerization of Some Vinyl Monomer. *Polym. Sci. U.S.S.R.* **1975,** *17*(8), 1919–1927. [Russ].
15. Trubnikov, A. V.; Gol'dfein, M. D.; Kozhevnikov, N. V.; Rafikov, E. A.; Stepukhovich, A. D.; Tomashchuk, V. I. Inhibition of Polymerization of Vinyl Monomers by Nitroxide and Iminoxyl Radicals. *Poly. Sci. U.S.S.R.* **1978,** *20*(11), 2751–2758. [Russ].
16. Trubnikov, A. V.; Gol'dfein, M. D.; Kozhevnikov, N. V.; Stepukhovich, A. D. Mechanism of the Inhibition of Polymerization of Vinyl Monomers Initiated by Benzoyl Peroxide by Stable Nitroxide Radicals. *Poly. Sci. U.S.S.R.* **1983,** *25*(10), 2497–2504. [Russ].
17. Rozantsev, E. G.; Gol'dfein, M. D.; Trubnikov, A. V. Stable Radicals and the Kinetics of the Radical Polymerisation of Vinyl Monomers. *Russ. Chem. Rev.* **1986,** *55*(11), 1070–1080. [Russ].
18. Rozantsev, E. G.; Goldfein, M. D.; Pulin, V. F. *Organic Paramagnetics*; Saratov Univ. Press: Saratov, 2000; 340 p. [Russ].
19. Rozantsev, E. G.; Goldfein, M. D. Some Conceptions of Organic Paramagnetics. *Oxid. Commun.* (Sofia). **2008,** *31*(2), 241–263.
20. Rozantsev, E. G.; Goldfein, M. D. Organic Paramagnetics as Antioxidants. *Polym. Res. J.* (Nova Science Publishers, Inc.: New York). **2008,** *2*(1), 5–28.
21. Goldfein, M. D.; Kozhevnikov, N. V.; Trubnikov, A. V. New Stabilizers for Acrylic Monomers. *Poly. Sci. U.S.S.R. Ser. B.* **1983,** *25*(4), 268–271. [Russ].
22. Kozhevnikov, N. V.; Goldfein, M. D. Inhibition of Methyl Methacrylate Polymerization in the Presence of Sulfuric Acid. *Poly. Sci. U.S.S.R. Ser. B.* **1988,** *30*(8), 613–616. [Russ].
23. Gol'dfein M. D.; Gladyshev, G. P. Kinetics and Mechanism of the Inhibited Polymerisation of Vinyl Monomers. *Russ. Chem. Rev.* **1988,** *57*(11), 1083–1097. [Russ].

24. Goldfein, M. D.; Kozhevnikov, N. V.; Trubnikov, A. V. Polymerization Inhibition During Methyl Methacrylate Synthesis. *Chem. Ind.* **1989,** *1,* 20–22. [Russ].
25. Kozhevnikov, N. V.; Zyubin, B. A.; Tsyganova, T. V.; Eisenberg, L. V. Inhibition of Acrylic Monomer Polymerization by Cupferron and Its Derivatives. *Izv. Vuzov. Chem. Chem. Eng.* **1989,** *32*(11), 98–102. [Russ].
26. Goldfein, M. D.; Gladyshev, G. P.; Trubnikov, A. V. Kinetics and Mechanism of the Inhibited Polymerization of Vinyl Monomers. *Polym. Yearb.* **1996,** *13,* 163–190. [Russ].
27. Kozhevnikov, N. V.; Goldfein, M. D.; Kozhevnikova, N. I. Macromolecular Chemistry and Environmental Problems. *J. Balk. Tribol. Assoc.* (Sofia). **2008,** *14*(4), 560–571.
28. Kozhevnikov, N. V.; Kozhevnikova, N. I.; Goldfein, M. D.. Solving Some Environmental Problems of Polymeric Chemistry. Proc. Saratov University. Ser. Chemistry. Biology Ecology. 2010. V. 10, No 2. P. 34-42. [Russ].
29. Goldfein, M. D.; Ivanov, A. V.; Kozhevnikov, N. V. *Fundamentals of General Ecology, Life Safety and Environment Protection*; Nova Science Publishers Inc.: New York, 2010; 251 p.
30. Kozhevnikov, N. V.; Goldfein, M. D.; Zyubin, B. A.; Trubnikov, A. V. Kinetic Features of the Emulsion Homo- and Copolymerization of Certain Water Soluble Polymers. *Poly. Sci. U.S.S.R.* **1991,** *33*(6), 1175–1183. [Russ].
31. Goldfein, M. D.; Kozhevnikov, N. V.; Trubnikov, A. V. Kinetics and Mechanism of the Processes of Formation of Polymer Emulsions Based on (Meth) Acrylates. Review. *Poly. Sci. U.S.S.R.* **1991,** *33*(10), 1907–1922. [Russ].
32. Kozhevnikov, N. V.; Goldfein, M. D. Kinetics of Emulsion Polymerization of Methyl Methacrylate and Its Copolymerization with Acryl- or Methacrylamide. *Vysokomol. Comm. A.* **1991,** *33*(11), 2255–2261. [Russ].
33. Kozhevnikov, N. V.; Zyubin, B. A.; Simontsev, D. V. Emulsion Copolymerization of Acrylic Monomers with Allyl Acrylate. *Polym. Sci. Rus.* **1992,** *34*(7), 46–51. [Russ].
34. Goldfein, M. D.; Kozhevnikov, N. V.; Trubnikov, A. V. Kinetics and Mechanism of Formation of Polymer Emulsions Based on (Methyl) Acrylates. *Polym. Yearb.* **1995,** *12,* 89–104. [Russ].
35. Kozhevnikov, N. V.; Zyubin, B. A.; Simontsev, D. V. Emulsion Copolymerization of Multicomponent Systems of Acrylic Monomers. *Polym. Sci. Rus.* **1995,** *37*(5), 758–763.
36. Kozhevnikov, N. V.; Goldfein, M. D.; Terekhina, N. V. Kinetics of the Emulsion Copolymerization of Butyl Acrylate with Water-Soluble Monomers in the Absence of Emulsifier. *Russ. Chem. Phys.* **1997,** *16*(12), 97–102. [Russ].
37. Kozhevnikov, N. V.; Goldfein, M. D.; Terekhina, N. V. Emulsion Copolymerization of Methyl Acrylate in the Presence and Absence of an Emulsifier. *Izv. Vuzov. Chem. Chem. Eng.* **1997,** *40*(3), 78–83. [Russ].
38. Kozhevnikov, N. V.; Terekhina, N. V.; Goldfein, M. D. Emulsion Polymerization of Methyl Methacrylate and Its Copolymerization with Hydrophilic Monomers. *Izv. Vuzov. Chem. Chem. Eng.* **1998,** *41*(4), 83–87. [Russ].
39. Kozhevnikov, N. V.; Goldfein, M. D.; Trubnikov, A. V. Emulsion Copolymerization of Butyl Acrylate with Water-soluble Monomers. *Inter. J. Polym. Mater.* **2000,** *46,* 95–105. [Russ].
40. Kozhevnikov, N. V.; Goldfein, M. D.; Trubnikov, A. V.; Kozhevnikova, N. I. Emulsion Copolymerization of (Meth) Acrylates: Characteristics of Kinetics and Mechanism. *J. Balk. Tribol. Assoc.* (Sofia). **2007,** *13*(3), 379–386.

41. Kozhevnikov, N. V.; Kozhevnikova, N. I.; Goldfein, M. D. Some Features of the Kinetics and Mechanism of Emulsion Polymerization of Methyl Acrylate. *Izv. Vuzov. Chem. Chem. Eng.* **2010,** *53*(2), 64–68. [Russ].
42. Kozhevnikov, N. V.; Goldfein, M. D.; Kozhevnikova, N. I. Kinetics and Mechanism of Emulsion Copolymerization of Methylacrylate with Some Hydrophilic Monomers. Proc. Saratov University. Ser. Chemistry. Biology. Ecology. 2012. V. 12, No 2. P. 3–8. [Russ].
43. Goldfein, M. D.; Kozhevnikov, N. V. Development of Environmentally Safe Technologies in Saratov State University. *Probl. Reg. Ecol.* **2005,** *4,* 92–95. [Russ].
44. Goldfein, M. D. Concepts of Chemical Physics as the Scientific Basis of Ensuring Life Safety. Proc. Saratov University. Ser. Chemistry. Biology. Ecology. 2009. V. 9, No 2. P. 79–83. [Russ].
45. Goldfein, M. D; Ivanov, A. V.; Kozhevnikov, N. V. *Life Safety and Environmentally-Economic Problems of Nature Usage: A Textbook*; RGTEU Univ. Press: Moscow, 2008; 405 p. [Russ].
46. Kozhevnikov, N. V.; Goldfein, M. D.; Kozhevnikova, N. I. One-Stage Synthesis of Polymer Flocculent on Acrylonitrile Basis. *J. Balk. Tribol. Assoc.* (Sofia). **2007,** *13*(4), 536–541.
47. Neumann, M. B. (Ed.). *Aging and Stabilization of Polymers*; Nauka: Moscow, 1964. [Russ].
48. Emanuel, N. M.; Buchachenko, A. L. *Chemical Physics of Aging and Stabilization of Polymers*; Nauka: Moscow, 1982. [Russ].
49. Emanuel, N. M.; Buchachenko, A. L. *Chemical Physics of Molecular Destruction and Stabilization of Polymers*; Nauka: Moscow, 1988. [Russ].
50. Zaikov, G. E. *Destruction and Stabilization of Polymers*; MITHM Press: Moscow, 1993. [Russ].
51. Emanuel, N. M.; Zaikov, G. E.; Krintzmann, V. A. *Chemical Kinetics and Chain Reactions*; Nauka: Moscow, 1989. [Russ].
52. Korshak, V. V. (Ed.). *Technology of Plastics*, 3rd ed.; Khimiya: Moscow, 1985. [Russ].
53. Ulyanov, V. M. *Polyvinylchloride*; Khimiya: Moscow, 1992. [Russ].
54. Kargin, V. A. (Ed.). *Russian Encyclopedia of Polymers*; Sovetskaya Entsiklopediya Publishers: Moscow, 1977; Vol. 3. [Russ].
55. Batrakov, V. G. *Modified Concretes*; Stroyizdat: Moscow, 1990. [Russ].
56. Storozhakov, S. Yu. Environment Protection from Pollution by Acetate-Containing Production Waste of Polymeric Materials. Doctoral Diss, 2000. [Russ].
57. Gembitsky, P. A. Synthesis of Methacide. *Russ. Chem. Ind.*, **1984,** *2,* 82. [Russ].
58. Gembitsky, P. A. A Method for Producing Polyalkylene Guminidine Based on the Higher Amine "Gembicide". RF patent No 2144,929 of 27.01.2009.
59. Tager, A. A. *Physical Chemistry of Polymers*; Khimiya: Moscow, 1968; 536 p. [Russ].
60. Biopolymers: Properties, Applications, Development Prospects. www.plastinfo.ru.
61. Biodegradable Polymeric Materials. www.unipack.ru.
62. Klinkov, A. S., et al.; *Disposal and Recycling of Polymeric Materials*; Tambov Technol. Univ. Press: Tambov, 2009. [Russ].
63. Klinkov, A. S., et al.; *Disposal and Recycling of Polymeric Materials. Textbook*; Tambov Technol. Univ. Press: Tambov, 2010. [Russ].
64. Klinkov, A. S., et al.; *Recycling and Disposal of Packaging*; Tambov Technol. Univ. Press: Tambov, 2010. [Russ].

CHAPTER 9

POLYMERS AND HUMAN ENVIRONMENT [1–10]

CONTENTS

9.1 Plastics Influence on Human Health 408

9.2 Classification of Polymeric Packaging Materials and Ways to Improve Their Environmental Purity 412

9.3 Common Factors Associated with the Influence of Polymer Packaging Materials on Human Health 416

9.4 Methods for Assessing the Harmful Effects of Polymers 418

9.5 Consumer Properties and Applications of Polymeric Materials in Medicine 423

9.6 Usage of Polymers in Livestock Facilities 425

Keywords 427

References 427

ABSTRACT

Polymeric materials are considered in the context of the human environment. The impact of plastics on human health is characterized. The problem of the ecological purity of packaging materials is posed. The methods of studying the harmful effects of polymers on human health are listed. The use of polymers in medicine and animal husbandry is touched upon.

9.1 PLASTICS INFLUENCE ON HUMAN HEALTH

Everyone enjoys relatively cheap and convenient polymer products and does not think of their influence on the human body. However, interesting data can be provided by a closer look at the technical process of manufacturing raw materials used to make various plastic products, especially utensils for food storage or consumption, and their chemical properties.

It was Leo Baekeland, a Belgian-born chemist living and working in America (New York), who invented the first synthetic plastic in 1907. From this invention, a new era of the mass use of this seemingly human-friendly material begins.

In the early 1950s, the businessman William Dart industrially produced the world's first plastic cup. Americans and, then, other nationalities quickly adapted to the new invention, constantly expanding the range of products and their purpose.

Today, it seems hard to imagine our life without the use of plastic utensils.

Semifinished products, a wide range of meat cuts, sausage, cheese, ham, and other food products, previously packed in plastic containers, which provide them long shelf life, have become the basis of urban life. Beverages bottled in different shapes and sizes are used as a convenient way of drinks' accessibility to the consumer.

Utensils are primarily made up of such common polymers which provide a beautiful appearance, the necessary strength and flexibility, so plastics are of such a high demand today.

You will hardly find an indication on any plastic product that the impact of external factors on the plastic such as heat and contact with liquid could make these comfortable cups and plates emit harmful carcinogenic compounds. Most polymers are very cheap to produce and selected by manufacturers for that quality. However, polymers, by their properties, belong to toxic substances and might provoke many human diseases of the central nervous system as well as of individual organs. In certain cases, there may be complications which may cause disability or even fatal cases.

9.1.1 CHEMICAL CONTAMINATION OF THE CONTENTS OF PLASTIC PACKAGING

People are exposed to chemicals not only while manufacturing but also while using plastic packaging, because some of its components contaminate food. Examples of food contamination with plastics have been reported for most types of plastics, including styrene from PS, plasticizers from polyvinylchloride (PVC), antioxidants from polyethylene, and acetaldehyde from polyethylene terephthalate (PET).

Styrene is most widespread, being the main raw material in the manufacture of plastic utensils. In the gaseous state, it irritates mucous membranes and provokes acute and chronic inflammatory processes.

Formaldehyde is a carcinogen which, by regular contact with the human body, causes side effects such as deterioration or loss of vision, liver dysfunction, and cirrhosis.

Dioxin is a persistent toxic substance which, due to its chemical structure, is very easily absorbed by the receptors of living organisms, modifying or even suppressing their functional activity. This compound decreases the body's immunity and leads to disruption of cell division, causes the development of cancer, reduces the reproductive function level, resulting in impotence and infertility.

Phosgene is a toxic substance which leads to pulmonary edema, alveolar malfunction, and suffocation.

Bisphenol A leads to hormonal abnormalities.

At the same time, it should be emphasized that, in addition to the chemicals released from plastic products, the latter themselves are direct polluters of the human environment, leading to significant health deterioration. In this regard, it is advisable to give examples of such most widely used polymers.

Polyvinylchloride (PVC). Food packaging, plastic films, containers for toiletries, cosmetics, rubber protection against gnawing in cots, linoleum, pacifiers, shower curtains, toys, water pipes, garden hoses, car upholstery, and inflatable pools. It is used for packing liquids for washing windows, edible vegetable oils; for making jars for packing loose food and all sorts of dietary fat. Only this plastic is almost unrecyclable. Its use may cause cancer, birth defects, genetic changes, chronic bronchitis, ulcers, skin diseases, deafness, visual disturbances, digestive disorders, and liver dysfunction. Moreover, there is evidence that the carcinogenic vinyl chloride contained therein may penetrate into food and, further, into the human body. Many additives highly toxic to humans are used for PVC production: phthalates, heavy metals, and so forth. Moreover, the processes of production, use, and disposal of PVC are accompanied by the formation of large quantities of dioxins (most dangerous poisons) and other highly toxic compounds.

Phthalates. Plasticized vinyl products made with phthalates, including vinyl clothing, emulsion paints, shoes, inks, inedible toys and children's products, food packaging and films, vinyl floor coverings, bags and tubes for blood, containers and components, surgical gloves, breathing tubes, laboratory equipment of general purpose, inhalation masks, and many other medical devices may cause endocrine disorders associated with the development of asthma and pathology of the reproductive system. Medical waste with PVC and phthalates, permanently burned, causes harm to human health due to the release of dioxins and mercury, causing cancer, birth defects, hormonal changes, decreased sperm count, infertility, endometriosis, and a weakened immune system.

Polycarbonate with Bisphenol A. It is used to manufacture bottles for water. Researchers associate low contact doses of Bisphenol A with cancer, impaired immune function, early puberty onset, obesity, diabetes, and hyperactivity, among other problems.

Polystyrene. It is used to make a large number of food containers for meat, fish, cheese, yogurt, and soap; containers in the form of a clam shell, plates from foam and hard plastics, transparent containers for cookies, packaging for peanuts, foam plastic packaging, cassette cases, CD boxes, disposable covers, thermal insulation materials, ice molds, wall tiles, paint trays, disposable cups for hot drinks, and toys. It may irritate eyes, nose, and throat; causes dizziness and fainting; contaminates food and is deposited in the body's fat. There is an increased risk of lymph and hematopoietic cancer for workers.

Polyethylene. Soda water bottles, carpets, chewing gum, coffee stirrers, cups for drinks, food containers and films, plastic packaging with heat-sealable film, kitchen utensils, plastic bags, flexible bottles, and toys. There is evidence of a certain degree of its carcinogenicity.

Polyester. Bedding, clothing, diapers, food packaging, tampons, and upholstery. It may cause irritation of the eyes and respiratory tract and a skin rash.

Urea-formaldehyde (urea-methanal). Chipboards, plywoods, building insulation materials, finishing fabrics. Formaldehyde is assumed to be a carcinogen, the cause of birth defects and genetic changes. Formaldehyde inhalation causes coughing, swelling of the throat, watery eyes, breathing problems, headaches, rash, and chronic fatigue.

Polyurethane foam. Cushions, mattresses, and pillows. It contributes to bronchitis, cough, problems with skin and eyesight. Under certain conditions, it releases diisocyanate, leading to serious lungs problems.

Acrylic. Clothing, blankets, carpets made of acrylic fibers, adhesives, contact lenses, dentures, floor wax, cooking equipment, disposable diapers, sanitary napkins, and paints. Breathing difficulty, diarrhea, nausea and vomiting, weakness, headache, and chronic fatigue.

Tetrafluoroethylene. Nonstick-coated cookware, irons, ironing board covers, sanitary engineering and tools. It irritates eyes, nose, and throat, may cause breathing difficulties.

Polyethylene terephthalate. It is used for producing packages (bottles, cans, boxes, etc.) when filling soft drinks, juices, water, and so forth. This material can be also found in various kinds of containers for powders, powdery food products, and so forth. It is recyclable and reusable.

High-density polyethylene (HDPE). It is used to make pots and packages for milk and water; bottles for bleach, shampoo, detergents, and cleaners; plastic bags, cans for engine and other oils, and so forth. It is recyclable and reusable. It has a negative impact on the environment as indicated above.

Low-pressure polyethylene. It is used in the manufacture of plastic bags, bendable plastic packaging, and some plastic bottles. Well recyclable and reusable.

Polypropylene. Bottle caps, CDs, syrup and ketchup bottles, yogurt cups, and packaging for photo films.

9.2 CLASSIFICATION OF POLYMERIC PACKAGING MATERIALS AND WAYS TO IMPROVE THEIR ENVIRONMENTAL PURITY

As the technology of packaging materials develops, the functions of packaging expand. Packaging is converted from an inert, indifferent barrier between food products and the environment to a factor of production, since it enables one:

1. to directly change the composition of a product; in this case, biologically active materials with immobilized enzymes (the additive tightly held in the matrix of a polymeric material) are used to make packaging;
2. to protect food from microbial spoilage, thereby prolonging its shelf life; for example, the shelf life of sausage products in an "active" shell increases by two to three times;
3. to create an optimal gas medium within the shell, which is widely used for food storage in a modified and controlled environment; the use of such packaging for retail sale is impractical because of its rather high price, but the method of storage of fruits and vegetables in large bags with a little window made of a selectively permeable material is widely used in the West; the fruits so stored stay fresh longer, their packaging is repaid by eliminating the causes of spoilage and shrinkage of the goods;
4. to adjust the temperature in processing foods under microwave heating (e.g., using metallized polymeric materials); a product in a metallized packaging can be heated up to 200°C and above in a microwave oven; in this case, most heat is generated in the coating, and the product is roasted as on a pan, which is unattainable during common microwaving.

Such packages are rightfully called active. This trend is of great interest, since the introduction of an additive to the polymer shell matrix rather than to food allows prolonging the effect of this additive, by adjusting its rate of mass transfer into the food product. This provides the required additive concentration gradient on the surface of the protective shell in direct contact with the food product.

An important advantage of active packaging is that, due to immobilization of the additives, their migration into the food product is minimized (or optimally adjustable) as, according to recent data, many food supplements pose a particular threat to health. For example, the well-known citric acid, often introduced into the composition of products, despite its apparent harmlessness, may be harmful with excessive consumption. Of course, such an effect is not noticeable in a normal diet, but for a certain group of people whose diet is mainly composed of canned and semifinished products, such supplements may be really harmful.

The body of some people does not digest milk, which is genetically caused by the lack of the enzyme destructing milk sugar (lactose). Introduction of lactase, the enzyme hydrolyzing lactose, into the polymer base of the packaging material allows obtaining a dietary product (lactose-free milk).

The high content of cholesterol (usually and incorrectly called cholesterin) in human plasma is usually associated with an increased risk of cardiovascular disease. Appointment of special preparations to the patient can be a way to reduce the cholesterol level in the plasma. At the same time, if necessary, it is possible to effectively reduce the cholesterol content in milk and milk products by using cholesterol reductase immobilized in the polymeric packaging material in direct contact with liquid food.

To protect food products from the adverse effects of pathogenic organisms and toxic products of their life, bactericidal packaging materials have been using in recent years. The use of antimicrobial protection systems based on hygienically safe latexes (aqueous dispersions of synthetic polymers) exemplifies the implementation of this method. A method of protecting meat products and cheeses has been developed by designing a latex composition of an original formula based on environmentally friendly aqueous systems containing antimicrobial additives and the subsequent formation of coatings directly on food. The proposed method has a relatively easy technical solution: applying multilayer multifunctional coatings onto the surface of the product precludes the use of high temperatures, which sometimes adversely affects the product properties. This provides dense and widespread fitting to the surface of the product, which guarantees no microcavities (areas of the potential development of undesirable microorganisms). Domestic original drugs, dehydroacetic acid salts with a broad spectrum of action on various microflora (yeasts, fungi, actinomycetes, etc.), and complexes of these additives in combination with special controls of vital microbial cells (they mainly protect the surface of the packaged product mostly susceptible to infection) are used as antimicrobial additives.

Protective coatings formed directly on the surface of foodstuff (unripe cheeses, sausages, delicatessen, and conventional meat products) feature antimicrobial activity, provide a loss reduction in the net weight, for example, about 2% for cheese for its ripening period and environmental security, accelerate the biochemical processes of maturation, improve working conditions for the care of cheese due to elimination of the stage of cleaning and repackaging, and reduce the effects of toxicants on the product and service personnel.

In unfavorable environmental conditions, food may become a source and carrier of potentially dangerous chemical and biologically active compounds produced by microorganisms.

Compounds hazardous to health are contained in raw materials and food products at various stages of their technological processing, packing, storage, and marketing. At the same time, the safety and quality of food are one of the main conditions to ensure health. In Russia, in recent years, due to the sharp decline in food production and the outflow of agricultural raw materials to low-power companies, the risk of microbial contamination increased and, consequently, the quality of food products deteriorated.

To reduce the moisture content inside the package, special absorbers are introduced into the polymeric material, generally, mineral ones (e.g., zeolites, permutites, etc.). The process of moisture absorption can be accompanied by microbial growth inhibition.

In recent years, enzymatic supplements are introduced into polymeric packaging materials. Of particular interest and social significance are the development and use of biologically active packaging with enzymes immobilized inside the polymeric material. Such materials are able to regulate the composition and biological and organoleptic (taste, consistence, color, and smell) value of food, to accelerate the technological processes of getting finished products. In Russia, this area is in its infancy.

Studies have shown that certain enzymes immobilized on a polymeric support, while retaining their biological activity (by 70–80%), are able to acquire some new properties. For example, for materials with immobilized enzymes, extended operating temperature and pH ranges are characteristic, which very positively influence the rate of technological processes (hydrolysis) of biological substrates (proteins, fats, and carbohydrates). Free enzymes and their mixtures are known as expensive preparations, mostly imported to Russia. Production testing of new materials with immobilized enzymes in processing plants of the agro-industrial complex has shown their reusability. For example, a biologically active polymeric material (BAPM) with immobilized pepsin (the enzyme to split proteins) sustains over 90 production cycles during cold fermentation of milk before cooking the curd.

Therefore, the application of BAPMs allows reducing the consumption of enzymes and enzyme mixtures by two to three times. At the same time, the use of BAPMs improves the quality of finished products (the product grade increases by 20–30%) and more efficient processing of food raw materials is achieved (the completeness of use of food raw materials increases by 50–80%).

The use of such "active" shells as edible coatings is very promising. Natural polymers (polysaccharides) are the film-forming base in this case. Starch and cellulose derivatives are most widely used. The properties of these polymers are truly unique: having a good film-forming ability (edible films), they are widely used as components of foodstuff, for example, as structure-directing agents (thickeners) in the pasty dairy, confectionery, fruits, and vegetables. Films based on cellulose derivatives (e.g., carboxymethyl cellulose and its sodium salt) and modified starches (e.g., carboxymethyl starch, CMS) protect food products from weight loss (by reducing the rate of moisture evaporation) and provide a certain barrier for the penetration of oxygen and other substances from outside, thereby slowing any processes causing spoilage (fat oxidation, protein denaturalization, etc.). The edible films based on natural polymers have high sorption capacity, which determines their beneficial physiological effects. For example, when ingested, these substances adsorb and remove metal ions, radionuclides (radioactive decay products), and other harmful compounds, thus acting as detoxicants.

By incorporating special additives (flavors, dyes) into a polymer shell, one can adjust the aromatic taste properties of the food product itself in an edible film. Therefore, the "active" edible shell can alter the consumer's sensory perception of the product, which is particularly important when taking food of therapeutic and prophylactic action, for example, food with reduced fat, sucrose, supplemented with vegetable (e.g., soy) protein. In addition, the ability of an edible film to hold a variety of compounds allows enriching foodstuff with minerals, vitamins, trace elements, complexes, and so forth, compensating the deficit of some food components necessary for the human.

The coatings on quick-frozen meat products are examples of the use of edible films based on natural polymers. It should be recalled that deep cooling is recognized as the best way to store meat (beef is stored for 1 year at −18°C and up to 2 years at −30°C). However, unpackaged frozen meat loses 1–3% of its weight during storage (due to moisture freezing) and undergoes negative quality changes. The formation of carboxymethyl cellulose-based coatings on block-frozen meat significantly reduces the effect of these factors. Furthermore, these coatings eliminate environmental pollution with the used packaging waste, because further processing is carried out with the meat coating. Similar edible coatings have been designed and used to protect meat and meat–vegetable semifinished products manufactured by frozen food factories.

Thus, even a brief overview of the capabilities and prospects of the use of "active" packaging in food technologies shows that the future belongs to these types of packaging materials and the current century will be the century of "active" packaging.

9.3 COMMON FACTORS ASSOCIATED WITH THE INFLUENCE OF POLYMER PACKAGING MATERIALS ON HUMAN HEALTH

Polymeric materials such as plastics used for making various products, including packaging, contain compounds which, in the course of their operation, are systematically released to the environment, including food. This is accompanied by not only pollution but also violation of the ecological balance between the environment and human health. This impact starts from the moment of the synthesis of high-molecular-weight compounds and continues by producing a variety of household and industrial products, many of which are intended for packaging.

As shown above, a large number of ingredients are involved in the formation of polymers: monomers, initiators, inhibitors, solvents, emulsifiers, plasticizers, fillers, and so forth.

The harmful effect of polymers is primarily determined by the amount of the contained migratory monomers which tend to have high toxicity, including carcinogenicity and other harmful properties (allergy, mutagenicity, etc.). This may be due to many monomers containing functionally active chemical groups, very reactive and aggressive biologically. In some cases, the toxicity of monomers is determined by the presence of contaminants due to poor purification. Such impurities may, even in small (trace) amounts, give characteristic bad odor to the product and drinking water, which is unacceptable for any packaging material. Initiators (or catalysts) are substances that alter the rate of chemical reactions by forming an intermediate complex with the reactants; they either remain (initiators) or do not remain (catalysts) in the final product. Catalysts are typically alkali and alkaline earth metals, mineral salts, bases, or acids. The presence of catalyst residues in a polymer material is judged by its ash content. Polymerization initiators are peroxides, persulfates, and alkyl compounds of metals, that is, very aggressive compounds requiring careful purification of the resulting polymers. Mercaptans are used as polymer chain length regulators, methyl or isopropyl alcohol is used as solvent; they are very hazardous compounds and require careful purification or washing. Stabilizers, antioxidants, and aging inhibitors are introduced into a polymeric composition in order to prevent degradation (decomposition) when processing into articles and during their operation. They are used in small quantities (from a fraction of a percent to a few percent), more often not more than 3%. They are connected mechanically to the base polymer and therefore easily migrate to the surface of the polymeric material and, further, to the media contacting therewith (air, water, and food). Amines, phenols, esters of various acids, and other compounds are most commonly used as stabilizers (see above), whose toxicity is well studied. Plasticizers are introduced into polymeric composition in amounts of 10% or more to facilitate processing into products and to achieve optimal process conditions. Plasticizers are typically low-molecular-weight or high-molecular-weight compounds (even polymers) which do not form chemical compounds with the main product. A plasticizer, mainly low-molecular-weight compound, should easy migrate to the surface of the material, so the esters of fatty (phosphoric, phthalic, adipic, and sebacic) acids are often used with low vapor pressures and high boiling points. Plasticizers have a good ability to dissolve in fats and oils, so the plasticizer migrating to the surface can easily move into fat-containing foodstuff, which is always in a significant amount in the modern man's daily diet. Furthermore, the presence of a plasticizer in plastics greatly facilitates migration of other small molecules, which are often more toxic than the plasticizer itself. Dyes and pigments are used for coloring plastics. They have the ability to volatilize in significant amounts into the environment. To prevent this, such inorganic and organic compounds should be chosen in making packaging which have no ability to dissolve in the polymer and are thus immobile. Fillers are an integral part of any polymer composition and their content may reach 90%. They are introduced in order to reduce the material consumption of the polymer (saving) and for imparting certain properties to the products obtained. Low-molecular-weight and high-molecular weight compounds are used as fillers.

Thus, the problem of the right choice of a packaging material for a particular food product is not more rhetorical because polymeric materials used for packaging represent multicomponent systems comprising substances harmful to the human body. During prolonged contact of the packaging with the product, all these components may migrate into the product and, further, into the human stomach. The consequences of such migration, unfortunately, may occur quite quickly or after a long time. In order for everyone to feel safe, (s)he needs to know about the influence of packaging components on human physiology. This is particularly important for the right choice of packaging for products that are extractants for low-molecular-weight compounds, such as oily products.

9.4 METHODS FOR ASSESSING THE HARMFUL EFFECTS OF POLYMERS

In this context, the sanitary and toxicological requirements for packaging materials are most important. Material testing should be carried out with a mandatory assessment of the biological activity of those chemical substances which can migrate into food. One of the main tasks of biological tests of such substances is establishing possible long-term effects on the human body. And the results of these studies should have a decisive influence on the hygienic regulation of the packaging material for a specific product. What should a packaging material be from the standpoint of hygiene?

The hygienic requirements for polymer packaging in contact with food are determined by the following factors:

Toxicity. The formulation of a polymeric packaging material should not include substances of high toxicity.

Cumulative properties, including specific impacts and consequences on the human body (carcinogenic, mutagenic, allergenic, etc.).

The packaging material should be ***chemically inert*** with respect to the product (it should not change the organoleptic properties of the product or release substances at doses exceeding their permissible levels). Sanitary-hygienic tests of new packaging materials are multistage.

Consider several main stages of assessing the harmful effects of polymers.

Sensory evaluation. Preliminary information on the possibility to use a certain packaging material for contact with food can be obtained sufficiently quickly on the basis of its physicochemical properties: the solubility in various media, volatility, odor, and color. Such rapid assessment allows, using taste, smell, appearance, consistency, and uniformity, to evaluate the possibility of any undue influence of the packaging material on food. Packaging polymeric and combined materials as well as the food product itself can be the objects of organoleptic evaluation. To provide the necessary objectivity of such an assessment, scientifically developed standards of its implementation are used, including the method of blind tasting, the availability of the necessary qualifications of tasters, their quantitative composition, as well as modern methods of data processing. The result of sensory tests is evaluated in "numbers" in accordance with the corresponding GOST or TU for the polymer material recommended for food packaging.

Sanitary-chemical tests. The main danger when using plastic packaging in direct contact with food is due to the low-molecular-weight compounds contained therein that can be released into the environment and migrate into the packed product. Therefore, sanitary-chemical tests occupy an important place in complex hygienic studies. They allow one to assess the nature and quantity of chemical substances released from the polymer composite material into a model environment or the product. The objects of this research are also monomers, catalysts, polymerization initiators and accelerators, and, of course, technological additives (stabilizers, plasticizers, dyes, fillers, and others).

Sanitary-chemical tests are carried out by chemical-analytical methods, evaluating the integral (total) and specific (individual) migration of foreign substances into food. Their presence in food is most commonly determined in artificial media simulating the nature of the product. Foodstuffs themselves are not suitable for such tests because they are complex systems where it is difficult or even impossible to analyze trace amounts of certain chemical compounds in their composition. The media used for sanitary and chemical analyses are called "modeling."

In each country, individual modeling media, extraction conditions, and related regulatory documentation are developed and implemented to conduct such tests, since no common international unified methodology for conducting sanitary-chemical studies of packaging materials is agreed upon.

The ***maximum permissible concentration*** (***MPC***) or the ***maximum allowable value of the integral migration of the substance*** is a criterion for assessing the quality of tests. Such data on packaging materials are included in the "List of materials, products, equipment examined in the Scientific and Practical Center of Hygienic Examination of the State Committee of Russian Federation and approved for contact with food and liquids." It is issued by Ministry of Health of the Russian Federation as a reference book.

Toxicological assessment on animals. The final stage of hygienic examination of packaging materials in contact with food is toxicological tests. They are conducted on animals (rats, guinea pigs, and monkeys) by administration of solutions of the migrating substances into their bodies. To evaluate the toxicity of a substance, two basic criteria in accordance with the scale of toxicity are used:

1. The median lethal dose LD_{50} is the dose causing the death of at least 50% of the tested animals when administered intramuscularly during a certain observation time, for example, 90 days, measured in milligrams or grams per kilogram of animal body weight (mg/kg); in other words, the dose required to kill half the members of a tested population after a specified test duration;
2. The median lethal concentration LC_{50} is a similar parameter evaluated when administered in a gaseous state via the respiratory tract of an animal, measured in weight parts of vapors per air volume unit (mg/m^3).

From these parameters, the cumulation coefficient C_{cum} is calculated, which is the ratio of the total dose causing the death of 50% of animals on multiple administration (**LD_{50}(m)**) to the dose causing the death of 50% of animals on a single exposure **LD_{50}(1)**, that is,

$$C_{cum} = LD_{50}\,(m) / LD_{50}\,(1).$$

Depending on the obtained values of these parameters, the degree of toxicity of hazardous substances and their threshold of migration from packaging into food are determined. Studies of toxic substances contained in a packaging material or food enable to find the basic criterion of toxicological assessment, namely, the permissible migration quantity (PMQ). This quantity is the ratio of the maximum allowable daily dose of a given substance, D_m (mg/kg), after the animal to human conversion, to the amount of food consumed by a person on average per day (usually 2–3 kg) V, that is,

$$\mathrm{PMQ} = D_m / V.$$

PMQ is a hygienic standard for evaluating harmful effects and should ensure human health safety during indefinitely prolonged contact of a packaging containing the substance with food. The *PMQ* criterion takes into account not only toxicity and cumulation but also other effects on the human body, such as allergenicity, mutagenicity, teratogenicity, and embryotoxicity.

According to the hazardous nature of the pathogenic influence upon the human body from the source, auxiliary and other compounds, all polymeric packing materials can be divided into two main groups with regard to their biological activity and the degree of migration from the polymer.

Acceptable ones. The use of this group of materials is allowed for the production of plastic packaging. Chemical compounds in these materials do not alter the organoleptic characteristics of food products inside the package, which has been proven by years of research. This group includes the majority of the compounds used in the preparation of polymers: monomers, plasticizers, fillers, stabilizers, dyes, and other additives, according to their DCM.

Unacceptable ones. Their use is not permitted for plastic packaging. The group of harmful compounds includes those which have high toxicity or other adverse effects on the body and represent a considerable danger in case of migration to the environment. For example, consider some regulations on the most toxic monomers and heavy metal compounds.

Stringent requirements for physiological harmlessness are applied to the plastic packaging for fat-containing foods, which have a high extraction capacity for low-molecular-weight compounds. To characterize the stability of a packaging material to the action of oils and fats, two main indicators are used, namely, fat resistance and fat penetration, to evaluate and rightly choose the necessary packaging material for a given product in accordance with the requirements of GOST. The range of greaseproof materials is not very wide. These include PVC, PS, and copolymers thereof (polyvinylidene chloride (PVDC), and impact-resistant plastics based on PS, acrylonitrile, and other elastoplastics used in manufacturing semirigid containers). Recently, the group of these polymers was supplemented by PET and some other polymers more harmless to the environment.

In the last quarter of the 20th century, the residual monomer content of vinyl chloride, acrylonitrile, and styrene in polymers was reduced significantly. This was due to changes in the technology for their synthesis, processing, and better purification of the monomers used and processing additives. These monomers were then the subject of intense scrutiny by the health authorities in many countries because of the alleged carcinogenic effects of vinyl chloride and acrylonitrile on the human body, as well as the high toxicity of styrene.

Studies and adopted regulations ensure the safety of the use of such food packaging and allow establishing the unsuitability of containers made of acrylonitrile copolymers for packing alcohol. The polymers used to package bread should be heat resistant, as bread is often packaged while warm.

The right choice of a packaging material for a particular food product is very important. In general, only the brand of polymer raw materials approved by the Russian Ministry of Health should be used for manufacturing materials and articles intended for use in contact with food. The permitted polymeric raw materials are the raw materials having passed the necessary assessment, including toxicological tests, the allowed brands of polymer raw materials consistent with the relevant health services.

It should be noted that the introduction of additive substances (stabilizers, antioxidants, plasticizers, fillers, etc.) into the polymeric raw materials intended for making articles in contact with food is allowed, if only such substances fall into the IV (low hazardous substances) or III (moderately hazardous substances) hazard classes according to GOST 12.1.007-76, that is, nontoxic substances.

The physicochemical properties of a particular polymeric material are another important problem. For example, HDPE is not a greaseproof material, so the articles made of it are not suitable for packaging products with high (>30%) fat. The use of permitted raw materials guarantees compliance with the requirements for finished products by their sanitary-epidemiological safety; however, this condition is necessary but not sufficient.

During the processing of polymeric materials, the formation and accumulation of thermo-oxidative degradation products capable of going into the contacting food and having an adverse effect on the health indicators of the product may proceed under adverse factors (too high temperature of processing, etc.). Therefore, testing of finished packaging materials and products for hygiene indicators is necessary.

9.5 CONSUMER PROPERTIES AND APPLICATIONS OF POLYMERIC MATERIALS IN MEDICINE

The wide usage of polymeric materials in medicine is primarily determined by their unique physicochemical properties (often referred to as "irreplaceable substitutes"), as well as high consumer properties (especially the lower cost compared to products from metals and their alloys), and the ability to be easily processed to manufacture disposable products. The usage of polymers to make medical equipment allows the serial production of tools, health products, specialty dishes, and various types of packaging for medicines (Table 9.1).

TABLE 9.1 Properties and Main Applications of Polymeric Materials in Medicine

No	Name	Consumer properties	Applications
1	Cellulose	Readily hydrolyzable. Forms flexible films	Semipermeable membranes for hemodialysis, packaging materials, spectacle frames
2	Natural rubber	High elasticity. The glass transition temperature −70°C	Tubular products: catheters, probes, tubes for blood transfusion, etc. Intermediates for latex and rubber (thin-walled products)
3	Polyethylene	Frost resistance −80°C. Sterilization by ionizing radiation. Resistant to disinfectant solutions	Packaging materials, adhesive tapes, catheters, drainage systems, irrigation devices; supporting plates for semipermeable membranes for hemodialysis, hemo-oxygenators; connecting elements; tube syringes, medicine droppers; glassware

Contd...

No	Name	Consumer properties	Applications
4	Polypropylene	High physico-mechanical and thermal properties. Resistant to mineral and vegetable oils, retains shape to 150°C. Can be sterilized by steam under pressure	Disposable syringes, parts and components for hemodialysis and oxygenation equipment, packaging materials, vascular prosthesis
5	Polystyrene	High chemical resistance, low water absorption, easily glued, insoluble in aliphatic hydrocarbons	Labware, housing and other structural elements of instruments and devices, disposable syringes, parts of medical instruments
б	Polyvinyl chloride	Sparingly soluble in ketones, esters, chlorinated hydrocarbons; resistant to moisture, acids, alkalis, salt solutions	Catheters, probes, bougies, drainage devices, systems for taking and transfusion of blood. Blood-conducting pipes, eyeglass frames, etc.
7	Fluoropolymer	Exceeds platinum, quartz, graphite, and all synthetic materials by chemical resistance. Does not cause blood clotting. The operating temperature range from −196 to 130–190°C	Cardiovascular catheters, intravenous cannulae, parts and components of apparatus for extrarenal blood purification, circulatory support. Laboratory glassware, health products, blood vessels, tape plastic ligaments and tendons, etc.
8	Polyamides	Abrasion resistance, tensile strength	Details and components of medical devices and instruments, laboratory glassware, eyeglass frames, transitional cannulae, surgical thread (Capron, nylon)
9	Poly-4-methyl-1-pentene	Lightweight, transparent, durable. Withstands 400 sterilization cycles (up to 104°C at steam sterilization)	Anticorrosion coatings for chemical apparatus and pipes. Syringes, parts and components of inhalers, housing elements for direct measurement sensors of blood pressure, laboratory glassware (flasks, pipettes, adapters, etc.)
10	Polyethylene terephthalate (polyester)	Resistant to the action of microorganisms	Clamshell of medical instruments, surgical thread, synthetic blood vessels, the initial material for implants
11	Polyacrylates and polymethacrylates	Solid polymers with melting point of 190°C, pass 42% of light, resistant to water, alkalis, oils, fats, and carbohydrates	Optics endoscopes, structural elements for other medical instruments and devices, contact lenses, spectacle lenses, drip systems for blood transfusion, durable glass

No	Name	Consumer properties	Applications
12	Polycarbonates	Impact and temperature resistance. Increased transparency. Products made of polycarbonate can be repeatedly steam sterilized under pressure at 132°C, but the physical and mechanical properties of the polymer significantly deteriorate after 100 cycles. Low-toxic polymer	Reusable syringes, parts and components of medical devices and instruments, laboratory glassware, prosthetic and orthopedic products
13	Polyurethane	Resistant to hydrolysis. Melting point 150–200°C	Parts and components for extracorporeal circulation equipment, intra-aortic balloon catheters
14	Polyorganosiloxanes	Cause no blood clotting. No smell. Physiologically inert	Oxygenator membranes, urinary catheters, parts of syringes, catheters, tubes, parts and components of medical devices and instruments, artificial heart valves

The polymers to be used in reconstructive surgery should have the following properties:

1. Physiological harmlessness;
2. No toxicity and carcinogenic properties;
3. Minimal irritant action on the tissues in contact with the polymer, and so forth.

In addition, specific applications of polymers in tissue and organ prosthetics impose a variety of stringent requirements on the range of their physicochemical and mechanical properties.

Such polymers are intended for permanent or temporary replacement of lost or damaged tissues or organs in a living organism.

9.6 USAGE OF POLYMERS IN LIVESTOCK FACILITIES

Materials made from waste polymers (obtained either in the major industries or as secondary resources) have good prospects in the construction of livestock buildings. In particular, the production waste of obsolete rubber products (transport tires, conveyor belts, etc.) is used in road surfaces, deck floors in livestock farms, and the production of walling. As has been shown, under certain conditions, polymers may release toxic substances of various kinds of exposure (general toxic, irritant, teratogenic, and mutagenic) into the environment (air, water, and food). Some of them have a negative influence on the reproductive function of animals. The toxicity of the polymers used is primarily determined by the nature of the released monomer (styrene, chloroprene, vinyl chloride, urethane, etc.). However, auxiliary components such as catalysts (acids, alkalis), initiators (oxygen, organic and inorganic peroxides, azo compounds), emulsifiers, stabilizers, dyes, and so forth may be toxic as well. The monomers having functional groups affecting the skin and mucous membranes, the liver, exhibiting allergenic effects, and inducing carcinogenesis are most harmful. Plasticizers introduced into some polymeric materials are low-molecular-weight organic compounds used as solvents for resins. They usually form no chemical bonds with the polymer and easily migrate to the surface of the material, especially at elevated temperatures and mechanical stresses. Dibutyl phthalate, dioctyl phthalate, ethyl benzene, and carbon monoxide are the most toxic substances released from PVC-made materials. The constructional materials based on polyurethanes may release diisocyanate, triethylamine, and so forth into the environment, and materials with the addition of epoxy resin release ethylchlorohydrin, ethyl diamine, acetone, styrene, toluene, and diphenylpropane. Fiberglass (glass-reinforced plastic, GRP, or glass fiber-reinforced plastic, GFRP) and concretes with the addition of polyester resins release ethylene glycol, ethyl benzene, styrene, toluene, maleic anhydride, isopropylbenzene, methyl acrylate, and formaldehyde into the air and water environments.

In addition, special studies have found quite a strong sensitizing effect of polymeric materials in direct contact with the body of animals (dog collars, wicker beds, etc.). A sensitized animal body is an unstable (in a reactive aspect) biological complex which easily exhibits an increased sensitivity not only to a specific chemical reagent but also to many other allergens of different origin. Ultimately, the animal's immune system is weakened, its sensitivity to abiotic environmental factors increases, which leads to diseases with various symptoms, whose manifestations depend on the contact paths in the body and the sensitizer concentration.

KEYWORDS

- **polymer**
- **human environment**
- **human health**
- **packaging**
- **environmental cleanliness**
- **medicine**
- **animal husbandry**

REFERENCES

1. Ivanova, T. I.; Rozantsev, E. G. Active Packaging as Reality and Perspective of the 21st Century. *J. "Paket"*. **2000,** *1*. http://www.kursiv.ru/paket/archive/02/active.html.
2. Lyubeshkina, E. K. Packaging with Additional Functions. *J. "Paket"*. **2000,** *4*. http://www.kursiv.ru/paket/archive/05/special1.html.
3. Lyubeshkina, E. K., Fedotova O. V. For secure links with packaging material. *J. "Paket"*. **2004,** *5*. http://www.kursiv.ru/kursivnew/paket_magazine/archive/28/46.php.
4. Golub, O. V.; Vasilyeva, S. B. *Packaging and Storage of Food*; Kemerovo Technol. Food Inst.: Kemerovo, 2005. [Russ].
5. Protasov, V. F. *Ecology, Health, and Environmental Protection in Russia. Textbook for High School*; Finansy i Statistika: Moscow, 2000. [Russ].
6. Knop, A. N. *Phenolic Resins and Materials on Their Basis*; Khimiya: Moscow, 1983. [Russ].
7. Andrianov, G. A.; Ponomaryov, Y. E. *Foam Plastics Based on Phenol-Formaldehyde Polymers*; Rostov Univ. Press: Rostov, 1987. [Russ].
8. Dontsov, A. A., Lozovik, G. Ya.; Nowicka, S. P. *Chlorinated Polymers*; Khimiya: Moscow, 1979. [Russ].
9. Levin, V. S., et al.; USSR patent No 351, 583. [Russ].
10. Levin, V. S., et al.; Ed. Testimonies. USSR patent No 322, 274. [Russ].
11. Kamalov, R. A. Polymeric Materials as a Factor in Sensitization of Animal Organisms. Proc. Conf. Orel. 2000. [Russ].
12. Kamalov, R. A. Polymeric Materials in Water Supply Systems for Animals. *Vet. Med.* **2005,** *2,* 7–8. [Russ].

INDEX

A

Absorption spectroscopy, 156
Acetaldehyde, 361, 409
Acetic acid, 333, 361, 362, 369
Acoustic energy, 101
Acquired immunodeficiency syndrome, 197
Acrylamide, 346
Acrylic, 337, 338, 411
Acrylonitrile, 112, 241, 296, 299, 300, 303, 310, 336, 422
  monomer, 112
Acryloxyethyl maleate, 344
Actinomycetes, 413
Activation energy, 299, 306, 308, 310, 322, 337–380
Active atoms, 24
Adaptive function, 167
Adiabatic, 101, 302
Algorithmic approach, 22
Alkyl radicals, 234, 292–294, 301
Allomones, 165, 166, 167
Allyl acrylate, 344
Alonso's theory, 219
Aminoquinone, 307, 308
Ammonium persulfate, 343, 344
Analytic geometry, 90
Animal husbandry, 408, 427
Anthropic principle, 36, 81
Anthropo-ecologic crisis, 13
Antiferromagnetic materials, 25
Antigen–antibody interaction mechanism, 313
Antileukemic activity, 325
Antioxidants, 230, 284, 288, 291–293, 312, 325, 340, 349, 352, 409, 417, 422
Antroposociogenesis, 36
Aqueous solution of sodium thiocyanate, 336
Aromatic azyls, 256
Aroxyls, 256, 259, 299
Arrow of time notion, 35
Artificial intelligence, 45, 46
Artificial leather, 355, 387
Artificial neural network, 46
Astronomical events, 141
Astronomo-physical concepts, 118
Astronomy, 8, 19, 29, 30
Atmosphere composition role, 126
Atomic absorption spectrometry, 150
Attenuated total reflection, 157
Attractive function, 166
Auto-disorganization process, 72
Automated flowing immunoanalysis, 155
Autowaves, 26
Avalanches, 55, 59
Average extinction rate, 142
Avogadro's number, 257
Azine-bis-3-ethylbenzothiazoline-6-sulfonic acid, 161
Azo-*bis*-cyanopentanol, 336
Azo-*bis*-cyanovaleryanic acid, 336
Azo-*bis*-dimethylethyl amidoxime, 336
Azobisnitrils, 232
Azo-*iso*-butyric acid dinitrile, 297, 336

B

Bagdasarian-Bamford's equation, 305
Banfield's radical, 300
Baryonic matter, 66–70, 72, 73
Basic structures, 35, 89
  algebraic structures, 89
  order structures, 89
  topological structures, 89

Benard cells, 23, 58
Benard's experiments, 25
Benthic and planar landforms, 54
Benzimidazole, 287
Benzothiazole, 287
Benzoxazole, 287
Benzoyl peroxide, 307, 309, 339, 357
Benzoylhydroperoxide, 232
Benzoylperoxide, 232
Benzyl paradox, 234
Bifunctional chelate complexes, 156
Big Bang, 36, 48, 65–67, 74, 77, 78
Big Bang cosmological expansion, 48
Big Crunch, 72, 75
Bioaffinity, 156
  antibodies, 156
  nucleic acids, 156
  receptors, 156
Bioanalogs, 46
Biocatalytic enzymes, 156
Biodegradable polymers, 368, 370, 375
Biodegradation of polymers, 330, 402
Biodetoxication of aromatic hydrocarbons, 160
Biodiversity, 118, 143, 187
Biological evolution, 22, 23, 32, 100, 110, 112, 115
  Darwinian basic laws, 23
Biosensor components, 156
Biosensor technologies, 118, 151, 159, 187
Biosphere, 7, 8, 11, 14, 30, 32, 42, 53, 118, 128, 131, 145–149, 168, 178, 205
  consumers, 11
  decomposers, 11
  producers, 11
Biospheric studies, 118, 187
Bisphenol A, 409, 410
Black holes, 67, 68, 71, 72, 73, 80, 81
  capsule, 73
Black information, 80
Blood circulation system, 169, 170
Blood-generating system, 170
Blow-up regime, 206, 209, 216
Boltzmann formula, 105
Boltzmann's constant, 257
Bonner–Fisher type, 272
Bottle-to-bottle scheme, 399
Bourbaki's definition of mathematics, 92
Bourbaki's structural-mathematical approach, 90
Bradford's technique, 161
Breakthrough of natural dams, 55
Brick-red crystals, 238
Bulk acoustic, 158
Bulk polymerization, 356
Bull plasma albumin, 156
Butadiyn-*bis*2,2,6,6-tetramethyl-1-oxyl-4-oxi-4-piperidyl, 273
Butyl acrylate, 337
Butylenimine, 251

C

Calcium, 169
Caledonian orogeny, 127
Calendering, 390
Cambrian period, 132
Cancer radiotherapy, 325
Caoutchouc, 353, 402
Carbon monoxide, 358, 361, 426
Carbon–carbon bonds, 401
Carbonyldiimidazole, 154
Carboxylic acid salts, 231
Caspian Sea, 55
Catastrophic events, 55, 60
Category-theoretic approach, 88
Category-theoretic concept, 89
Cause-and-effect relationships, 20, 178
Central nervous system, 169, 170, 409
Chemical structure theory, 253
Chemiluminescence, 154
Chemomediators, 165–167
Chlorofluorocarbons, 345
Cholera, 194, 195
Chromatomass spectrometry, 239
Chromophore group, 290
Civilization development strategy, 14
Classical equilibrium thermodynamics, 100, 115
Clausius principle, 104
Clausius–Clapeyron's equation, 54

Coastal riptide systems, 58
Cobalt, 168–170, 339
Cognition problem of dark matter, 79
Coherent structures, 18
Colloid systems, 322
Columbus' voyage, 174
Computer technology, 42, 43, 46, 177
Conceptual shelves, 65
Conformational wave, 315
Convection theory, 58
Cosmic microwave background, 51
Cosmology, 51, 65, 69–71, 76
Cretaceous Period, 128, 129
Critical micelle concentration, 319, 320
Crushing worn tires and tubes, 400
Cryosphere, 24
Crystal lattice, 27, 258
Cultural-historical systems, 106
Cumene peroxide, 310
Cumulative properties, 418
Cumylhidroperoxide, 232
Cumylperoxide, 232
Curie law, 259
Curie's expression, 256
Curie–Weiss law, 257, 260–264, 267, 268, 271, 273
Cybernetics, 7, 8, 44, 45, 70, 93, 177
Cyclingprocess, 62

D

Dacron™ fibers, 397
Dark energy, 51, 53, 63, 64–70, 72, 74–81
  vacuum of space, 51
Dark mass, 65–67, 69
Dark matter, 18, 63–67, 69, 70, 73–75, 77–80, 82
  energy, 18, 82
Dark substance, 64, 65
Day of Judgment, 209
Debris flows, 55
Debye's equation, 325
Debye's theory, 323
Deductive analysis, 91
Degradation process, 72, 160, 393
Degree of evolutionability, 69
Degrees of freedom, 207
Detergent–water systems, 322
Detoxification, 118, 187
Diethylenetriamine pentaacetic acid, 156
Dimethoxidiphenylnitroxyl, 307
Dimethyldi-(*n*-β- naphthylaminophenoxy) silane, 340
Dimethylformamide, 312, 336
Dioxin, 373, 409
Diphenyl2-picrylhydrazyl, 296
Dipole–dipole interaction, 323
Dirty systems, 274
Disease spread problem, 194
Diversity, 2, 34, 38, 47, 57, 71–74, 80, 124, 128, 129, 132, 133, 135, 142–147, 173, 186, 347, 394
Donor–acceptor complex, 307
Donor–acceptor pair, 272
Dynamic system, 148
Dynamo, 178

E

Earth relief role, 124
Earth's crust, 57, 58, 331
Earth's magnetic field, 129
Earthquakes, 55, 60, 62, 63
Ecological education, 12
Ecotoxicants, 150, 157, 159
  atomic absorption spectrometry, 150
  gas chromatography, 150
  mass spectrometry, 150
Einstein's gravitational equations, 48
Einstein's mass–energy relationship, 33
Einstein's theoretical construction, 3
Electric and magnetic energy, 101
Electrochemiluminescence, 157
Electromagnetic field, 24, 159
Electromagnetic radiation, 24
Electron affinity, 308, 309, 311
Electron paramagnetic resonance, 234, 336
Electron transport particles, 316, 317
Electron-donor solvent, 312
  dimethylformamide, 312

Elodea, 160–165
canadensis, 160, 164
sprouts, 160
Empirio-criticism, 6, 7
criticism of empirics, 7
Emulsion polymerization, 342, 356
Endometabolites, 165, 166
Entropy, 23, 26, 35, 75, 101, 103–108
Environment, 11, 23, 26–28, 37, 40, 63, 69, 71, 100–114, 124, 128–130, 136, 146, 149–151, 159, 165–167, 171, 177, 180, 185, 186, 313, 317, 322, 330–335, 341, 343, 354, 355, 361–365, 369, 373, 375, 402, 411, 412, 416–422, 426
Environmental cleanliness, 427
Environmental crisis, 39
Environmentally friendly, 330, 332, 334, 364, 402, 413
Enzyme-labeled antibodies, 152
horseradish peroxidase, 152
laccase, 152
Enzyme-linked immunosorbent assay, 151
Epidemic curve, 197
Ernst Julius Öpik's hypothesis, 122
Ethoxyquin, 288
Ethyl acrylate, 344
Ethylenediaminetetraacetic acid, 156
Ethylenimine, 251
Etymologically, 172
Evanescent wave biosensors, 157
Evolution, 6, 18, 21–23, 26–36, 53, 56, 59, 63–88, 108–115, 118, 122–128, 131, 136, 145–149, 167, 187, 196, 202, 206, 209, 351
biological evolution, 109
Charles Darwin's evolutionary theory, 110
modern Darwinism, 110
Evolution models, 115
Evolution of species, 118, 131, 187
Evolutionary process, 36, 68, 141, 144, 145
Evolutionary youth degree, 114
Evolutionism, 30, 31, 36
Exchange process, 105
Exometabolites, 165, 166
External energy, 101

F

Ferromagnet, 25
Ferromagnetic exchange interaction, 258
Ferromagnetic materials, 25
Ferromagnetic-type interchain interactions, 261
Fiber optic biosensors, 157
First law of thermodynamics, 104
Flocculant, 346
Flowing immunoanalysis, 155
Flow-injection analysis, 155
Flow-injection liposome immunoanalysis, 155
Fluid system, 56
stone substrate, 56
Fluorene, 160, 163, 164
Fluorescein isothiocyanate, 157
Fluorescence microbead immunosorbent assays, 153
Forest-steppe zone, 55
Formaldehyde, 361, 409, 411, 426
Fossil record, 124, 131, 132
Four-element system evolution, 11
Framework of civilizations, 3
Fremy's diamagnetic salt, 277
Fremy's salt, 277, 278
Friedman's theory, 51
Friedmann's open model, 68
Friedrich Nölke's hypothesis, 123

G

Galactic geochronological scale, 53
Gas chromatography, 150, 161
Genetic information storage, 113
role of thermodynamic factor, 113
Geneva principles, 251
Geocatastrophes, 60, 61
autogeocatastrophes, 60, 61
bifurcationgeocatastrophes, 60, 61
geocatastrophes, 60, 61
Geocratic type, 123

Geodynamics, 55, 56
Geologists, 57, 58
Geomorphological system, 60
Geomorphology, 58, 60
Geophysics, 48, 55, 57, 242, 277
Geosystems, 18, 54–58, 60, 82
Gerontologic value, 114
Giant luminosity, 80
  quasars, 80
Gibbs function, 109, 113, 114
Gibbs thermodynamic potential, 103
Glaciers-ocean-atmosphere-land system, 59
Gladyshev's theory, 100, 108, 115
Global climate models, 118, 129, 187
  astronomo-physical, 118
  fluctuations in solar activity, 129
  galactic mass extinction, 129
  geologo-geographical, 118, 119
  impact hypothesis, 129
Global evolution, 36, 65, 69, 70, 72, 74–78, 81
Global evolutionary process, 18
  scientific apparatus, 18
Globalization, 14, 117, 118, 149, 172–175, 181, 185–187, 194, 204, 224
Glutaraldehyde, 154, 158, 366
Glycidyl ethers, 351
Glycolysis, 376
Gödel theorem, 95
Gomberg's method, 232
Gomberg's radical, 296
Gravitational models, 218
Gray holes, 73
Gray transformation, 73
Great glaciations, 127
Greenhouse effect, 126, 127, 331
Grinding effect, 401
Growth dynamics of World urbanization, 213
Growth tectonics, 56
  tectonics of the Earth's core, 56

H

Hägerstrand's process, 221
Hair, 169–171
Heisenberg's
  exchange Hamiltonian, 257
  linear model, 261
  linear system, 263
  one-dimensional model, 271
    ferromagnetic interaction, 271
Helmholtz free energy, 103
Heterocyclic nitroxide, 240, 306
Heterogeneities, 71, 72, 73
Heterogeneoussystems, 31, 71, 100, 101, 109, 112, 151, 152, 341
Hevea latex sap, 332
High-density polyethylene, 359, 411
High-efficiency liquid chromatography, 154
High-impact PS, 391, 392
Homo sapiens, 148, 205
Homogeneous, 23–28, 66, 71–74, 77, 79, 100, 151, 152, 207, 244–246, 341, 374, 381
Hubble constant, 48, 50
Hubble's law, 48
Human environment, 146, 355, 364, 408, 409, 427
Human health, 150, 151, 155, 159, 168, 331, 359, 365, 408, 410, 416, 421, 427
Human sciences, 2–5
Humanitarian culture, 2
Humanitarian systems, 106
Humanities, 2, 3, 4, 14, 35, 216, 330
Hydrazyls, 296, 298
Hydrocarbon polymers, 290
Hydrocarbon radicals, 295
Hydrodynamics, 18
Hydrogen peroxide, 158, 232, 233
Hydroperoxide decomposition process, 292
Hydroquinone, 340, 341, 357
Hydrosphere, 24, 53, 59, 118, 119, 126, 145, 146
Hypophysis, 170

I

Iminoxyl free radicals application, 313

Imminoxyl radicals, 239
Immune system, 169–171, 410, 426
Immunoaffinity chromatography, 154
Immunochemical techniques, 151, 187
Immunochemical test systems, 151
Immunoenzymatic analysis, 152
Immunosensors, 158
  electrochemicalimmunosensors, 158
    amperometry, 158
    conductometry, 158
    potentiometry, 158
  opticalimmunosensors, 158
    absorption, 158
    fluorescence, 158
    luminescence, 158
  opticoelectronicimmunosensors, 158
  piezoelectricimmunosensors, 158
Impact hypothesis, 129
Indicator function, 167
Information,
  crisis, 39
  explosion, 39
  hunger, 39
  industry, 42
Infrared spectroscopy, 239, 336, 351
Insecticides, 153, 157, 158
Intensification process, 42
Intensity of extinction, 133–136
Internal energy, 101, 103, 104
  heated bodies, 101
Internetization, 47
  connectedness, 47
Interspecific interactions, 130
Intervening opportunities, 219, 220, 221
Intranuclear energy, 101
Intratomic energy, 101
Ionic polymerization, 308
Isentropic, 101
Isobaric, 101
Isochoric, 101
Isomerization, 240, 312
Isothermalprocess, 101

K

Kantian cosmogony, 35
Kant–Laplacecosmogonic hypothesis, 6, 29
  planetary system, 29
Kidneys, 169
Kinetic effects, 308
Kinetic phase transitions, 27

L

Landslides, 55, 59
Law of conservation of energy, 49
Law of temporary hierarchies, 108
L-dihydroxyphenylalanine, 161
Le Chatelier's principle, 148
Light-addressable potentiometric sensor, 157
Linear function, 21
Linear low-density polyethylene, 359
Linearity and symmetry principles, 107
Liposome immunoaggregation, 155
Lipps–Signor effect, 132
Liquid–crystal transition, 27
Living organisms, 106, 127, 146, 160, 171, 332, 334, 409
Logistic curve, 201, 202, 203
Logistic equation, 194, 198, 199, 201, 202, 204, 212, 213, 224
Log-log scale, 137–140, 144, 212
Low-density polyethylene, 359
Lunisolar tides, 63

M

Machinery, 11
  link between society and nature, 11
Macroscopic inhomogeneities, 27
Magnesium, 169
Magnetic properties, 230, 245, 259, 271, 274, 325, 387
Malthus' equation, 201
Malthus' model, 199
Man–biosphere relation, 8
Man–biosphere–cosmos harmony, 7
Man-machine hybrid information processing systems, 43
Man-to-nature relation, 101
Marxist tradition, 7

Mass extinction, 128, 129, 136, 140, 141, 143
extinction events, 129
mass extinctions-phantoms, 129
true mass extinctions, 129
Mass spectrometry, 150, 155, 239
Mathematical experiment, 93, 94
Mathematical modeling, 86, 93, 97
Mathematics, 86–97, 177
Mathematization of science, 94–97
knowledgemathematization, 95
Kurt Gödel's theorems, 95
leads, problems, limits, 94, 95
McConnell's model, 258
Mean information field, 222
Mechanical energy, 101
Mechanical recycling, 376
Mediatization, 47
Medicine, 167, 195, 359, 408, 423, 427
Medium-density polyethylene, 359
Megaecology, 330
Melt flow index, 380
Metal–chelate complexes, 156
Methacrylic acid, 242, 298, 302, 303, 337, 340, 341, 343
Methanolysis, 376, 397
Methyl acrylate, 310, 337, 426
Methyl methacrylate, 296, 298, 300, 306, 310, 311, 339, 393
Micelle–solvent interface, 322
Microelementoses classification, 168
endogenic, 168
exogenic, 168
iatrogenic, 168
technogenic, 168
Migration, 36, 146, 182, 194, 204, 205, 216–224, 354, 387, 412, 417–421
models, 217, 223
patterns, 216, 218
Milankovitch's
astronomical cycles, 62, 122
astronomo-climatic concept, 120
astronomo-climatic cycles, 121
paradigm, 118, 121
radiation curve, 121
Minerals, 187, 242, 415
Modern equilibrium thermodynamics concepts, 108
Gladyshev's theory, 108
Modern natural science, 82, 86, 146
Modern nonequilibrium thermodynamics, 115
Molecular and kinetic theories, 102
Monophenols, 160
Morphism, 91
Mozart or Beethoven music, 3
Velazquez or Dali paintings, 3
Muscular tissue, 169
Mutagenic effect, 363
Mycobacterium tuberculosis, 365
Mycotoxicants, 151
Mycotoxins, 150
Mystery of the living, 2

N

Naphthalene, 160, 163, 164, 165
Natural philosophy, 4, 5, 64
Natural rubber, 241
Natural sciences, 2–14, 21, 22, 32, 35, 57, 86, 88, 91, 93, 97, 107, 273
Negentropic pyramid, 73
Neoformations, 73
Neopositivism, 8
Neuman–Rozantsev's reaction, 313
Newton's law, 77, 219
Newtonian gravitational constant, 49
gravitational constant, 49
N-hydroxysuccinimide, 154
Nicolas Bourbaki's concept, 89
Nicotinamide adenine dinucleotide's reduced form, 316
Nitroxide nitroxyl triradical, 274, 275
Nitroxyl paramagnetic concentration, 306
Nitroxyl radicals, 292
Nitroxyl–nitroxide radical, 300
Nitroxyls, 256, 268, 292, 293, 295, 300–303, 307
Nitroxyls (or nitroxide nitroxyls), 268
N-nitroso-*N*-phenylhydroxylamine, 341
Nonequilibrium open systems, 25, 26

Nonlinear
dynamics, 18, 57, 58, 82, 128, 195, 198, 204, 216
effect, 54, 56, 118
function, 21, 54
hydrodynamic equations, 54
oscillations, 21
science, 18, 21, 22, 82, 147
systems, 21, 209
trigger system, 55
Nonnonlinearity, 21
Nonplanar configuration, 235
Noosphere, 47, 53, 118, 145, 146, 148–150, 187
Greeknoos, 149
education, 13
orientation, 13
science, 13
transformations, 13
Novorotor units, 401
Nuclear reactions, 80
Nylon fibers, 396

O

Occupational exposure limit, 367
Ocean bottom uplifts role, 125
Offensive function, 166
Oligourethanacrylate, 299, 303
Ontological framework, 64
Optical quantum generators, 312
Opticoelectronic biosensors, 158
Organic acids, 166, 297, 361
Organic magneton, 274
Organic paramagnets, 237, 249–252, 258–260, 278, 325
Oscillating motion of the atoms, 101
Osseous tissue, 169, 170, 171
Oxotetraphenylpropanol, 247
Ozone-friendly azeotrope mixture, 345

P

Packaging, 330, 355, 358, 359, 368–374, 386, 397, 399, 408–424, 427
Pancreas, 169, 170, 171
Paradigm, 9
Paramagnetic probe method, 316, 317, 322, 325
Parizer–Parra–Popl methods, 234
Penrose's theorem, 72
Permetrinic acid, 153
Peroxides,
as initiators, 302
Personal computers, 40, 44
Pesticides, 150, 151, 152, 154, 157
Phanerozoic
age, 140
biota, 132
geological age, 142
geological periods, 133, 134
Phase-resolved fluoroimmunoassay, 153
Phenanthrene, 160, 163, 164, 165
Phenols, 157, 166, 291–294, 340, 349–353, 417
Phenomenological model, 194, 224
Earth's population growth, 204
Phenylmercuric bromide diphenylmercury, 231
Pheromones, 165, 166, 167
Philosophy and science relations, 4
criticism of empirics theory, 7
drama of ideas, 5
handmaiden of theology, 6
materialism and empirio-criticism, 6
philosophical and methodological apparatus, 5
philosophical knowledge, 5
Dialectics of Nature, 6
Phosgene, 409
Phosphorus, 169
Phthalates, 410
Piperidole-based stabilizers, 284
Planar configuration, 235
Planck's theoretical construction, 3
Planet's shells, 53
lobal systems, 53
geospheres, 53
Planetary cosmic force, 148
Plasma physics, 18
Polarization fluoroimmunoassay, 153

Pollution, 147, 159, 177, 178, 184, 185, 242, 275, 330–332, 336, 341, 354, 362, 364, 373, 374, 394, 402, 416
Polyacrylamide, 346
Polyacrylonitrile, 112, 337
Polyalkylene guanidines, 364, 402
Polyamides, 284, 402
Polycarbonate with Bisphenol A, 410
Polychlorinated biphenyls, 150, 151, 155, 157
Polycyclic aromatic hydrocarbons, 150, 151
Polyester, 411
Polyethylene, 275, 276, 286, 288, 289, 318, 330, 359, 372, 402, 409, 411, 423, 424
Polyethylene terephthalate, 409
Polyhexamethylene guanidine, 365
Polymer, 112, 276, 285, 290–295, 304, 308–311, 330, 333–362, 367–369, 372, 374, 376–401, 408, 412–419, 421, 422, 425–427
Polymerate, 346
Polymeric materials, 241, 330, 334, 412, 418, 423, 426
Polymerization kinetics, 304, 307, 325
Polymers aging, 402
Polynucleic acids, 32, 113, 114
Polyolefins, 242, 284, 286, 288, 302, 348, 349, 359, 373, 377, 380–385, 393, 402
Polypropylene, 286, 295, 330, 348, 372, 411, 424
Polysaccharides, 32, 110, 332, 415
Polystyrene, 410, 424
Polyurethane foam, 411
Polyurethanes, 284–286, 376, 426
Polyvinyl chloride, 284, 285, 289, 329, 354–359, 372, 381, 386–390, 393, 402, 409, 410, 422, 426
Polyvinylidene chloride, 284, 422
Popper's theory of scientific knowledge, 9
Population growth, 182, 194, 199–201, 204, 205–212, 215, 216, 224
Positive heuristics, 9
Positivism, 8, 9
  Lat.positivus, 8
Positivist philosophy, 5
Potential energy of gravity, 101
Power station, 399
Power-law distribution, 139, 144
Predtechensky hypothesis, 123
Principle of,
  falsifiability, 9
    Lat. falsusand facio, 9
  verifiability, 8
    Lat. verifacare, 8
    neopositivism, 8
Problem of,
  spread of diseases, 194, 224
  recovery and recycling of polymeric materials, 371
Production of,
  high-density polyethylene, 360
  linear low-density polyethylene, 360
  low-density polyethylene, 359
  medium-pressure polyethylene, 360
  polymeric materials, 362
Propylenimine, 251
Prostate gland, 170
Protective function, 166
Proteins, 32, 110, 113, 154, 167, 169–171, 314, 332, 334, 366, 414, 415
Protobiogeocenoses, 146
Proto-evolution, 70, 75
Protoevolutionary changes, 72
Pseudomonas aeruginosa, 365

Q

Quartz crystal microbalance, 157
Quartz optical fiber, 158
Quasi-closed, 103, 108, 109
Quasi-equilibrium, 103, 107, 147
Quaternary ammonium compounds, 365
Quaternary glaciation, 125
Quaternary period, 120–122, 125, 126
  Milankovitch's curve, 122

R

Radiant energy, 101, 126

Radioactivity, 58
Raman spectroscopy, 239
Raoul's law, 248
Rapid movement of glaciers, 55
Recombination–molecular dimer, 232
Recycling of,
  polyamides, 394
  PVC, 386, 389
  rubber products, 399
Regalconverter, 391
Regal process, 390
Regulation of interactions, 166
Resonator, 24
  oscillating system, 24
Restraint of competitors, 166
Retardation phenomenon, 55
Reynolds number, 61
Rhythm of geophysical processes, 61
River basins, 58, 59
Rubber, 241, 288, 332, 353, 362, 367, 386, 388, 390, 393, 398–402, 409, 423, 425

S

Sanitary-chemical tests, 419
Science and philosophy, 2, 5, 6, 14
Science of sciences, 5
Scientific and technical information, 40
Scientific revolution, 3
Scientific thinking methods, 3
Sea-level fluctuations, 130
Second law of thermodynamics, 104–106
Secondary LDPE, 377, 379
Secondary polyethylene, 378
Seismic energy, 62, 63
Selenium, 168, 171, 172
Self-developing system, 147
Self-organization, 18, 20, 23–28, 36, 53, 56, 58, 68–70, 73, 78, 205
Self-organizing systems, 26, 31
Self-oscillations, 27, 59, 61
Self-reflection, 91
Self-similar solution, 208, 209
Self-sufficiency, 92, 94
Sensory evaluation, 419
Sepkoski database, 132
Set-theoretic approach, 88–90
Settheoretic stage, 90
Sinitsyn's hypothesis, 123
Skin, 169–171, 367, 410, 411, 426
Social information, 38, 41, 45
Social-economic systems, 106
Socioeconomic process, 42
Socio-natural laws features, 10
Sodium dodecyl, 320
Solar activity cycles, 121
Solar radiation, 101, 118, 120, 122–124, 126, 330, 393
Solid-phase extraction, 153
Spatial rhythmic structures, 58
Special symbolic deductive systems, 92
Spectrophotometry, 154, 155
Spin clamp method, 239
Spin–spin interaction, 259
Stabilization of,
  caoutchouc and rubber, 352
  monomers, 339
  polyamides, 351
  polymers, 353, 402
Stable radicals, 230, 253, 259, 300, 303, 304, 307, 308, 312, 313, 325
Stationary process, 207, 353
Sterically hindered amines, 284, 291
Stone chronicle, 61
Storms, 60
Structuralization, 90
Structure theory, 253, 254
Styrene, 296, 298, 300, 303, 306, 310, 311, 339, 372, 393, 409, 422, 426
Substance energy technologies, 37, 46
Sunspots, 62
Supercritical fluid extraction, 153
Supernova explosion, 129, 130
Super-pluralistic society, 47
Surface acoustic wave, 157
Surface plasmon resonance, 157
Surface transverse wave, 157
Suspension polymerization, 356
Symmetry breaking, 25

Synergetics, 18, 24, 26, 31, 38, 56, 57, 68, 77, 81, 216, 349
Syringaldazine, 161
System-temporal integrity principle, 81

T

Tacoma Narrows Bridge destruction, 61
T-butylhydroperoxide, 232
Technical sciences, 3–5, 13
Tetracyanoethylene, 270
Tetracyanoquinodimethane, 308–310, 342
  anion radical, 309
Tetracyan-*p*-quinodimethane, 270
Tetrafluoroethylene, 411
Tetramethyl-4-hydroxypiperidino-1-oxyl, 263
Tetramethyl-4-oxopiperidyl-1-oxyl azine, 268
Tetramethyl-4-oxypiperidyl-1-oxyl phosphite, 268
Thermal energy, 100, 101, 103, 106, 247, 266
Thermal-oxidative process, 353
Thermodynamic
  equilibrium, 28, 103, 106, 107, 108, 233, 262, 296
  evolution theory, 110
  factor, 113, 238
  properties, 101
  rates, 107
Thermodynamics, 22, 23, 26, 35, 56, 68, 81, 100–111
  equilibrium process, 102
  irreversible, 102
  nonequilibrium process, 102
  reversible process, 102
  systems, 100, 103, 105, 108
    adiabatic systems, 105
    closed systems, 105
    heterogeneous, 100
    homogeneous, 100
    isolated systems, 105
    open systems, 105
Thermostat, 111, 112
Third culture, 3, 4
Threshold limit value, 361
Thyroid gland, 167, 169–171
Tidal waves, 129
Topological structures, 89, 90
Tornadoes, 60
Total internal reflection fluorescence, 157
Toxicity, 418
Toxicological assessment on animals, 420
Trans-1,2-cyclohexyldiethylenetriamine pentaacetic acid, 156
Transducer, 146, 157–159
Transitional process, 72
Trans-Mediterranean Renewable Energy Cooperation, 180
Transnational corporations, 174
Triglycerides, 248
Triplet magnetic exciton, 266
Tsunamis, 60
Turbulence, 18, 25, 54
Tween–water system, 322
Typhoons, 24, 60
  hurricanes, 60
Tyrosinase, 157, 160–165
Tert-butylperoxide, 232

U

Ultrafine interaction, 235
Ultraviolet, 127, 129, 231, 291, 333, 336
  light, 352, 369
  radiation, 127, 298, 349
Universals (global), 31
  antigravitation properties, 65
  antigravitation, 65, 76
  evolutionism, 18, 29–31, 36, 64, 82, 145
  Latin Universalis, 31
  problems, 3
Universum, 30, 36, 63, 91
Upper respiratory tract, 195
Urbanization, 194, 213, 215, 224
Urea-formaldehyde, 411
Utensils, 408

V

Vapor–air mixture, 54
Vegetative peroxidases, 160
Verdazyl radicals, 237
  tetrazyls, 237
Verdazyls, 256, 298, 299
Verhulst's equation, 201, 202
Verhulst's logistic model, 200
Vinyl acetate, 296, 298, 337
Vinyl chloride polymerization, 355
Vinyl monomers, 230, 296, 298, 300, 303, 304, 308, 309, 312, 325, 342
Vitamin D deficiency, 169
Volcanic,
  eruptions, 55, 60, 62, 63, 124, 130
  islands, 59
  phenomena role, 124
Volcanism, 124, 129
  hypothesis, 130
Volcanoes, 59, 126
Vulcanization grid, 401

W

Warning function, 167
Waste rubber diaphragms, 399
Water–detergent system, 320
Water–Tween system, 321
Weiss constant, 260–263, 268, 270
White holes, 73
  naked singularity, 73

X

Xenobiotics, 151, 154, 155, 160–165, 242
  screening, 151
X-ray diffraction analysis, 274
X-ray fluorescent spectroscopy, 156

Z

Ziegler–Natta catalysts, 360, 361
Zinc, 168, 169, 170, 339